Jack Ward Thomas

THE JOURNALS OF A

FOREST SERVICE CHIEF

Jack Ward Thomas

THE JOURNALS OF A
FOREST SERVICE CHIEF

Edited by Harold K. Steen

FOREST HISTORY SOCIETY Durham

in association with

THE UNIVERSITY OF WASHINGTON PRESS

Seattle and London

Jack Ward Thomas: The Journals of a Forest Service Chief
has been published with support from the U.S. Forest
Service History Program and the Lynn W. Day
Endowment for Publications in Forest History.

Copyright © 2004 by the Forest History Society

The Forest History Society
701 Wm. Vickers Avenue
Durham, North Carolina 27701
919–682–9319
www.foresthistory.org

University of Washington Press
P.O. Box 50096
Seattle, WA 98145-5096
www.washington.edu/uwpress

Library of Congress Cataloging-in-Publication Data is
available from the Library of Congress
ISBN 0-295-98398-1

CONTENTS

Introduction   3

1990   14

1991   38

1992   44

1993   54

1994   75

1995   145

1996   276

Index   401

Jack Ward Thomas

THE JOURNALS OF A
FOREST SERVICE CHIEF

INTRODUCTION

In 1989, Dr. Jack Ward Thomas was arguably the most prestigious of the nearly two thousand research scientists employed by the U.S. Forest Service. Thus, it was no surprise when Chief Dale Robertson "invited" Thomas to be chairman of an interagency committee of scientists to devise a scientifically credible management strategy for the northern spotted owl. Too, it was no surprise that the report would turn out to be highly controversial, because implementation of its recommendations would have a direct and significant impact on the economy of the Pacific Northwest. It is no exaggeration to say that the spotted owl provided Thomas with a political prominence comparable to that he had held so long in science. In 1993, during the early months of President Clinton's first term, Chief Robertson, portrayed as a Reagan appointee, was assigned to other duties and chose to resign. Thomas was named his successor.

Thomas's tenure as chief had a rough beginning. Under Civil Service regulations, he was not eligible to head an agency: he did not meet the qualifications for membership in the Senior Executive Service. Oddly, he had been one of the first Forest Service employees to be recommended for such training but had refused on the basis that he had no interest in becoming a senior executive and preferred to remain a scientist. But the Clinton administration was committed to the appointment, no doubt in large part because it would symbolize a clean break from the Reagan-Bush era of focus on commodity production from the public lands. It was no small matter that Thomas was a scientist and wildlife biologist and would be the first chief who was not a forester (although Chief Max Peterson had been a civil engineer), a highly symbolic shift. The only option was for Thomas to be a political appointee, the common practice for other federal agency heads but anathema to the tradition-proud Forest Service. In addition to the usual pressures of leading a large agency, during his tenure Thomas carried the burden—some of it self-imposed—of not having a "legitimate" appointment by Forest Service standards.

The pace of a chief's life is hectic: externally imposed priorities are insensitive to ongoing demands on time. Thomas the scientist faced the additional challenge of being abruptly immersed in politics, Washington, D.C.–style. He was appalled by those who seemingly casually proposed to place advancement of the administration's agenda ahead of scientific fact and apparently even existing law. And because he did not consider himself a political partisan, he was disgusted by his brutal treatment during partisan congressional hearings. Thomas was uncomfortable in his role, and within two years he submitted his resignation. By threatening to replace him with an "outsider," the secretary of Agriculture persuaded him to stay on for another year—until after the November 1996 election. In December 1996 Thomas retired and became Boone & Crockett Professor of Wildlife Conservation at the University of Montana.

Jack Ward Thomas was born in Fort Worth, Texas, on September 7, 1934. He grew up in a small town and characterizes himself as having rural, southern, Depression-era roots. He hunted and fished on nearby private lands—there were no public lands, such as national forests, in his part of Texas. Later, when he encountered federal lands where one could walk without "permission," he became devoted to the notion that public lands were a very special part of the American heritage. While chief, when defending the integrity of the national forests against proposals to transfer portions to state ownership or sell them to the private sector, he spoke passionately of his earlier experiences and pledged that the land would remain federal. He would also assert that the overriding purpose of the management of the national forests—and probably other public lands—had become, unintentionally, the preservation and protection of biodiversity.

Thomas entered the Agricultural and Mechanical College of Texas (later Texas A&M University) with plans to study veterinary medicine. After learning that there was a field called wildlife management, he switched majors. At that time, the university was an all-male institution with a required and substantial military training program. Thomas found that he enjoyed the discipline that is part of the military experience and intended to make it his career upon graduation, with a regular commission in the Air Force. But the need for new officers was declining about that time, and his entry into active duty was delayed several times. When the call did come, he failed the physical because of an old football injury. He therefore remained with the Texas Parks and Wildlife Department, where he had worked since graduation. His affection for and interest in the military remains, and from time to time in the journals that follow are phrases— "load my last clip, fix bayonets and charge"—that are reminiscent of his earlier training and goals.

He stayed with Texas Parks and Wildlife for ten years, but with his growing family, he eventually "starved out" and took advantage of an opportunity to move to Morgantown, West Virginia, in 1966 to become a research wildlife biologist with the U.S. Forest Service. While there, he earned a master's degree in wildlife biology at West Virginia University. Perhaps the most important lesson that he brought with him from his Texas experience was that wildlife management was "90 percent about people and 10 percent about animals"—animals and their habitats would be managed only to the extent that the people would support it. But education in wildlife management, he observed, focused primarily on animals: the managers had to learn about people on their own.

In 1969, Thomas moved to Massachusetts as principal research wildlife biologist assigned to a branch of the Forest Service's Northeastern Experiment Station, in Amherst. He also completed the Ph.D. program at the University of Massachusetts, where he specialized in forestry and land-use planning. This apparent shift in interest was really more a blending of his belief that effective wildlife management needed to be centered within a broader, human context, at a time that the Forest Service itself was beginning its slow shift along similar lines.

The Thomases were happy in Amherst, but in 1974 he accepted his "dream assignment" as chief research wildlife biologist in La Grande, Oregon, at a branch of the Pacific Northwest Forest and Range Experiment Station. Wildlife had been part of the range research program, but now the station director wanted to bring the study of wildlife management into its own. Thomas was seen as just the person who could bring the needed talent to the task. He very much enjoyed his new and expanded responsibilities, and his career and scientific prominence grew accordingly. Then in

1989, Forest Service Chief Dale Robertson and three other agency heads persuaded him to head an interagency committee to develop a management strategy for the northern spotted owl, whose primary habitat was the old-growth forests of the Pacific Northwest. This assignment, which he accepted with reluctance, would become central to profound change—for Thomas, for the Forest Service, and for forestry itself.

In June 1986, Thomas had begun keeping a journal, off and on. By the time of his spotted owl responsibilities, he was recording his thoughts more frequently. Later he jotted, "This will be a journal of 'random thoughts.' My purpose . . . is unknown to me, but I feel a compulsion to begin. Perhaps it will serve as a tickler of memory for the book I intend to write, but of course never will." It is history's good fortune not only that Thomas maintained a journal but also that he is a superb writer: he captures the moment with clarity, grace, and passion. By December 1997 he had filled 2,906 handwritten pages that printed out to nearly a thousand. The extract that follows ends in November 1996 and represents about 25 percent of the total, focusing on the spotted owl project and his time as chief of the Forest Service. The other 75 percent comprises mainly his wilderness and hunting essays, personal matters, and additional examples of his involvement with the owl issue and day-to-day events while chief. The "discarded" majority holds the potential of a full-sized book of his wilderness and wildlife entries, plus a modest volume on the spotted owl.

The extensive cutting has inevitably produced gaps in entry dates. However, some gaps already existed. For example, there is a thirteen-month lacuna following the May 14, 1992, entry, the period that included the election and inauguration of President Clinton, with all of its significance to Thomas's career. In the full journals there are only two entries from this period, and since neither is related to the shift in political leadership, they were cut. Thus, nothing crucial or relevant to his time as chief is missing from the narrative, and the reader can rest assured that every effort has been made to retain continuity.

Assistant Secretary of Agriculture James Lyons, who appears prominently throughout Thomas's journals, had traveled to La Grande to persuade him to accept an appointment as chief. Thomas hesitated in part because his wife, Meg, was dying of cancer, and moving to Washington, D.C., would be difficult for her. Lyons spent two days with the Thomases, both as an advocate and as a friend. Their association had begun many years earlier. When Lyons held a staff position with the Society of American Foresters, he had arranged for Thomas to participate on a committee to develop a policy for old-growth timber. The committee report would go beyond conventional projections of harvest schedules to note that old-growth provided a "special habitat, a unique attribute of diversity in the successional spectrum, and a thing of great beauty and even spiritual significance."

By the time of Thomas's spotted owl assignment, Lyons was chief of staff to the Democratic majority on the House Agriculture Committee. It was from that vantage that Lyons watched the Bush administration try to come to grips with the Thomas committee recommendations to preserve the owl, including a failed effort to oust Chief Robertson for apparently being too accepting of a parallel decline in the forest economy of the Pacific Northwest. A short time later, the Clinton administration took a new look at the controversy and, among other things, appointed Lyons assistant secretary of Agriculture with responsibility for the Forest Service.

During the presidential campaign, Clinton had promised a "timber summit" in Portland to find a way through the legal and political stalemate; by court order, logging on public lands within the range of the spotted owl had been suspended. Thomas made a presentation to President Clinton, Vice-President Gore, and assorted Cabinet members at that meeting. He warned them that the overriding objective of the national forests had evolved into a necessity to maintain biodiversity. Although he pleaded with the president to either confirm that reality as desirable or state a new focus, the mission of the Forest Service was never clarified.

Prior to the Portland summit, Lyons approached Thomas to assemble a team of scientists to prepare options to protect the owl but also to allow some logging to resume. Thomas reluctantly agreed and formed the Forest Ecosystem Management and Assessment Team (FEMAT), which produced an array of options for the president to consider; eventually Option 9 was selected. Option 9 as described by FEMAT was "morphed" by subsequent teams dealing with the requisite environmental impact statement and record of decision into something more resembling Option 1, which they called the "green dream."

Then, Robertson was out as chief, and Lyons began persuading Thomas to be his successor. Thomas remembers that it was Meg who made the decision and overruled his objections. She died shortly after they moved to Washington. His loneliness, coupled with the stresses of being chief, prompted Thomas to turn to writing even more extensively in his journals as therapy, and history is the beneficiary.

The Thomas journals portray a range of complex and controversial issues. For example, a pulp mill in Sitka, Alaska—a mainstay of the local economy—had operated under a fifty-year contract with the Forest Service to provide pulpwood. By the 1990s, the company faced expensive capital investment and had closed the mill but was studying the feasibility of retooling it. At the same time, the Forest Service saw that it was unlikely that it could produce the promised amount of wood over the long term because of environmental constraints that had appeared since the contract had been signed. In the meantime, another mill shut down for other reasons, terminating all employees and removing all its equipment. The two signatories faced off across the table and accused each other of contract default, setting the stage for litigation that continues at this writing. The journals lead us day by day through legal and political maneuvers by the Alaska congressional delegation, an indecisive secretary of Agriculture, and lawyers from the Department of Justice.

Another issue, and one that received wide media coverage and produced much rhetoric from interest groups, was the Emergency Salvage Timber Sale Program, a six-page addition to Public Law 104-19, commonly called the Rescission Act of July 27, 1995. Thomas uses both "salvage rider" and "rescission bill" to refer to the salvage program. This law provided the Forest Service with a degree of immunity from certain environmental constraints as it removed dead and dying trees from "unhealthy" forests. Thomas ordered that environmental laws continue to be obeyed. The means of removal would be via timber sales to the forest products industry. The stage was well set for controversy, and an often-exasperated Chief Thomas had to cope with full-frontal hyperbole by environmentalists and industry supporters.

Law enforcement had vexed Chief Robertson as well. Timber theft has been around as long as there have been trees and people who wanted them, but more recently the widespread growing of marijuana on national forests had made hard crime a fact of life. The conventional forest ranger was ill-prepared to deal with the situation, so the Forest Service created a Division of Law Enforcement, staffed by professionally trained law enforcement officers. For an agency committed by tradition and law to practicing multiple use—balancing recreation, timber, wildlife, and the myriad other values Americans place on their national forest lands—the "inflexibility" of the law officers brought internal conflict: cops wanted to report to and be evaluated by cops, not forest rangers. If an officer was seen as being too lenient on timber theft, a whistleblower would cry that the agency was condoning crime—a story the press loved. When Thomas, supported by both the Office of Inspector General and the FBI, disbanded an investigative team because it had turned up little new information, he was accordingly accused of a coverup. Then there were bombings and burnings of ranger stations with accompanying threat to life. At times like these, his past life as a scientist in La Grande must have looked good indeed.

Most former Forest Service chiefs recount few dealings with the Department of the Interior, except the occasional meetings to report to each other on related land management programs. In contrast, Thomas writes of a series of meetings with the secretary of the Interior, and more frequently with the director of the Bureau of Land Management. The latter relationship is less surprising in that the acting director of BLM was, like Thomas, a biologist and a long-term Forest Service employee, and they had a special empathy. Eventually Mike Dombeck would succeed Thomas as Forest Service chief.

Three issues that the Forest Service directly shared with BLM were the spotted owl, grazing on public lands, and management of the Columbia River basin. The Forest Service and BLM were moving ahead with the recommendation of the Interagency Scientific Committee report on the spotted owl, but BLM Director Cy Jamison insisted that his agency would find a better way—that is, achieve higher levels of timber harvest and protect spotted owl habitat. This was Thomas's first real experience with blatant subordination of scientific fact to political expediency, and it left a sour taste. But Jamison's efforts ended with the fiasco of a "God Squad" hearing, and he left with the Bush administration. After Dombeck was moved up for a long stint as acting director, the agency fell in line with an option derived from FEMAT's work.

President Clinton appointed Bruce Babbitt secretary of the Interior, and the new secretary quickly began the "War on the West," as it was dubbed in the press. Babbitt insisted that grazing cattle and sheep on the public lands required much more careful regulation. Thomas generally agreed, but the aggressive secretary seemed to be encroaching on the national forests, which were administered under the secretary of Agriculture. This conflict of authority is the topic of many journal entries.

Because some species of salmon were listed as threatened or endangered by the National Marine Fisheries Service, the agencies responsible for land management in the Columbia River's vast basin were obliged to comply with the Endangered Species Act requirement to protect habitat. The Forest Service and BLM are only two of the federal agencies with Columbia basin ownership, and once again Thomas and Dombeck, with teams of scientists and practitioners, worked within an ever-smaller window of opportunity to head off a "train wreck" that could make the conflict over the spotted owl pale in comparison.

Another controversy involved the New World Mine, on the Gallatin National Forest near Yellowstone National Park. The park superintendent, who as part of the National Park Service was an Interior employee, directly engaged the Forest Service—an Agriculture agency—in open conflict, a fairly untypical historical event. The mine had begun operation a century before and had lain inactive for many years. Recent shifts in the economy had prompted the new owners, a Canadian corporation, to request appropriate permits from the Gallatin National Forest to reopen it. Following both the law and standard procedures, the forest supervisor commissioned preparation of an environmental impact statement. At the same time, the Yellowstone superintendent began a vigorous campaign to discredit the Forest Service's efforts, claiming that the park itself would be placed in jeopardy.

Enter the White House. The political strategy was to buy out the mine— initially by trading off national forest lands—to show the administration's commitment to protecting the environment. Chief Thomas found himself ordered—he refused—to intervene in the impact statement process so that no evidence could appear that might derail the purchase, which had become a high-profile environmental cause for the administration. President Clinton himself participated in a ceremony at Yellowstone, where he announced the plan to purchase the New World Mine. Though disgusted that a national forest operation would be staged in a national park, Thomas admitted that the president's charm pretty much disarmed the disgruntled Forest Service attendees, who had been scapegoated for a bungled political operation and then relegated to being an audience for a Park Service show.

Thomas continually explained to a range of audiences, including congressional committees, that the Forest Service was hamstrung by conflicting and overlapping statutes. Added to that were court decisions that reflected plaintiffs' skill in selecting under which part of which statute to bring suit, a process certainly not geared to sorting out disparities of legal language. Too, the chief believed strongly that the regulatory agencies (the Environmental Protection Agency, the Fish and Wildlife Service, and the National Marine Fisheries Service) should in some way share the responsibilities of the management agencies (the Forest Service and the Bureau of Land Management) to achieve the desired level of stewardship over the land. Although he found sympathetic ears for his arguments, the situation remains the same.

Chief Thomas had been repeatedly promised that his status as a political appointee, necessary because he was not a member of the Senior Executive Service, would be remedied. But it was not, and it added to his desire to resign. Ironically, a decade earlier while he was at La Grande, Experiment Station Director Robert Buckman had tried to persuade Thomas to take the necessary training assignments in order to qualify for SES. Buckman insisted that Thomas was "one of the guys who could be chief of the Forest Service." Thomas refused: he did not want to be chief, and he believed that the SES produced managers when the Forest Service needed leaders.

There are some highly emotional accounts—ones that can still bring a tightening of the throat even after many readings—of the deaths of fourteen firefighters in Colorado. Thomas writes of the immediate aftermath and subsequent meetings with grieving families and transports the reader directly into the chief's shoes: you are there, you can see the people, and his sadness becomes your own.

Interspersed with the many other important issues described in the Thomas journals are descriptions of personal relationships. Often it is these relationships that determined outcomes more than the merits of an issue. The journals reveal how he interacted with his political bosses in the Department of Agriculture, and how he dealt with congressional partisans.

A branch of government that usually is in the background of Forest Service activities is the Department of Justice, which alone provides the agency its representation in court. In detail, Thomas shows us just how much legal strategies—totally at the discretion of the Department of Justice— have shaped agency policies. Budget hearings are another challenge, as we see when a member of the Alaska congressional delegation, angered by the pulp mill situation, is also on the Appropriations Committee. As Thomas noted in his journal, "Things simply don't work the way that students are taught in natural resources policy classes—not even close. And there is simply no way that scholars of the subject can understand the *ad hoc* processes that go on within only loosely defined boundaries." The reader will be able to watch and learn as, time and again, *ad hoc* processes yield the environmental policies of the Clinton administration's first term.

Although Thomas continued his journals after he was no longer with the Forest Service, and continues them to this day, the following extract ends with his last day as chief. After a series of retirement ceremonies during his final weeks, including the traditional praise and regrets for his leaving, the bureaucracy had the final say. "When the movers were gone, Jim Long came for my security debriefing. Then Sue Addington checked in my property, took my Forest Service key, my office key, my identification card, my government credit card, and my telephone credit card, and removed my access to the computer." A prosaic finish to a lustrous Forest Service career.

Many thanks are due, first and foremost, to Jack Ward Thomas for keeping a marvelous journal and then for making it available for publication. He also carefully read the manuscript and corrected errors of fact. Footnotes and italicized interpolations giving background information are his, added as the manuscript was prepared for press. A closely related project was an in-depth interview with Thomas, which along with many photos from his personal collection is deposited with the Forest History Society. At the Forest History Society itself, Steve Anderson, Cheryl Oakes, Michele Justice, and Carol Severance provided their typically high level of technical and moral support. Al Sample, president of the Pinchot Institute for Conservation, offered valuable insights into ways to improve the manuscript. Jerry Williams, national historian for the Forest Service, provided technical information. Eddie Brannon, director of Grey Towers National

Historic Landmark, brought the Thomas journals to my attention and strongly advocated their publication. Karl Perry, Chuck Meslow, Nancy Herbert, and Mike Ferris, all of the Forest Service, supplied photos. Sally Atwater cheerfully and efficiently brought editorial order to the text. And computer whiz Billie Plehn helped ward off the too-frequent hexes by cyber-gremlins, who show no compassion for innocent editors.

1990

By early 1986, it was becoming evident that the northern spotted owl was closely associated with old-growth forests of the Pacific Northwest, primarily west of the Cascade Range in Washington, Oregon, and northern California. The old-growth forests in private ownership had been largely logged, and such forests on the national forests and the Oregon and California Railroad lands managed by the Bureau of Land Management were being logged at a steady rate, causing fragmentation of the remaining old-growth.

Regulations governing the management of the national forests contained a phrase requiring that all native and desirable nonnative vertebrate species be maintained in a viable state and well distributed within planning areas. The intention was to preclude the necessity to list any such species under the requirements of the Endangered Species Act, whose purpose is the preservation of ecosystems on which threatened or endangered species depend.

The Forest Service made several attempts to come to grips with the issue. Since every one of those efforts projected some reductions in the annual timber cut, however, resistance from the timber industry, labor unions, and elected officials aborted them.

By 1989, the issue could no longer be avoided. Petitions for the U.S. Fish and Wildlife Service to list the northern spotted owl as threatened under the provisions of the Endangered Species Act were deemed imminent and likely to be successful. The agency heads of the Forest Service, U.S. Fish and Wildlife Service, Bureau of Land Management, and the National Park Service chartered the Interagency Scientific Committee to Address the Conservation of the Northern Spotted Owl. The charter for this committee—the ISC for short—was incorporated into law in Section 318 of Public Law 101-121 in

Northern spotted owl. USDA Forest Service photo, by Tom Iraci.

October 1989. The committee was directed to develop "a scientifically credible conservation strategy for the northern spotted owl."

The results of the ISC's work set off a chain of events that was to dramatically change the approaches to the management of the federal lands in the Pacific Northwest. As the senior research wildlife biologist for the Forest Service, I was appointed to select the committee and serve as team leader.
—JWT, 2003

10 April 1990, Portland, Oregon

This will be a journal of "random thoughts." My purpose in this journal is unknown to me, but I feel a compulsion to begin. Perhaps it will serve as a tickler of memory for the book I intend to write, but of course never will. I think, maybe, it is for my children and their children. Maybe it will let them know who I am (or was) and what was important to me.

I have just finished six months of work leading a team of sixteen scientists in developing a "scientifically credible" conservation strategy for the northern spotted owl. The work was focused, prolonged, and intense. Six months of gathering, synthesizing, and sifting evidence. Six months of rational thought. Six months of formulating and testing constructs and hypotheses. Six months of the conscious and methodical stifling of feeling and emotion and instinct.

In the last two months of the job there was no respite, no days off, nothing but work and sleep with more work and less sleep with the approach of the deadlines. When the end came, we felt that we had done a good job and had arrived at the only possible place that would satisfy our mission. We finished without exultation, but with exhaustion, and a profound feeling that something had changed forever for each of us, for the agencies that employ us, and for the management of the public lands.

In the last weeks of our deliberations someone hung a print over the coffeepot. It was a scene of a small girl and her dog proceeding down a path through the woods. She had taken a fork in the trail. The sign at the fork pointing in the direction she had taken said, "The Rest of Your Life," and the sign pointing the other way said, "No Longer an Option." Atop the signpost was an owl peering at the little girl and her dog. The picture was titled "Never Look Back." The night we finished the report at 3 A.M., I stared at that picture for over an hour with a strange sense of foreboding. My life would never be the same again.

What lies ahead is unknown but it will be different. What we have done will be praised by some (quietly, I suspect) and condemned (loudly, I suspect) by both environmentalists and industrialists. What we have done may well be the trigger that sets off the conservation debate from the last half of the twentieth century. The outcome will, quite likely, determine if there is any chance for sustaining biodiversity. If we cannot do so in North America, is there a chance anywhere?

The focus will be on me as the "Thomas Report"[1] is evaluated, praised, and castigated. This is most ironic, as I have no firsthand experience with spotted owls. But the issue is more than spotted owls and timber—it always has been—and down deep everybody knows it.

The faces of the thousands of people who will be hurt by this haunt my sleep and my private thoughts. A congressman asked me, sarcastically, dur-

---

1. Jack Ward Thomas, Eric D. Forsman, Joseph B. Lint, E. Charles Meslow, Barry R. Noon, and Jared Verner. 1990. *A conservation strategy for the northern spotted owl.* Portland, OR: USDA Forest Service, Bureau of Land Management, U.S. Fish and Wildlife Service, and National Park Service.

ing the hearings last week, "Have you ever thought about the people who will be hurt by what you have recommended?" I answered, "Every goddamned night from one to five in the morning. But that doesn't change the biology of the situation."

Most of the politicians and the agency administrators know that we have been overcutting our public lands for years, but they never wanted to face the music. Every decision was made with the purpose of forestalling the moment of truth. Now, through a bizarre path of law and regulation, the northern spotted owl and a team of obscure biologists have brought the festering truth to a head. The question is, Will we, collectively, acknowledge the disease and lance the boil, or ignore it all once again?

If we cast our eyes away, one more time, it cannot be for long. We are, simply and irrevocably, at the end of an era. The timber that makes up most of the remaining old-growth is not left by accident but by design. The best timber on the best sites at the lower elevations and on relatively flat ground with the most stable soils has already been logged. It is not an issue about spotted owls and timber, and it never has been.

Yet the environmentalists will base their arguments on spotted owls because there are regulations that require the Forest Service to maintain viable populations of vertebrates well distributed in the planning areas. The Endangered Species Act lies ahead in the trail like a cocked and ready grizzly bear trap, only thinly covered by pine needles. It is their best weapon.

Those who have much at stake in the continued cutting of old-growth at as near current rates as possible—both in industry and labor unions, and inside the Forest Service and the Bureau of Land Management—will also emphasize that the issue is spotted owl versus jobs and prosperity (for a little while, anyway). This ploy is attractive because it trivializes the issue and avoids the real issue of faulty projections of timber supply, and the *real* issues of overcutting and subsidy.

With the strain of the owl assignment mostly behind me, I called my wife last night. We have not been together for three weeks. Really, it has been longer than that, for her last two visits to Portland found me monomaniacally consumed with the final draft of our committee report on a conservation strategy for the northern spotted owl.

She asked me how things were going and I told her what I thought. She told me what she felt. She always does. I, like most scientists and natural resources management professionals, think too much and feel too little. She insisted that I feel and share and to let her feel and share with me. I couldn't. It is too frightening.

17 April 1990, [location unknown]

I am teaching a short-course for beginning wildlife biologists in the Forest Service. There is obviously a great deal of strain in the ranks as the Forest Service comes to grips with the recommended strategy for conservation of the northern spotted owl. One of the speakers, sensing this tension, asks a question that strikes me as insightful. "Is it possible to solve a value-driven problem through technical means?" If the "technical solution" does not correspond with the outcome desired by groups with values, would it be accepted merely because it is technically (even "scientifically") credible? The answer is no, and the immediate response will be to argue the technical merits of the solution and/or to attack the skill, integrity, or bias of those who have provided the technical analysis.

When the stakes are as high as they are in the case of the conservation of the northern spotted owl, extremists on both sides of the issue will attack with equal vigor and vehemence. These attacks will take on an aura of "scientific" objectivity and demands for a reevaluation and a committee to evaluate the work of the previous committee. There will be, in the meantime, a call for more research to clarify points that are less certain than would be desirable in a more perfect world.

I wonder, in my more cynical moments, if the real purpose for the hired guns on both sides is not the continuation of the conflict. For so long as the combat rages, there is a need for the hired guns, and more attention and resources flow to their legions. The lawyers on both sides sit on the sidelines like large cats licking their whiskers waiting for the chance to finish off the wounded. These gladiators, the hired guns and the lawyers, always win, regardless of the outcome. They always will so long as there are true believers to bring forth the resources to continue the skirmishes.

Is it possible, then, to provide technical solutions to value-driven problems? Perhaps, but only when the politicians are forced to take an issue unto themselves and make a decision. At that point, they may embrace the "scientifically credible solution" as a way out that relieves them, at least partially, of the responsibility by laying it at the feet of the scientists and the technicians.

The last day of our efforts to prepare a "scientifically credible" management strategy for the northern spotted owl finally came. We had struggled for six months and the end was at hand. Toward the end, there was over one straight month of effort, and the days became longer and longer as the deadline approached. During the last three days, each workday stretched to twenty hours as team members struggled with their individual assignments. Then, suddenly, on the last morning, we were finished. There was no feeling of exultation—no feeling at all. Rather, numbness set in. We left the office to walk to breakfast and David Wilcove looked up. "The swallows are back," he said. "When we started, they were just leaving."

Where had the six months gone? The six months ran together in my mind in a jumble. Six months in which I had been home only seven days. Six months in a city of concrete and glass and people everywhere. Six months in exile from the Blue Mountains. Now the swallows were back and we were free of our burden. It felt good.

But then came the chilling realization that now the consequences of our action would descend upon us and follow us all the rest of our days. We had done our job and we felt we had done it well. Yet what we had done signified the end of an era, and with the end of that era would come economic dislocation and human suffering.

Watching the swallows moving north, my feelings change, suddenly, to a feeling of sadness. Our team's time together has come to an end. We will break apart today and the magic of our bonding will dissipate. But it will always remain in my memory.

We have worked so long and so hard and so close for so long. Each team member contributed the best he had to offer. Each subordinated ego to the team effort. No one lost his temper or patience under the strain— professionalism personified.

As I walked behind the group, lines from *Henry V* came to mind. For a time we have indeed been "we few, we happy few, we band of brothers." No matter what happens now, that will have made it all worthwhile.

The swallows are back! I quicken my pace to catch up with my brothers. I am content.

23 May 1990, Washington, D.C.

I testified today before the Senate Public Lands Subcommittee as part of a panel that included John Turner, director of the Fish and Wildlife Service; Cyrus "Cy" Jamison, director of the Bureau of Land Management; Dale Robertson, chief of the U.S. Forest Service; and Doug Houston, research biologist, National Park Service. The subject was the "Jack Ward Thomas Report," its credibility and social and economic impacts.

The chairman, Dale Bumpers of Arkansas, called for opening statements and recognized Malcolm Wallop of Wyoming. The good senator launched into a vituperative tirade that lasted some ten minutes. He insulted me personally; attacked the "science" of the Interagency Scientific Committee report; suggested that all the members of the ISC be fired for insubordination in refusing to follow the charter given us to consider an array of alternatives; accused us of being rabid environmentalists bent on the destruction of the American way of life; plus some other selected slurs and insults.

He was followed by Senator John McClure of Idaho, who said the same things but in a slightly milder tone. Then, Wallop and McClure left the room without having either the nerve or the class to participate in the give-and-take of questioning.

My impression was that most of the people in the hearing room were embarrassed by the performance. If Wallop and McClure thought they were helping the timber industry position, I think they were sadly mistaken. I have never seen such a tasteless, overbearing, and bullying performance in my life.

Halfway through Wallop's spitting up of bile, I felt my shock and anger leave me and a feeling of pure amusement set in. Every organization needs a clown prince and Senator Wallop certainly fills that role. The really amusing part is that I don't think he means to be funny and does not know he is well recognized as a clown and an embarrassment to the Senate.

Senator Timothy Wirth of Colorado made a statement thanking the ISC team for its work, and recognizing the quality of that work, and took his leave. Senator Mark Hatfield of Oregon also thanked us for our work and described the magnitude of the social and economic consequences of the report. His dignity and grace shone like a bright light compared with the darkness of manner of Wallop and McClure. Here was a senator who is a player. A senator from one of the most affected states, who still had the style and class to behave as a gentlemen and as a senator. I was proud that he was the senator from my state, whatever his opinion and wherever he comes down on the spotted owl issue.

Senator Bumpers opened the questioning and seemed to have little idea where he was going, except to make the point several times that the projected impacts on employment were exaggerated by adding the potential effects of the Thomas Report to losses already anticipated under the preferred alternatives in the forest plans. He turned the hearing over to Senator Hatfield and left the room.

Senator Brock Adams of Washington, who was not a member of the committee, had sat in for a while and then left. Now we were down to the only real player. Hatfield was alone behind the elaborate curved dais that towered above the witness table and audience. I suspect there is some method in that design. I suppose it could be intimidating, as it is intended to be.

Hatfield began to ask questions that had been provided, by invitation, by the timber industry. The questions had been given to me the day before, and I had had time to consult with my colleagues on the answers. We went through only six questions, and he shifted to quizzing the agency heads about process, anticipated action and impacts, and addressed no further questions to me. In his closing remarks, he made a poignant speech about the people who were going to be hurt as the timber supply from federal lands declined. He pleaded that this reduction be carried out in a way that caused the least human suffering.

After the hearing closed, he walked around the dais to visit for a few minutes. I asked him why he had not continued down the list of questions prepared by industry. He said to do so would have been pointless. It was obvious that no answers would emerge that would weaken the credibility of our report. And so my testimony ended.

The heads of the Fish and Wildlife Service, BLM, and the Forest Service had spent a considerable amount of time maneuvering to put the responsibility in the pocket of the Fish and Wildlife Service director by saying if the subspecies is listed as threatened by the Fish and Wildlife Service, then the agency "has the lead" to prepare an agreeable "recovery plan." Turner did not flinch, saying, in effect, that they would carry out the law to protect fish and wildlife and that the Thomas plan would be a good starting point for a recovery plan. Then Robertson spoke up, saying that he agreed and reminding the committee that the National Forest Management Act was a "pretty tough act" and that the FS has to come up with a new strategy for spotted owl management. And it all keeps coming back to the strategy developed by the ISC team.

For the first time it really came home to me that our plan had a very significant chance of being adopted. The chief was actually going to take the lead. Good—at last. But it came to me equally quickly that it may cost him, and maybe me, our jobs.

Jamison frankly announced that BLM would not follow the strategy proposed; they would find a "better way" (I think that means a way to hold timber harvest at present levels). Then, of course, they will leave it up to the Fish and Wildlife Service to decide if that better way is acceptable. It will not be. But Cy will be able to say he tried and was overruled by the Fish and Wildlife Service. Nice political move. Not very courageous, but a nice move.

Today, I saw the best and the worst in people: senators who behaved like thugs and senators with style and grace, agency heads dedicated to obeying the letter and spirit of the law, and an agency head dedicated to evading the spirit of the law. I cannot speak to his motivations.

Friday, [nd] June 1990, La Grande, Oregon

When my secretary laid this day's mail on my desk, she, knowing my habit of ignoring mail until the "right time," pointed out one letter. It was a letter signed by Senator Dale Bumpers in behalf of the Public Lands Subcommittee, asking for detailed answers to 176 questions concerning the ISC's proposed conservation strategy to be added "to the record" of the recent hearing before the Senate. While I was pondering the questions, she came back into the office and stated, "They just said on the radio that the Fish and Wildlife Service will announce the decision on whether to list the northern spotted owl as threatened early next week." It seems there is little doubt that the owl will be so designated, and that will again shift public and political attention to the question of what the required recovery plan will be. That, in turn, means the media will again devote attention, intense attention, to the ISC report.

Friday, [nd] June 1990, Washington, D.C.

The Bush administration, having been sold a bill of goods by Senator Slade Gorton of Washington, announced that a Cabinet-level task force would reexamine the "spotted owl situation" and derive a "balanced" solution that will save the spotted owl and cause less economic and social disruption than would result from adoption of the Thomas Report. Of course, the people on that task force are far too important as Cabinet officers and other such prestigious folk to have time, or patience, or expertise, to deal with the situation. Therefore, there is a second-line team of lesser political appointees assigned to do the job. It is our (Chuck Meslow, Barry Noon, and me) job to brief the second team on the ISC report and deal with their questions as they try to get sorted out where they are going. The second team is chaired by Jim Moseley, the brand-new assistant secretary of Agriculture, who supervises the Forest Service and the Soil Conservation Service. As I finish my briefing and begin to answer questions, it becomes very obvious very fast that these folks haven't got a clue. We would feel sorry for them if the situation were not (simultaneously) so funny, so ridiculous, and so serious in terms of political ramifications. It is almost impossible to take this seriously. Yet we dare not do otherwise.

It is easy to sense that these players have already learned that they are engaged in a huge political mistake, a blunder of significant proportions. It is a classic lose-lose situation. There are no answers that will not make enemies in all directions. The professionals, the bureaucrats, and the eggheads were beautifully positioned to take all the heat, and the White House stepped in.

After the briefing, you could see it in the eyes of the brighter group members that they realized two things: (1) We were not the "government hacks" they had been led by industry to expect; and (2) in the end, if our committee does not support their solution, it will go to court, and it will not stand.

The second team obviously contains, as a primary component, lawyers and economists, whose minds run in very different channels than scientists'. The lawyers look for ways to slide around the Endangered Species Act and the economists want to do rigorous risk analyses to tweak the ISC report and judge the gain in timber supply against increased risk for the owl.

You can watch their jaw muscles twitch as they see their dreams of such a neat, impeccable process slip away as a real alternative. Their faces fall even further when it sinks in that the ISC report is already a compromise that takes the unprecedented course of prescribing a strategy that will give up the habitat of 50 percent or more of the known population at the time of listing, and that any strategy with less impact is apt to increase the loss even further, and perhaps dramatically.

After I finished my briefing, Noon began a presentation on why we cannot perform the rigorous risk analysis that they had wished for. Chief Robertson and the man from the White House (who nobody is supposed to know is from the White House) moved from where they were to chairs behind me so that they could see the screen on which Barry was showing slides. He leaned over to the chief and whispered, "We are on the wrong side of this issue." Ten minutes later, he whispered again, "These guys are heavy hitters; they know their stuff."

Moseley, as chairman, interrupted Barry and said we would have to continue at another time. The task force is going to the Pacific Northwest tomorrow to get a firsthand look, to talk to industry representatives, workers, environmentalists, and politicians. Now, that ought to be extremely informative. At least the responses will be totally predictable and they won't have to absorb anything they don't already know. When in doubt, form another committee to review the committee.

The field trip will take a week. Then, several members of the task force (second string) must be on vacation for ten days. That will take them to the first of August. They then have three weeks to prepare their final report.

It suddenly dawns on them that, perhaps, the ISC is essential to the process, and Moseley says that we should be prepared to be back in Washington the first two weeks in August. No certain dates, but be on standby.

One of the lawyers resumes his questions about why we can't just raise owls in captivity and put them where we want them. We tell him that is not the problem and won't be if we don't rip up the habitat that is left until it becomes a problem.

Well, he wants to know, what's the problem if we can solve the problem we create by raising and transporting owls? I explain that such is not in keeping with the Endangered Species Act, not to mention the National Forest Management Act. He says that we are assuming that we are confined by law when they may be able to change the law to suit the situation.

The Interagency Scientific Committee in the chief's conference room, Washington, D.C. From left: Eric D. Forsman, E. Charles Meslow, Barry R. Noon, Jared Verner, Jack Ward Thomas, Joseph B. Lint, July 1990. USDA Forest Service photo, courtesy of Nancy G. Herbert.

Well, we say, it is unproven technology, and we can't be certain that it would work. In fact, when the survival rate for juvenile owls in the Roseburg District of BLM is 0.20, and the survival rate of subadults is 0.50 and "tame owls" survived at the same rate (which is hugely optimistic), only 10 of every 100 released birds would survive to be adults. Next, the birds, to have any chance, have to be "hacked" on site. This species is not independent upon fledgling: they are fed by the parents until September. This means the hack has to cover four to four and a half months. There is no telling where the birds will go when they disperse.

Well, the lawyer says, "We can just move owls."

It is time for us to leave and we are glad to do so. This ought to be interesting as it all plays out over the next few weeks.

[nd], Washington, D.C.

This time it is testimony before Congressman Bruce Vento's National Parks and Public Lands Subcommittee on the merits of several (five, to be exact) bills concerning resolution of the old-growth/spotted owl issue. There are several panels to be heard in today's hearings. Panel one is composed of Dale Robertson, chief of the Forest Service; Cy Jamison, director of BLM; and a representative of F&WS. Their statements are throwaways. The questions are what everyone is waiting for. Congressman DeFazio goes after Robertson—I think more to be antagonistic than anything else—and ends up pushing his bill demanding alternatives to the Thomas Report. I don't know if he knows his bill is not going anywhere and he is just posturing, or he really thinks that there are alternatives that will stand up to scientific scrutiny and court testing and will cost less in terms of timber supply. There are alternatives that will stand up to testing, but they will all cost more than the ISC strategy. And there are alternatives that will cost fewer jobs but won't stand scrutiny.

If industry does not quit picking at the ISC report, they may reap the whirlwind and get protection of every known owl and every newly discovered pair. I don't think the environmentalists will stand still for any further reduction in owl numbers, and they shouldn't. The environmentalists will go with the ISC report, at least the mainstream will, and it will stand up in court. I don't think they will go along with anything ISC does not bless. It is the best deal they can come up with and they had best take it while it is still on the table. Delay and more research will only make things worse for their cause. That is the history of the situation and it is likely to continue.

Jerry Franklin, the top-ranked Forest Service forest ecologist, and I are next to testify. Jerry has an opening statement that emphasizes the need to think beyond old-growth and spotted owls to broader concerns of ecosystems and biodiversity. I have no prepared testimony and announce that I am present at the request of the chairman to answer any questions that the committee might have. There are several questions that give the dynamic duo the chance to plug old-growth as an ecosystem and to proclaim the issue as being more than spotted owls and old-growth.

Franklin is given a question about "new forestry," and he suddenly goes too far and promises too much. He expresses his opinion that silviculturists can produce the forest condition that will produce owls and whatever other creatures may be associated with old-growth. Jerry cautions that we don't know how this will influence the amount of timber available. As he talks, I watch the eyes of the committee; they are willing to buy anything that may give them a reprieve from the hot seat of having to face up to lower cuts. "New forestry"—whatever that is—may well cost more than the habitat conservation areas and the 50-11-40 rule.[2] But they are willing to grasp that seed of hope without analysis and without even knowing what new forestry is. It is amazing—but perhaps no more amazing than the willingness to embrace the Jamison Plan put forward by the BLM director without knowing what the plan is or what its long-term consequences are likely to be. The only thing they do know is that they won't have to face reality for another two years. As Scarlett O'Hara so aptly said when faced with a tough situation, "I won't think about that now. I'll think about that tomorrow."

[nd], airborne, en route from Washington, D.C., to Oregon

I have just finished two weeks in Washington, D.C., with the task force working group and am on the way back to Oregon. These flights are some of the rare times for introspection. The two weeks opened with yet another review of our committee's work. This time, four more wildlife scientists have been commissioned to review the ISC's "Conservation of the Northern Spotted Owl" report. I am allowed to sit in as Drs. James G. Teer, director of the Welder Wildlife Foundation; Richard Lancia, professor, North Carolina State University and editor of the *Journal of Wildlife Management;* Malcolm Hunter, professor, University of Maine; and James Karr, professor, Virginia Tech, go through their critique. Unanimously they give the report high marks and caution the working group about further compromises. It is clear to them, and they make it clear to the working group, that the compromises with the welfare of a threatened species made by the ISC are unprecedented in giving up perhaps 50 percent or more of the population, and no further "balance" can be either justified or tolerated.

Several members of the working group come at the reviewers with ques-

---

2. Fifty percent of the area between habitat conservation areas (HCAs) consists of trees with an average diameter at breast height (dbh) of 11 inches, with a 40 percent canopy closure. HCAs were areas designated by ISC for protection of extant old-growth and for development of old-growth over time.—*JWT*

tions aimed at "weaknesses" in the ISC report. Each attack provides the reviewers opportunity to make the report look even better.

At this point, I am ready to turn over the ISC's answers to the 176 questions presented to the ISC by the Senate Subcommittee on Public Lands for written response. Those looking for weaknesses in the report through this mechanism will be sorely disappointed. In fact, these questions, so carefully prepared by lawyers and hired guns for the timber industry, are what the ISC began to think of as batting practice pitches delivered at half speed and right down the middle, giving the batter a chance to hit them out of the park. We did.

Just when we were ready to leave, Assistant Secretary Moseley handed me a list of questions and answers prepared by the Northwest Timber Association and given to the working group. Moseley asked me to respond to these comments as soon as possible and, when these answers were prepared, the remainder of the core team could return home. I was to stay for the next week.

The next day, the core team prepared the responses to the materials presented by the Northwest Timber Association. NTA obviously never expected the ISC to get a chance to respond to these comments. Not only are the statements easy to refute, they give us a chance to reveal this particular set of hired guns for exactly what they are. The questions and answers are easy to brand as obfuscatory at best and outright lies at worst. But all in all, the ability to question the ISC is an advantage to them. It may produce an inconsistent answer, some amount of disagreement, a flare of temper, and an intemperate response that reveals "unscientific bias" on the part of even one member of the ISC. But so far at least, the constant bombardment of questions and accusations by the hired guns has worked against them.

It is obvious that the industry types are outlobbying the environmental side, which has presented no questions. Each day we run into Mark Rey or some other hired gun from industry in the USDA building or the Forest Service building or the halls of Congress. I fear that in the end, all of this persistent effort will pay off and we will lose the war, having won every battle of that war. It seems that the environmental side simply cannot bring itself to support a bunch of government biologists. Besides, they have their own agendas and their own hired guns. I suspect sometimes that hired guns savor the combat more than they desire victory. Industry's hired guns survive by saving money for their keepers. The environmentalist's hired guns are not as well served by small victories as they are by glorious defeats in courageous battles—in the mode of midgets versus Titans.

The core team, having provided the answers to NTA's statements, departs the Washington scene to make some attempt at dealing with their real jobs—doing research. We all long for that.

[nd], La Grande

As the days have gone by since my return from Washington, the bits and pieces of news that come to me from Rob Walcott of the policy shop of the Council on Environmental Quality and Deputy Chief James Overbay of the Forest Service give every reason to hope for a decision essentially in line with the ISC recommendations. The two representatives on the working group from Interior continue to hold out for a higher cut and creation of the biggest possible "train wreck," thereby forcing a jeopardy opinion at some point by F&WS over some decision by FS or BLM, and creating a circumstance where the God Squad[3] can be called into action. This scenario assumes the God Squad will put the economy (code word "jobs") above the welfare of a threatened species. This view, however, is much in the minority position. The Interior Department representatives are likely afraid that a decision in favor of the ISC strategy for the national forests will leave BLM hanging out all by themselves with the Jamison Plan.

If that were to occur, BLM Director Cy Jamison and, in turn, Interior Secretary Lujan will have gone from heroes to goats, having made themselves and the administration look both weak on the environment and politically inept at the same time—a double screwup. Cy may end up not being governor of Montana after all.

The working group, evidently, is still struggling with finding some way to embrace the Thomas Report but depart from it enough to avoid appearing to have given in to a bunch of biologists.

Things are looking good enough that it seems an appropriate time to pack up the horses and head into the Eagle Cap Wilderness. That seems a good place to be when the task force makes its announcement. Once the

---

3. Slang for a committee authorized by the Endangered Species Act to be convened when it is necessary to determine whether actions required to save a threatened or endangered species have too high a social or economic cost.—*Ed.*

announcement is made, the press will be on me like a kestrel on a June bug for my comments on this or my reaction to that. The press is clever at weaseling out comments you never meant to make. And a flub at this stage could undo our carefully maintained role of dispassionate scientists. In order to cover all the bases, I sent a computer message to the ISC suggesting they make the following comment: "Our committee said in the final report that we have proposed and now it is up to others, elected and appointed officials, to dispose. They have now done their job as we have done ours. It was a tough decision. How that decision will fare is up to Congress and the courts. If there are other questions, they should be put to the chairman of the ISC."

30 August 1990, [in camp, Eagle Cap Wilderness Area, Oregon]

On the eve of my departure for the Eagle Caps, I was working late in my office clearing my desk and leaving instructions for work to be done in my absence. At 7:00 P.M. Pacific time (10:00 P.M. Eastern time), Undersecretary Jim Moseley called. I could tell from his voice, which was tired and dispirited, that the call was not good news. Moseley wanted me to hear the news from him and not from some newshound with the freshly leaked story.

The working group has recommended to Secretary of Agriculture Clayton Yeutter to go with the ISC report with an allowable cut of 3.0 billion board feet [bbf] for Forest Service lands in 1991, scaling down to 2.7 to 2.6 in subsequent years. Secretary Yeutter was in agreement. However, when they took it to the White House, something (Moseley doesn't know what) went wrong. The president's chief of staff, John Sununu, was in the Soviet Union giving advice on how to organize the Soviet premier's office. So Sununu (who has been perceived as the big, bad "booger" who will eat everybody alive if they go with the ISC report) was not present. Interior Secretary Lujan was evidently the stumbling block, saying things like "no bunch of biologists are going to determine policy for the United States government." That is understandable—he will look very bad if the ISC report is adopted now, after he let Jamison convince him that there was "new or better science" or "other experts" who had devised a "better way." To adopt the ISC report now is to have to eat those words, and he simply doesn't have the stomach for it.

It now looks as if the train wreck proponents have carried the day. The timber cut level being proposed is an annual cut of 3.7 to 4.2 bbf and you sacrifice the number of habitat conservation areas necessary to hold the cut level. They will ask Congress for "sufficiency language" to preclude the environmentalists from challenging the decision in the courts. That will, if Congress approves, have the effect of declaring that whatever is prescribed will, de facto, provide adequate protection to the spotted owl and that is that—problem solved.

I had listened quietly to this point and now I began to speak quietly and calmly though my chest was tight. I said that we would expect to see our committee in front of Congress within ten days and there was no way we could support that decision. At that point, we would have to follow the dictates of our profession, which would lead us into direct conflict with the administration. Truly, we must say in the manner of Martin Luther, *"Heir stehe Ich. Ich kannicht auder."*

Thursday, 6 September 1990, airborne, en route from La Grande to Washington, D.C.

I am off to a meeting in Salt Lake City on elk-cattle competition. I have no real desire to go, but in today's competitive climate for research dollars, I dare not be absent. On the other hand, it will be good to talk elk and let spotted owls take care of themselves for a while. Also I am looking forward to seeing colleagues I have not seen for some time.

As I am strolling across the Salt Lake City airport, there is a most unwelcome voice calling "Jack Ward Thomas to a white paging telephone." My secretary is on the line. "Please call Assistant Secretary Moseley as soon as possible" is the message.

Moseley says that Secretary Yeutter has met with White House Chief of Staff Sununu about the spotted owl decision. Things have changed since last Thursday. It now looks as if the task force will buy the ISC plan with just enough modification to claim a change in the plan. But Secretary Yeutter has a few ideas for modification to the ISC strategy that he wants to discuss with me to see what the consequences are and whether the ISC can live with those changes. Moseley asks if I can be in Washington in the morning. The request is very polite but it is an unmistakable order. Of course, I will be in Washington as soon as I can.

When I call my secretary back, the arrangements have already been made and tickets await me at the Delta counter. The connection time is short and there is no time to call Moseley and confirm that I am on the way. But once airborne over Utah, I place a call and let his secretary know. She tells me to stand by while she places a call to him on his car phone to let him know we are scheduled to meet in the morning.

As I walk back to my seat on Delta Flight 931 at 500 miles per hour at 31,000 feet above Utah on the way to Washington, D.C., I wonder to myself if it is possible that only yesterday I was traveling over Burger Pass in the Eagle Cap Wilderness and seeing very little to distinguish now and 1885. Somehow, I think I might have liked 1885 better.

Friday, 7 September 1990, Washington, D.C.

I am in Chief Robertson's office at 7:30 A.M. He tells me that we are expected at Assistant Secretary Moseley's office at 8:30 A.M. and things are apt to be a bit more hectic than expected. There is a "leaked" story in the *Washington Post* this morning that quotes a "high administration official" as saying the task force has "reluctantly agreed" to go with the ISC strategy, as there was no rational alternative between the requirements of the Endangered Species Act and the ISC report. The headline trumpets that the decision has been made and that 20,000 jobs will be lost in the timber industry and associated small towns by the year 2000.

Then Robertson is on the phone with Moseley, and I can tell he is not a happy camper. When he hangs up, he says that presidential Chief of Staff Sununu is "furious about the leak and is blaming the Forest Service." Robertson is concerned and a little hot at the same time. It is obvious that the leak did not come from the Forest Service, as there are details there that even Robertson did not know. Now, Deputy Chief Jim Overbay and Associate Chief George Leonard arrive in the office, and a discussion begins as to who leaked the story and what the motivation of the leaker was.

Robertson and I set out for Moseley's office. The chief is tall and long and lean and walks fast in any case. Now he is agitated and walks even faster. I am almost between a fast walk and a trot to keep up. I give him a quote from my favorite one-liner philosopher, Satchel Paige: "Never hurry, it jangles the nerves." He doesn't smile and he doesn't slow down, either.

Moseley is waiting for us in front of his secretary's desk. We shake hands and he turns immediately to Robertson and says, "We've got a firestorm. Sununu is livid. Hatfield feels like he has been double-crossed. Gorton is hot."

Robertson assured Moseley that the leak did not come from the Forest Service, and Moseley assured him that he believed that and Agriculture Secretary Yeutter believed that but Sununu was convinced otherwise. We left immediately for Secretary Yeutter's office. Yeutter, in contrast to others I had seen this morning, was calm and unflustered and went straight to the point.

He briefed me on what had transpired over the past several days. Last Thursday, things looked very bad for the ISC report. Secretary Lujan is hanging tough that the working group must come up with an alternative to the ISC strategy that will save the owl and minimize the job losses. Unless the working group, and in turn the task force, does that, they have failed in their assignment. He strongly praised the innovativeness of BLM Director Cy Jamison and recommended the Jamison Plan as an alternative. Lujan attacked me, wanting to know "what makes Thomas the assigned guru of spotted owls, anyway?" He wanted a cut of 3.5 to 4.0 bbf.

Then, sometime over the weekend, things changed again. Yeutter thinks this was related to Senator Hatfield's agreeing to a cut level of 3.0 bbf in fiscal year 1991. At any rate, Sununu seems ready to accept the ISC strategy in principle but wants to squeeze a few hundred thousand more board feet to perhaps 3.4 bbf or more for fiscal year 1991.

After explaining some technical points, I then gave the secretary the ISC's bottom line. "Mr. Secretary, the ISC cannot accept any weakening in the proposal that we have put forward. We have already taken the unprecedented step of accepting a probable reduction of 40 percent or more in numbers of a species officially listed by F&WS as threatened. I believe that the majority of the concerned scientific community and the mainline environmental groups will accept the ISC strategy because it is sound in concept and because of the reputation of the scientists that put the strategy forward. We have walked a tight line. What we have recommended approaches the absolute limit of what we consider necessary for owls to have a good chance of survival over a hundred-year period. We simply cannot and will not go any further. Our professional ethics and the standards of our profession will not allow it."

To his everlasting credit, Secretary Yeutter said, "Above all else, you and your committee must remain loyal to your personal honor and your professional ethics. I expect that and will support that. You and your committee have my respect. You have earned that."

Yeutter, Moseley, and Robertson agreed that I could go home this afternoon if I could arrange a flight. However, when I arrived at my hotel to check out, there was a message that I should return to Overbay's office and stand by if I were needed. Secretary Yeutter is scheduled to visit with Sununu at 4:00 P.M. The meeting, in the end, was delayed until next week.

Watching politicians and political appointees make forestry policy is not a pretty sight, nor does it instill confidence. It may well be that it is similar to the old line about making sausage: You will enjoy it more if you don't watch it being made.

[nd], La Grande

Don Knowles, deputy undersecretary of the Interior, called me today to inform me that Marvin Plenert, regional director of F&WS, will head the recovery team for the northern spotted owl. Knowles will serve full-time as the team coordinator and Interior Secretary Lujan's "policy adviser" to the recovery team. The governors of Washington, Oregon, and California have been asked to submit nominees. Similar requests have been made of undersecretaries of Interior and Agriculture and agency heads of the Forest Service, Park Service, Bureau of Land Management, and Fish and Wildlife Service. It seems likely that many—probably most—of the team members will not be biologists because it will be possible for the team to consider "economic and social factors" that could not, by law, be considered in the decision to list the owl as an officially recognized threatened species.

The announced time frame is to have a draft of the recovery plan by December 1991. That will then be subject to review and revision, then presented in public hearings, then revised, and adopted by December 1992. Of course, the timber sales put forward in 1991 and 1992 will be allowed to go forward, which means business as usual through 1993 and the 1992 elections will be safely past. If the issue is kept alive that long, it will be possible to wring many millions in additional campaign contributions from industry and labor groups before the final decision.

I told Knowles that I sincerely hoped that they would be able to come up with a "plan at least as equally scientifically credible as the ISC plan" (their stated goal). But I added that I did not think that possible. If the outcome was something the ISC could not attest to, it seemed likely that it would fail when challenged in court.

Knowles seemed to agree, but I don't think he cares. The operative political purpose is to draw things out, get by the election, pay off the politicians' debts (with public money and timber), and then come up with a plan that will not work and which will be challenged in court. That could draw out the process another one to two years—i.e., through 1994–1995—and then, after being overruled in court, the outcome can be blamed on the scientists and the courts.

I can't believe a recovery plan essentially formed under the direction of a nonbiologist political appointee and staffed with handpicked folks can have any credibility. The process will be interesting to watch.

10 September 1990, Kahneeta Resort,
Warm Springs Indian Reservation, Oregon

Finally, a big enough issue will arise, often from very obscure situations, which will force a quantum change in the agency's management. With that change will come weakening of old alliances with established constituencies and strengthening of new alliances.

If this does not occur at exactly the opportune moment, it's professional death to the leaders who suffer premature cognition, or worse yet, who embrace new missions and constituencies prematurely, and professional death to those who cling too long to the past. Now, I believe, is a magic moment when the stars of politics, public opinion, and circumstances make change possible. I believe that the chief of the Forest Service knows the moment is at hand. He has come to the moment reluctantly, and justifiably enough is afraid. First, he must survive for his own sake and for the sake of the Forest Service. He must, above all, hand off the agency to the next in the line directly descended from Gifford Pinchot. Right or wrong, that is his first duty.

He serves under a conservative president with even more conservative operatives with an orientation to economic criteria for judging success. He cannot lead from the front—except in a very strange way. He can, and is, allowing more leadership from below. The establishment of the ISC was such a move; the establishment of the "New Perspectives in Forestry" is another. He seems to embrace the magic gingerly, as a hungry coyote sniffs a porcupine. Perhaps he does as much as he can—perhaps as much as anyone can do and survive. And survive he must, in my opinion. My experience with spotted owl issues for the past nine months has given me the opportunity to view the performance of a political appointee as head of a natural resource management agency. Watching Cy Jamison play the game so as to please superiors and enhance his own political stature has led me to appreciate the necessity of maintaining a professional natural resources person as Forest Service head. There must be dedication first to land stewardship, and to personal ambition at some lesser level. It is a fine line, indeed.

*After the God Squad's pronouncement ended BLM Director Jamison's
attempts to evade Judge Dwyer's injunction on further timber sales from
old-growth stands within the range of the northern spotted owl, the chairman of the House Committee on Agriculture, Kiki de la Garza of Texas,
sought a solution through congressional action. Working through the committee's chief of staff, James Lyons, de la Garza appointed a team to develop*

The Gang-of-Four at a Yale Forest Forum on the Northwest Forest Plan, October 23, 2001. From left: Norm Johnson, John Gordon, Jerry Franklin, Jack Ward Thomas, and Jim Sedell (one of the two additional members of the panel). Photo courtesy of Global Institute of Sustainable Forestry.

an array of management options from which Congress might choose to
impose a legislated solution to the impasse. This team was charged with
developing an array of alternatives and deriving estimates of the impact
of each option on timber yield and associated jobs as well as ecological
viability.

Officially, the team was the Scientific Panel on Late-Successional Forest
Ecosystems, but spokesmen for the forest industry quickly dubbed it the
"Gang-of-Four," after a group of dissident communists. The name stuck.
Team members were K. Norman Johnson, forester-economist, Oregon State
University; Jerry Franklin, forest ecologist, University of Washington; John
Gordon, forester, Yale; and I.

As the team was leaving an initial meeting with the congressional committee, Congressman Volkmer of Missouri called out, "Don't let us get
surprised by some damn fish!" James Sedell and Gordon Reeves, research
fisheries biologists from the Forest Service, were added to the team to ensure
compliance with Volkmer's concerns. Although the team referred to themselves as the "Gang-of-Four Plus Two," that name did not catch on.

No viable options to the general scheme that had underlain the strategy
laid out by the ISC emerged, but the Gang-of-Four nevertheless delivered
a broad array of alternatives to the Agriculture Committee for its consideration. Hearings were held. Sensing the political volatility of the issue, the
committee forwarded no legislative proposals. It, too, had decided to defer
to the upcoming presidential election.

—JWT, 2003

1991

Monday, [nd] October 1991, airborne,
en route from La Grande to Washington, D.C.

As I write this, I am on my way to testify before a joint session of sub-committees of the House Committees on Agriculture and Interior. The subject will be the delivery of the report by the "Gang-of-Four" (Norman Johnson, Oregon State; Jerry Franklin, University of Washington; John Gordon, Yale; and me) on alternatives for management of late-successional forests in the Pacific Northwest. I do not look forward to the next few days. The timber industry and its allies seem to be becoming more and more desperate as they see their power and their chances of holding on to sub-stantial timber cuts from the public lands decline.

Their strategy seems to be changing, from one of debate over facts intended to obfuscate the issue to one of an effort to vilify and intimidate any scientist who dares to surface ideas counter to their aims and welfare. The latest instance of this attack is a report from a Portland law firm enti-tled "A Façade of Science." This report was fabricated by careful selection of bits and pieces of testimony given by members of the ISC in depositions related to two court cases. The ISC team members could never understand why we were forced into giving depositions related to court cases in which there was no intention to call us as witnesses. Now we know the reason.

At first, only selected bits and pieces of carefully selected testimony were quoted out of context in news releases and letters to the law firm's clients. We assumed the purpose was to try to convince the clients that they were on to something and the clients should pay for even more efforts, or to intimidate us, or both.

But now the full treatment emerges: "proof" of conspiracy and "sloppy science," built on careful, out-of-context replies to hour after hour of questions put forward by industry attorneys, while our government lawyers sat by and watched without comment or much interest that we could readily discern.

Strangely, I feel no anger, and certainly I do not feel intimidated. The report strikes me as pure slime, and I will see to it that the document receives the widest possible distribution. They have simply gone too far this time, and my bet is that the report will hurt their cause—and seriously. The timber industry and timber workers are badly served by their legal representatives and those who represent their political interests. As a leading professional environmentalist told me yesterday, ". . . as much damage as industry's hired guns do to themselves and their cause, we need to figure out a way to get more money to them. They are turning out to be our secret weapon—obnoxious, bullying and inept, and they never learn from a mistake."

I talked to three congressmen for the Pacific Northwest delegation yesterday who are trying diligently to bring some semblance of order out of the present stalemate that faces public land forest management in western Oregon and Washington. These are men known to be sympathetic to the timber industry and to their constituents who are employed in the wood products business. They were flatly told by industry that any "fix" from Congress must entail a timber sale level of 3 billion board feet from Region 6 of the Forest Service, or they want no solution at all, preferring what they call a train wreck—their play being the hope that such a train wreck will produce such a political and economic disaster that surely the Endangered Species Act and other constraining legislation will be overturned and the good old days will return.

The congressmen were appalled at industry's cavalier attitude— in particular toward the welfare of its workers—and its absolute misreading of the political situation. Each of them said that, essentially, the industry had given up its place at the negotiating table. Industry representatives appear to have committed another blunder—likely the most serious yet. But maybe they know something not obvious to the congressmen or to me.

As I read over the Gang-of-Four report in preparation for tomorrow's grilling, I feel good about our work but saddened by having to be the messenger who points out to Congress and the people of the United States that the Forest Service has been living with the myth of potential timber harvest that is dramatically exaggerated, and that was clearly understood, at least at the national forest and BLM district level. I know that various forest supervisors have tried to call attention to that fact and failed. Then, like good soldiers, they shut up and tried their best to "get the cut out." Should they have done more? What was their duty? The situation could not stand in the long run.

The agency made a Faustian bargain that tied the agency's budget and mission to the cutting and growing of timber in fulfillment of Gifford Pinchot's promise that the national forests would eventually make money. Times have changed far faster than the Forest Service, whose progress toward satisfaction of its multiple-use mandate has been impeded by the dukes from the timber states who occupy positions of power in the Senate and House. But now, their power seems to be crumbling, for myriad reasons, as questions concerning the management of our national forests become more and more national and less and less regional in scope.

I can see the lights of Washington below now as the plane follows the Potomac River toward National Airport. I do not look forward to tomorrow. It is getting more and more difficult to appear calm and professional for hour after hour in the witness chair. I am approaching physical, mental, and emotional exhaustion. Bill Brown says that everyone has his breaking point. I know that mine is somewhere nearby. But I think, not this time, not this time.

Tuesday, [nd] October 1991, Washington, D.C.

The Gang-of-Four (Franklin, Johnson, Gordon, and Thomas), plus our advisers on fisheries matters (Gordon Reeves and James Sedell), are all present and accounted for and start the day with a visit with Congressman Bruce Vento, chairman of the Public Lands subcommittee that deals with forestry matters, and his aide, James Bradley. The meeting is more of an exchange of pleasantries than anything else in preparation for our appearance before a joint committee meeting of Interior and Agriculture this afternoon.

Vento informs us that Congressman Norm Dicks of Washington is trying for another one-year fix in the budget negotiation between the Senate and House committees that are trying to reconcile the budget for Interior and Related Agencies. That fix involves placing the ISC strategy into law with "release language" to allow timber sales in keeping with that strategy—i.e., such sales could not be legally challenged.

Industry is lukewarm and won't go along unless it is guaranteed a cut target of 3 billion board feet for the Forest Service's Region 6. That is totally out of the realm of possibility in technical terms alone. If industry can't get a promise of 3 bbf, they prefer the train wreck.

Vento and other House members think that such a strategy is unwise for the industry in that it has alienated nearly all the members of the Pacific Northwest delegation. Further, he doesn't believe the train wreck will be as dramatic a political plague as does industry. The debate has simply dragged out too long, and there are too many other train wrecks going on over the country—some of much greater magnitude—for this particular economic and social dislocation to have much impact at the national level.

Conversely, the environmentalists now firmly believe that they can get more than the reservations of old-growth forests proposed under the ISC strategy, perhaps much, much more. Probably 30 to 50 percent of the timber sales made under such a one-year fix would go into areas suggested by the Gang-of-Four for protection. They aren't willing to live with that, and so the chances for Mr. Dicks to succeed don't appear to be good.

Congressman Les AuCoin called last week about using the ISC strategy as a short-term fix. I suggested that it would not receive support from either side for the same reasons mentioned above. My skill in developing political prognosis is improving rapidly, a somehow disturbing recognition.

About midmorning we met with the chairman of the House Agricultural Committee, Kiki de la Garza, for a review of the situation. He reviewed a news release with us in which our conclusion that the projected annual sale quantity was 15 to 20 percent too optimistic was emphasized. As members of the House and Agriculture committees move toward legislation to address the old-growth issue, it is strikingly obvious to them that the ASQ will drop significantly no matter which option they choose. They want to make it very clear that they, the Congress, are not responsible for the 15 to 20 percent of the decline caused by overestimates of ASQ in the new forest plans.

We warned the chairman that the Forest Service and the Bureau of Land Management will vehemently challenge our contention, knowing that agency credibility is on the line. However, it seems obvious to many congressional members that the ASQ numbers are simply too high and must be adjusted. That is accepted as a fact.

It is very discouraging how low the credibility of the agencies has declined with Congress. Yet our leaders continually tell the troops, and seem to believe it, how high our stock is with Congress and the interest groups. If the Forest Service can't recognize that the Gang-of-Four's assignment was a direct slap in the face of the Forest Service, the Bureau of Land Management, and the Fish and Wildlife Service, it is hopelessly mired in denial.

As we enter the hearing room at 12:30 P.M. for the 1:00 P.M. session, a man dressed in a logger's outfit is handing out news releases. The title is, "The Best Science Money Can Buy." The news release reads as if it is the reaction to what we have to say. How that could be legitimate is beyond me. But it bothers me that I am becoming inured to such pervasive dishonesty and lack of ethics and integrity.

The author of the release is accusing us of doing what we have done for money: science for sale to the highest bidder. How totally ludicrous. Each of us has put in hundreds of hours of unpaid overtime and will receive untold pain for our efforts. Announcing to the world that the emperor has no clothes, particularly when the emperor is either your employer or the biggest influence on your employer, is neither pleasant nor likely to be personally rewarding.

The hearings drag on for nearly five hours. I am not certain what the purpose is, as the questions are neither very challenging nor insightful. It is obvious in a number of exchanges that the purpose is for the congressman to ask a question or questions given them by someone they wish to accommodate for one reason or another, or to get a statement on the record to placate a particular interest group. It is abundantly clear that no minds are being changed as a result of the hearings.

Wednesday, [nd] October 1991, airborne, en route from Washington, D.C., to La Grande

This day began with breakfast with Dr. James Sweeny, wildlife ecologist for the National Forest Products Association. I saw him at the hearings yesterday for the first time since I and other members of the ISC blasted him and his organization for a very biased "analysis" of the basics of the Fish and Wildlife Service's report that recommended listing of the northern spotted owl as threatened under the Endangered Species Act. I think that he took the rebuke to heart and wanted to reestablish a working relationship.

Mark Rey, executive director of the American Forest Resource Alliance, joined us for breakfast. He is often referred to by some congressional aides as the Prince of Darkness. He is a very intelligent, well-educated, well-read man. He is a hard-fighting gladiator for the timber industry and its interests.

My next appointment is with the Gang-of-Four and Congresswoman Jolene Unsoeld of Washington. She wants our advice on several pieces of legislation she is pushing to relieve private landowners of the takings provisions (or did she use the word "take," incorrectly?) of the Endangered Species Act as it applies to spotted owls. She proposes to do this with legislation that declares "new forestry" (undefined) as fully suitable to provide and/or maintain spotted owls and excuses all private landowners who practice new forestry to be in full compliance and not subject to the takings provisions.

She seemed sorely disappointed when I said that there were simply no data to back that up, and the leading spotted owl biologists could be expected to refute that premise. Intuitively, the results of new forestry should produce habitat conditions more conducive to spotted owl habitat than current practices. But would the results be *suitable* habitat? Nobody knows.

My next appointment was a courtesy call on Chief Dale Robertson. I briefed him on my impressions of the hearings. He seemed very subdued and without his usual vigor and good humor. I suspect that he was not happy with what we had to say to Congress and is struggling to come up with how the Forest Service will reply in next week's hearings. On the other hand, I suspect that John Beuter, the deputy assistant secretary of Agriculture, will make that decision and handle the testimony. If so, that would be excellent, as it would be clear that a political appointee is making a political response.

1992

[nd] March 1992, North American Wildlife
and Natural Resources Conference, Charlotte, North Carolina

The Forest Service put on a first-class reception on the evening of the second day of the conference with the purpose of recognizing some of the agency's cooperators. Chief Robertson began his remarks with the statement, "We are primarily a forestry outfit, but we care about wildlife, too."

I felt like I had been punched in the stomach. Here was the absolute truth—inadvertent, perhaps, but the absolute truth. In spite of everything that has happened with the spotted owl, in spite of the fact that the Forest Service has become the premier wildlife research and management organization in the world, in spite of the occasion, the chief of the Forest Service would make such a declaration.

Several of the Forest Service's old "combat biologists" exchanged glances and shook their heads. We still are not a real multiple-use management agency. Even today, our leader says, with no embarrassment, in front of hundreds of wildlife management professionals, that "we are primarily a forestry outfit, but we care about wildlife, too." That statement could have been anticipated, and even accepted, twenty years ago, but not today.

22 April 1992, Portland, Oregon

I have just finished a day with three government lawyers, with colleagues Drs. Barry Noon, Eric Forsman, and Bruce Marcot from the Interagency Scientific Committee to Address the Conservation of the Northern Spotted Owl, and with Dr. Martin Raphael and Ms. Kim Mellon from the Forest Service team that performed the environmental impact statement and recommended the ISC strategy (from among five alternatives) as the management strategy to be followed by the Forest Service until a recovery plan is developed by the Fish and Wildlife Service. The latter six of us shared the distinction of having helped perform the risk assessment for the five alternatives considered under the EIS procedure for the long-term survival for the northern spotted owl.

Federal Judge William Dwyer lifted the injunction against timber sales in suitable spotted owl habitat on national forests last week. The environmental side in the case immediately announced their intent to sue the Forest Service over the supposed inadequacy of the ISC plan to preserve the spotted owl and the EIS team's "inadequate risk assessment." They have presented documents to Judge Dwyer and hope that he will grant a continuation of the injunction against timber sales in suitable owl habitat.

It is clear that they hope that a continuation of the injunction will keep the pressure on Congress to pass legislation setting up an "ancient forest reserve" that will permanently set aside designated lands from timber harvest or other management activities. Such reserves, if legally designated, would be millions of acres of de facto wilderness with a capital W. They fear, with justification, that the habitat conservation areas temporarily set aside for owls will, sooner or later, under one excuse or another, be manipulated to produce wood fiber.

The ISC strategy, in fact, allows that it may be possible to develop silvicultural treatments that may sustain owls in managed forests. Once there is proof of the viability and feasibility of such a management regime, the ISC strategy considered it possible for the system of habitat conservation areas to be abandoned.

The environmentalists' move is both technically unjustified and extremely risky in the political sense. I do not believe that this can come out any way but bad for the owl, for the retention of old-growth, and for the reputation of the environmentalists with the public and with Congress. Somehow, Congressman George Miller, chairman of the House Interior Committee, comes to mind. Several months ago, speaking of the ISC conservation strategy for the spotted owl, he told me, "Tell your environmental friends that they have just pulled off the great stagecoach robbery. But tell them to leave the watches and rings alone."

Here they are going back after the watches and rings. Good strategy? I don't think so, but maybe they know something I don't. There is more likely a combination of greed, zealotry, and ego at work. It could be a fateful mistake.

12 May 1992, Washington, D.C.

John Gordon (dean of Forestry, Yale), Jerry Franklin (professor, University of Washington), James Sedell (fisheries biologist, U.S. Forest Service), and I met this morning with Congressmen Miller of California (chairman of the Interior Committee), Vento (chairman of the Public Lands subcommittee), and DeFazio of Oregon. DeFazio is trying to get Vento and Miller to agree to a compromise on the bills being prepared by various House committees to deal with forestry in the Pacific Northwest.

DeFazio is stuck between a rock and a hard place in that his district is rather evenly split between environmentalists and those who depend directly or indirectly on the timber industry. Further, his district encompasses BLM lands and the Oregon Coast Range, which are the most severely overcut federal timberlands in Oregon. His mission, which he feels he cannot help but accept, is simply to figure some way to protect the northern spotted owl (or appear to) and hold the timber cut on the federal lands at as high a level as possible. In order to do that he wants to mix and match alternatives put forward in the Gang-of-Four report so that the impact on his district is lessened. His "innovative approach" is to create a new category of land classification dubbed the ecological reserve, in which a variety of successional stages will be maintained (which requires cutting) and stands will be treated so that desirable old-growth characteristics will occur more quickly (which, of course, requires cutting). When the maps he has prepared are examined, it is quickly apparent that the BLM lands are placed in this category along with other federal lands in the Coast Range.

The three congressmen asked for a candid appraisal and they got it. DeFazio did not like what he heard, that the BLM lands and the Oregon Coast Range were already sadly deficient in late-successional forests because of a combination of fires and accelerated timber cutting; that there are severe problems with anadromous fisheries habitat that would be exacerbated by his proposed course of management action; and that there was no free lunch. The management options are few. We explained that the BLM lands and the Oregon Coast Range were simply totally inappropriate places to try to conduct landscape-scale experiments. If such experiments are to be conducted—and they are surely needed—researchers should pick the best areas for that work, not a congressional committee trying to forge a compromise over what segment of the population is spared from the consequences of past overcutting.

We left that meeting for a hearing before the Senate Subcommittee on the Environment, chaired by Senator Baucus of Montana. The first panel of witnesses was composed of the four senators from Oregon (Hatfield and Packwood) and Washington (Gorton and Adams). The senators on the committee who were present were Baucus, Chafee of Rhode Island, Steve Symms of Idaho, and George Mitchell of Maine (the Senate majority leader). All the senators (save Symms, who is leaving the Senate and seemed only mildly interested) made statements. Each of the statements castigated the administration's chicanery in dealing with the spotted owl issue and decried the state of national forest management, particularly in the Pacific Northwest.

Then each of the senators from the Pacific Northwest had his say. Senator Hatfield, who seemed to have aged ten years since I last saw him, put forward a defense of his environmental record and claimed that the Endangered Species Act was only intended to apply to such actions as "building a dam or a bridge or a road." He said it with a straight face.

Senator Packwood warmed to the issue and essentially positioned himself as the foremost opponent of the ESA. He stated again and yet again that the cost of listing a species must be considered before the species is listed. Of course, knowing the cost at that point would require the preparation of a recovery plan prior to listing, which is obviously not possible due to funding constraints. The listing process is the filter by which species to be addressed by a recovery plan are selected. But it does indeed have a certain political appeal to the unsophisticated—i.e., "people should count for as much as some bird!" Not only will that play well with the timber industry and frightened timber workers in Oregon, it should be good for big contributions from outside Oregon.

Senator Gorton took the prize for misleading rhetoric. He seems too bright a man to be so confused. His obfuscation is coolly calculated.

Senator Adams has seen the light and knows the old days are gone and will never return. He knows, but his speech is his swan song. He is leaving the Senate under a cloud of charges of sexual harassment of employees.

He says there is no truth in the charges, but the lion will leave the pride without a fight. It is sad and too bad because he has had the ability to see anew and know that we must change our ways. Too bad. Too, too bad.

The senators leave, having talked much about things of which they have only superficial knowledge. But as Senator Packwood said, "This is not science, this is politics." And he is right: this is now pure politics.

The new panel is composed of Assistant Secretary Beuter, who oversees the Forest Service, and a deputy assistant secretary of the Interior named Edward Cassidy, who looks, walks, and talks like the political appointee he is. They say little, and very carefully at that. They do not wish to upstage the administration's show that will take place tomorrow—the God Squad decision, the release of the recovery plan, and the secretary of the Interior's alternative owl plan—already dubbed the extinction plan by the biologists who have seen it. The committee quickly bypasses the political appointees and calls Drs. Sedell and Meslow and me to the table. There are a number of questions, but basically they want to know the consequences of reduction of the owl's range and reduction in numbers of 75 percent. Answer: Bad. Consequences for potentially threatened runs of anadromous fish? Bad. The chances of success? Bad.

Then the committee affords a chance to discuss needed changes in the ESA. I seized the moment to expound on the need to prevent listings, and to consider communities and systems at landscape scale. I waxed poetic about the need to put the right people on the problem. Their eyes glazed over, they simply didn't get the point. Worse, they probably never will, as the ideas and the level of abstract thought required to move to that next level are complex and difficult to master. If these senators just don't get it, what chance do we have with the public? Not much, at least not much at the present time.

Now things begin to deteriorate as the senators' attention spans are exceeded. They struggle through the rest of the panels with neither attention nor interest. They are tired and bored, and it shows. Do these hearings really matter? Is information truly imparted? Or is it merely an elaborate dance to unheard music but in engraved patterns? Somehow, in watching the show, I can see in my mind's eye the mating dance of whooping cranes—the bows, the stretching of necks, the raucous cries, and the flapping of wings. All of these things are the preamble to the work being carried on in the offices of the aides to be brought at appropriate times to the senators' attention. A strange game and a fine dance and the only dance in town. It is the way that it is.

14 May 1992, Washington, D.C.

John Gordon, James Sedell, and I spent the morning working with James Lyons, staffer for Congressman de la Garza (chairman of the Agriculture Committee), and Tim DeCoster, aide to Congressman Volkmer (chairman, Subcommittee on Forests, Family Farms, and Energy), on potential resolution of differences on bills introduced by Congressmen Miller, Vento, Volkmer, and DeFazio. The problem still revolves around efforts by DeFazio, with support from Norm Dicks, to find some relief from the heavy impact of ISC and other plans on timber harvesting on the BLM lands and Coast Range and the Olympic Peninsula. We told them—again—that there was simply no free lunch in dealing with timber management in these areas. There is simply no room for compromise because of heavy past cutting and fires.

Even if there is a decision to step away from the owl habitat guides, there is an immediate problem with impacts on anadromous fish habitat in coastal streams. As these streams have no dams, they are potentially a key to saving salmon runs in the Pacific Northwest. Then there are the other twenty or so species associated with old-growth forests that are awaiting listing as threatened or endangered. DeFazio's aide seemed to have learned something since yesterday. He seems somewhat reconciled to having to face facts.

The discussion turned again to "departure"—cutting at rates above the sustained-yield rates. Several of the bills call for a sale level from national forests in Oregon and Washington of 3 billion board feet in the next year, compared with a sustained-yield level of 1.7 bbf. They visibly flinch when I tell them that anything above 1.2 bbf is a practical impossibility. This is a statement they do not want to hear and continue to insist it can be done. Jim Lyons, in particular, wants to know why the "can do" agency can't get up 3 bbf, intimating that the agency is simply dragging its feet. I explained that the field troops are simply worn out and demoralized. They have been ridden hard and put away wet for three straight years, while the politicians and political appointees fumble time and time again. Esprit de corps does matter, but this essence of the Forest Service has been exhausted. Elan is gone and will not return until a worthy mission again shines bright and whole.

I said that departure was simply refusing to face up to facts, another means of deficit spending—buy now and pay later. The timber industry, most politicians, and the Forest Service brass are still in a state of denial. Surely, they think, there must be some way to dodge the bullet or, at the very least, to delay the consequences until after the next elections in November.

As scientists are thrust forward into positions of power and influence in natural resource management, it is well that they be humble. For of all people, scientists should be acutely aware that we know so little and that there is no final truth. Scientists are not accustomed to power and are likely to use power poorly. If we are not careful, we may be like the Wizard of Oz that Dorothy discovered behind the curtain, manipulating an impressive show of smoke, mirrors, and images. There is much danger of hubris inherent in the developing situation.

And there are those who would and will turn such hubris to discredit scientists and any possible role for them in natural resource management decision making. Such people, particularly those in the timber industry, prefer deals made on the basis of "common sense," employment figures, political contributions, and profits. The ability to cut a deal for various levels of timber cut is much easier in a decision milieu that does not include science and scientists.

Our session with the congressmen ended shortly after noon. When Sedell and I arrived back in the Forest Service offices, the information was out on the God Squad's action on the question of allowing forty-four timber sales proposed by the Bureau of Land Management, which were declared by the Fish and Wildlife Service as likely to jeopardize the survival of the spotted owl, to go forward.

I am quite certain that the Forest Service's top leadership has not recognized that the power has shifted within the agency. Scientists, long considered insignificant or at least subordinate to the National Forest System, have now obtained power—though most of them don't even know it. This power exists in the fact that no proposed management can survive political and legal scrutiny that cannot muster support, or at least tacit approval, from scientists.

The decision, on a 5-2 vote, was to allow thirteen of the forty-four sales to proceed. This would seem, superficially, to be an incredibly stupid decision and one absolutely assured to please no one, deepen legal confusion, and be very difficult to rationalize. However, a closer reading of the decision revealed a very significant caveat. BLM cannot proceed with the thirteen timber sales until the agency adopts a "scientifically credible" management plan for the northern spotted owl. So far as I know, there are only two such plans in existence that will pass muster on scientific credibility—the ISC strategy and the Fish and Wildlife Service recovery plan (which is nothing more than a modified ISC strategy).

What this decision amounts to is that the God Squad gave away about a thousand acres of owl habitat to save face for Interior Secretary Lujan and BLM Director Jamison. Then they ordered BLM to come into compliance with either the ISC approach or the recovery plan. As I read and reread the caveat, I had a clear vision of Julie Andrews singing, "A little bit of sugar makes the medicine go down, makes the medicine go down, makes the medicine go down . . ." Lujan and Jamison are now hoisted on their own petard.

Meanwhile, the recovery plan was released in Portland with a mere assistant secretary of the Interior as master of ceremonies. The recovery plan is a good plan and the result of much effort. But as planned, the recovery plan release is completely overshadowed by the God Squad Persistence Plan, the top-billed vaudeville show in Washington, D.C.

Then there is the matter of timing. God Squad activities, the oft-delayed release of the recovery plan, and the Lujan alternative were timed for release five days before the statewide elections in Oregon and Washington. My God, what a coincidence! It all makes one proud to be an American.

After Cy Jamison, director of the Bureau of Land Management (with the consent of Secretary of Interior Manual Lujan) withdrew the Oregon and California Railroad lands under BLM management from adherence to the ISC strategy to protect the northern spotted owl, Judge William Dwyer shut down all logging of old-growth timber on federal lands within the range of the owl. In issuing his injunction, Judge Dwyer instructed the federal government to answer three questions germane to his decision to continue or lift the injunction. Would BLM's decision adversely influence the effective-

Portland timber summit, April 1993, with Vice-President Gore and President
Clinton. USDA Forest Service photo, by Tom Iraci.

*ness of the ISC strategy? Was there information that had become available
since the decision to adopt the ISC strategy that would influence the predicted outcomes? Would the execution of the BLM strategy likely lead to the
extirpation of any of the thirty-two species that the ISC team had identified
as closely associated with old-growth forests in the Pacific Northwest?*

*Jamison had moved to convene the Endangered Species Committee—
the God Squad—to consider allowing BLM to go forward with a number
of timber sales on the premise that the impact on the economy of the Pacific
Northwest and on the revenue to the counties concerned was so severe as to
allow the associated risk to a threatened species. BLM joined timber interests in essentially placing the ISC strategy on trial. Prior to my appearance
as a witness, Mark Rutzick, the attorney representing the timber industry,
announced that he would "defrock the high priest of the cult of biology." In
the end, the Endangered Species Committee in a "split the baby" decision
said a token number of sales could go forward, but thereafter BLM would
have to come into compliance with the ISC strategy. It was a humiliating
decision for Jamison and Lujan, in that the committee essentially upheld*

*the ISC strategy. The sales were never cut and the decision was made to
defer action until after the 1992 presidential elections.*

*In the meantime, Dale Robertson, chief of the Forest Service, appointed
a team to answer the judge's questions. The Scientific Assessment Team
comprised Martin G. Raphael, Eric D. Forsman, A. Grant Gunderson,
Richard S. Holthausen, Bruce G. Marcot, Gordon H. Reeves, James R. Sedell,
and David M. Solis, all of the Forest Service, and Robert G. Anthony of the
U.S. Fish and Wildlife Service; I was named team leader.*

*Because there were far more species (including plants and invertebrates)
that were associated with old-growth forests than the 32 identified in the
ISC report, we sought, and received, permission to expand the analysis. Our
report* was delivered in March 1993. The team found that because of the
absence of specificity in the "Jamison strategy," we could not predict the
efficacy of the ISC strategy without BLM's full participation. Further, no
new information had become available since the issuance of the ISC strategy
that would lead us to believe that it would not achieve its objectives if fully
implemented. We found 667 species (35 mammals, 38 birds, 21 reptiles and
amphibians, 149 invertebrates, 122 vascular plants, and 190 nonvascular
plants and fungi) associated with old-growth, 518 of them "closely associated." This included an analysis of 112 stocks of anadromous fish. We
concluded that 482 of the 518 closely associated species had a low risk of
extirpation and that information on the remainder made it impossible
to assay risk of extirpation.*

Judge Dwyer continued his injunction.
—JWT, 2003

---

* Jack Ward Thomas, Martin G. Raphael, Robert G. Anthony, Eric D. Forsman, A. Grant
Gunderson, Richard S. Holthausen, Bruce C. Marcot, Gordon H. Reeves, James R. Sedell, and
David M. Solis. 1993. *Viability assessment and management considerations for species associated
with late-successional and old-growth forests of the Pacific Northwest.* Portland, OR: USDA Forest Service, National Forest System, and Forest Service Research.

1993

In June 1993, while I was on a consulting job, along with Jerry Franklin of the University of Washington and Norm Johnson of Oregon State University, for the Makah Tribe on the Washington coast, I received a call that my wife, Margaret, had suddenly fallen ill and been diagnosed with cancer of the colon. I rushed home via charter aircraft in time to see Meg just before surgery. The news was bad: the cancer had already reached her liver and the prognosis was death in six to eighteen months. In the weeks afterward, we agreed to go on with our lives as we always had, for as long as we could.

The presidential election of 1992 was a three-way contest between President George H. W. Bush on the Republican ticket, Governor William Jefferson Clinton for the Democrats, and H. Ross Perot as the candidate of the Reform Party. The spotted owl issue played a prominent role in the election in Oregon and Washington. Bush and Perot talked of "owls versus jobs" and promised to reform the Endangered Species Act. Clinton was more cautious and merely promised a solution to the impasse. Bush and Perot split the conservative vote and whatever portion of the vote was predicated on the jobs-versus-owls issue. Clinton thus carried both Oregon and Washington and, at least partially as a result, won the presidency.

On April 2, 1993, as he had promised during the campaign, President Clinton convened a conference, which he chaired. A number of people associated with the issue over the previous decade were invited to testify, including me. I warned the president that the primary mission of the Forest Service had evolved through interacting laws and interpretations by the federal courts into the preservation of biodiversity. I asked him to clarify this issue by either affirming that de facto mission or specifying otherwise.

At the close of the conference, the president announced that he was nam-

ing a team and instructing them to provide, within ninety days, a series of options for his consideration to address the crisis presented by the ongoing federal court injunction on old-growth timber harvesting in the Northwest. I was assigned to head that group—the Forest Ecosystem Management Assessment Team.

FEMAT was to engage more than six hundred participants over the next ninety days in an effort that went on around the clock, seven days a week. In early July 1993, ten options* were delivered to the president, from very high emphasis on environmental concerns with a very limited timber yield to the exact opposite. Some team members jokingly characterized the array as ranging from the "green dream" to the "brown bomb."

President Clinton chose Option 9, a compromise between levels of environmental protection that would pass Judge Dwyer's examination; it included a "probable" sale quantity of about 1.2 billion board feet per year. At that point, the effort was turned over to another team to perform the required environmental impact statement and record of decision. In that process, "bells and whistles" were added that considerably altered the option, making it more closely resemble Option 1, the green dream.

—JWT, 2003

* Forest Ecosystem Management Assessment Team. 1993. *Forest ecosystem management: An ecological, economic, and social assessment.* Portland, OR: U.S. Department of Agriculture, Department of Interior, Department of Commerce, and Environmental Protection Agency.

Monday, 2 August 1993, en route to Washington, D.C.

I testify tomorrow before a hearing of the House Subcommittees on Agriculture, Interior, and Merchant Marine on the President's Forest Plan for the Pacific Northwest. With me are K. Norman Johnson of Oregon State University, who did the assessment for timber yield; Brian Greber of Oregon State University, who headed the economic assessment subgroup; James Sedell of the Forest Service, who was team leader for the aquatic team; and Roger Clark of the Forest Service, who was in charge of the social assessment.

This long-running game over the old-growth forest issue, which has grown to include fisheries and riparian habitats, may be drawing to a close. The president has chosen an option and is pressing forward with it.

In the past, the administration in power could not (or would not) come to grips with the issue. As Judge William Dwyer said in the Seattle Audubon case, there was a clearcut record of obfuscation, delay, and evasion of the law. Intended or not, this produced a train wreck, with the timber industry in dire straits and all timber sales in northern spotted owl habitat shut down by federal court order.

In previous efforts to produce a solution that would withstand technical and legal scrutiny, scientists were left to twist in the wind with no support from either congressional or administration politicians. The plans, without such support, became known as "the Jack Ward Thomas plan" or "the scientists' plan." They withstood both intensive peer review and technical scrutiny, but the politicians could not bring themselves to accept them and suggested other approaches developed by timber industry biologists or political appointees without appropriate technical credentials. None of those proposals survived critical review.

It was open season for politicians and timber industry representatives to attack the plans, the science that backed them up, and the integrity and motivations of the scientists. After the president's Forest Conference in Portland, in early April, the Forest Ecosystem Management Assessment Team was assigned to make yet another effort to devise a technically valid and politically acceptable plan to deal with ecosystem management, retention of an old-growth forest ecosystem, and obedience to the Endangered Species Act (i.e., recovery of the northern spotted owl and the marbled murrellet) and the regulations issued pursuant to the National Forest Management Act to "maintain viable populations of . . . vertebrates well-distributed over the planning area on Forest Service lands." That was a tall order.

Faced with ten options—none of which produced more than a third of the timber cut levels of the mid-1980s—the president chose Option 9. As he expected, he was condemned by the gladiators on both sides of the issue. Timber industry and labor gladiators predicted job losses in the hundreds of thousands and severe economic disruption on an international scale. The gladiators representing the environmentalists wailed that entire ecosystems were likely to collapse. Both sides attacked Option 9 before they had even seen it. Journals known for accuracy and verification began to publish rumors—most of which were grossly inaccurate.

The White House team that controlled the political aspects of the plan assured that the president was able to make a decision and announce it without the gladiators, or Congress, picking through and over the options. Inten-

Forest Ecosystem Management Assessment Team studying maps for Option 9 of the Forest Plan for the Pacific Northwest. *From left:* Jack Ward Thomas, Eric D. Forsman, Barry S. Mulder, Richard S. Holthausen, A. Grant Gunderson. Photo by E. Charles Meslow.

tional or serendipitous, this produced a situation whereby the kill-the-messenger activities of the past played poorly. It is much harder to attack "the president's plan" than to attack "the Thomas plan" or "the scientists' plan." It is difficult to say that the scientists did not follow instructions or have run amok when the giver of the instructions praises their effort.

Members of the Northwest congressional delegation split down the middle, with some castigating the scientists and some extolling their virtues. Speaker of the House Tom Foley (prodded by his aide, Nick Ashmore) came out vehemently against the plan but finally quieted down. Congress had demonstrated its impotence on this issue over several sessions and could do nothing more than wrangle and posture this time around either. Secretly, I think they are relieved at being taken out of the picture. Why not? There is a solution in place that was essentially the best they could hope for—they've known for over a year that 1.2 billion board feet was the approximate timber cut level—and they don't have to take a stand on a contentious issue.

3 August 1993, Washington, D.C.

Our little contingent of scientists meets in the assistant secretary's office prior to our appointment in the House for a hearing of the subcommittees on Agriculture, Interior, and Merchant Marine on the FEMAT report. When we arrive, an aide hands us the questions that have been prepared by the industry and environmental gladiators and given to their favorite congressmen for use at their discretion. We read over the questions with some trepidation. Past experience has shown us that these questions can be nasty, particularly when they catch you by surprise.

Having an assistant secretary (Jim Lyons) who is an exstaffer for the House Agriculture Committee, with two assistants who are also exstaffers for congressmen, gets you a break sometimes. Questions that can be devastating when asked in a courtroom atmosphere—particularly when asked in a well-thought-out sequence—can be turned on the questioner if the witness has time to think about the answer in advance.

The questions from the industry gladiators turn out to be very weak. Not at all up to their usual standards, either technically or in terms of political effect. The environmentalists don't even try. What does this mean? Is industry worn out? Giving up? What? Are the enviros O.K. with Option 9? Hard to say.

Once the hearings start, there are, perhaps, as many as three times more Democrats at the dais than Republicans. None of the Republican nasties are present save Bob Smith from Oregon, who is a rather weak adversary. He is smart enough but lacks passion and doesn't work very hard. However, he is making noises about running for governor of Oregon, and so it will be interesting to watch.

Jim Lyons and Tom Collier, chief of staff to Secretary of the Interior Bruce Babbitt, lead off the hearing with testimony. They don't pull any punches and lay the train wreck right at the feet of the Bush administration and the timber industry. Their testimony stated frankly that the two previous Republican administrations set out to evade the nation's environmental laws and were caught up by the federal courts time after time. They state that workers in the timber industry had been badly misused over and over and had suffered grievously as a result. They, obliquely, note that Congress had been unwilling to resolve this issue.

President Clinton had, if nothing else, stepped into this vacuum and made a decision. And the administration was moving forward with an administrative solution.

Several Democrats (Volkmer, Unsoeld, Vento, among others) spoke kindly and with appreciation for the work of the scientists. None spoke critically. Only Peter DeFazio of Oregon pursued any rigorous questioning, all of which was directed to Dr. Brian Greber over assessment of job losses. His obvious intent was to make those losses appear as dramatic as possible as the preamble to making the federal aid package to his congressional district as large as possible and his role as fighting for his constituents as visible as possible. He did well in the role he chose to play.

Not one congressman, not a single one—Republican or Democrat—used a single question provided by the timber industry gladiators! What does that mean? In fact, the hearing was a comparatively low-key affair.

The hearing sputtered to a stop at about 1:00 P.M., some three hours after it started and two hours earlier than anticipated. We were lucky this time. We can hope that the Senate hearings will go as well. Could we be lucky enough that the Senate will pass? I don't look forward to getting cussed out again by Senator Malcolm Wallop of Wyoming, who put on a tirade over the Interagency Scientific Committee report on spotted owls. Hatfield of Oregon might be a bit more hostile this time around. He has referred to me, privately, as "arrogant beyond belief" and as a megalomaniac.

As the meeting was ending, Jim Lyons asked me to meet with him tomorrow for about an hour. I suspect that the moment of truth is at hand. The assistant secretary has visited with me several times over the past two months about the possibility of my becoming chief of the Forest Service.

At first, I told him that I was flattered, but given Meg's battle with cancer, that such just wasn't possible. However, when I mentioned the conversation to Meg, she became the steel magnolia, and she informed me that she was not pleased that I had made this career decision without consulting her. We talked over the issue and she said that she wanted us to take the job.

We talked, forthrightly, that the odds were that she would not live long and that her death was likely to be drawn out and painful. I suggested that she deserved to live and die in La Grande among friends who care, rather than in Washington. She said that she would concentrate on living and let the dying take care of itself. She asked me, again forthrightly, if I was afraid for her or afraid of the challenge and the trials and tribulations of being chief at this moment in history.

"What will you say when in later years, you ask yourself if you could have made a difference and we didn't even try?" The steel magnolia speaks again from a core of strength that I lack sometimes. She said that I really had no choice. We have traveled a long journey together that is coming to an end. It simply is not possible to turn back now, not for either of us.

I agreed that if asked, I would serve. But there was always the hope that the mantle of leadership would pass to another—perhaps to one of six whom I suggested as better qualified and more deserving than I. Now, with the instructions to meet with Mr. Lyons on the morrow, there is a sense of dread and only a small hope that he will tell me that another has been chosen. Tonight will be a long and restless night.

Wednesday, 4 August 1993, Washington, D.C.

The first thing this morning there is a meeting of Assistant Secretary Lyons, Terri Gordon, secretary of natural resources for California, and me about a proposed symposium on the new information concerning the northern spotted owl in northern California. In the course of the discussion, Ms. Gordon asks Jim Lyons if I can be assigned to organize this effort. Lyons answered, "No. He will have other, more important things to do."

My heart fell. They have made up their minds as to the job of chief of the Forest Service. My mind races and I am only vaguely aware of the remainder of the conversation between Lyons and Gordon.

After Terri's departure, Lyons comes directly to the point. I am to be named chief and the announcement will be made on the 23rd of August. The conversation is amazingly brief. As I stand up to leave, we shake hands and he says, "By the way, you have been promoted to Grade ST-17 as a research scientist. I believe that you are the only scientist of that grade in the Forest Service. Congratulations."

Walking down the long corridor in the Department of Agriculture, I find myself elated over being promoted to Grade 17 as a research scientist: that is an achievement recognized by scientist colleagues. Somehow, that seems a more significant thing than being designated to be chief of the Forest Service. As I walk that corridor, I feel for Chief Robertson. This will be hard on him. He will not, probably, receive the credit he is due. He held the Forest Service together during very difficult times. He supported the ISC strategy, although in doing so he was almost fired. He instituted "new perspectives" and embraced ecosystem management. He is a decent man who, I believe, did his best. He will hand off the chief's job to another Forest Service professional, although not one of his choosing. I can only hope to do as well.

Monday, 9 August 1993, La Grande

Assistant Secretary Jim Lyons called today. There has been a snag with appointing me chief. Since I am not a member of the Senior Executive Service and have not been qualified for entry into that group, I am not eligible for the position as it now exists. Lyons, therefore, wants to convert the position to a Schedule C appointment—a political appointment. We discussed the matter for some time. Both of us have concerns about breaking with tradition and making the chief's job a political one. Lyons wants to move ahead with appointment of a new chief—immediately. I reiterated, over and over, that I did not want to become the first political appointee to hold that position. The tradition of a professional chief to head the agency is too important for the Forest Service over the longer term to make this move now, no matter how justified and desirable it seems at the moment.

The precedent is the significant thing. It makes little difference how pure the motives of this administration are or how well qualified the appointee. The ultimate result will be that some time in the future, a new president will, in order to pay off political debts or to follow his or her own agenda, fire and hire chiefs on that basis. I have witnessed, up close and personally, the consequences of compliant political hacks holding the reins of power in land management agencies. The result can be a disaster for the agency and for the natural resources that they manage. I will not be a party to such a change and tell Mr. Lyons that.

Jim is understanding but frustrated and agitated. There is the alternative of advertising the job for all who wish to apply and then selecting from the qualified candidates. But that too would be a break with tradition, and the selected candidate would come from outside the Forest Service. Besides, even if I was selected, the process would take two to four months. Then, it would be 120 days before I could make changes, which badly need to be made, in the ranks of the Senior Executive Service that directs the Forest Service.

After long discussion, Lyons still wants to go with a Schedule C appointment and put me in the job immediately. I maintain my position that such a move would not be good for the Forest Service over the longer term. Then he asks, "What if we convert the chief's job to a Schedule C position in any case—whether you take the job or not—and you are offered the position? Would you accept?" That is a tougher question and I am uncertain.

Once again, at the end of the conversation, I gently suggest that perhaps it would be better all around if a person in the Forest Service who was in the Senior Executive Service might be selected. Once again the suggestion was rejected. We will speak of this again next week.

Friday, 13 August 1993

In keeping with Friday the 13th, my mail contained the paperwork from the timber industry lawyers: the secretaries of Agriculture and Interior and I and the entire FEMAT are being sued. The basis of the suit is that FEMAT operations were conducted in a manner not in compliance with the Federal Advisory Committee Act. The brief was filled with page after page demanding documents and affidavits, and question after question concerning how FEMAT operated.

Well, I will deal with that when the time comes. The White House team set the rules and FEMAT did the technical work. I have learned to simply tell the truth in the briefest possible manner and then let the lawyers fight it out and the judges rule.

There are moments, however, when I sit at my desk answering interrogatories and sit hour after hour being deposed by lawyers and sit on the witness stand in court trying to tell the truth while not being discredited by the legal attack dogs—at such times it is sometimes difficult to remember that somewhere there are biologists in denim pants and work boots doing fieldwork far from the world of lawyers in their three-piece suits, shiny black shoes, crisply ironed white shirts, and fresh haircuts. I feel terribly out of place in this world—and the worst part is, they know it. A remark came back to me that one of the industry lawyers made. He said I reminded him of a grizzly bear bayed by dogs: angry, puzzled, and frightened.

Now when I am at the mercy of the lawyers, I keep the image of the cornered grizz' in mind, knowing that the smooth Harvard lawyer has never seen a bear swat the life out of a dog with one sweep of a paw. The thing about bear baiting is that sometimes you get the bear, but sometimes the bear gets you!

22 August 1993, Halifax, Nova Scotia

Assistant Secretary Jim Lyons called our hotel room last night to discuss the position of chief of the Forest Service. The time has come for a decision on both our parts. The situation has not changed since our last conversation: I do not want to accept a political appointment and Jim insists that it is critical that I be in position as chief as soon as possible. He makes it clear that the position will not go to any other person currently in the Forest Service if I do not accept. There is a long silence—yes or no? Far back in my mind is the nagging question—the ultimate question—that has always been there. What will you say on the *last* day? Will you say that you didn't have it in you to even try?

I have told every young professional whom I ever have counseled about duty and vocation and mission. I told my sons the same things, and they all believed me. Now, the prolonged silence on the other end of the phone screams at me my own words about sacred things: duty, vocation, mission.

The answer comes, at last, clear and sure, "I will do what you think best." It is done. Strangely, I feel sadness and not pride nor satisfaction. Here is yet another fork in the road with the path taken marked as "the rest of your life" and the other as "no longer an option." Jim Lyons acknowledges my answer quietly with a "thank you" and hangs up. It seems a solemn moment.

Then Meg, who is lying on the bed listening, begins to laugh. I find it difficult to discern humor in such a solemn moment and inquire as to the source of her amusement. "You didn't even ask how much the job pays. You didn't ask about moving expenses. You don't know if the job counts toward your retirement. Mr. Practical strikes again!"

Now, I begin to laugh as well. She is exactly right: these obvious, practical questions never crossed my mind as a consideration. No wonder she won't let me write checks. There is nothing like a touch of reality tinged with self-deprecating laughter to kill off a truly sanctimonious moment!

We are both still laughing as I call Jim Lyons back to ask some real-life questions.

After the conversation ends, Meg asks, "What did he say?"

"Well," I respond, "he wanted to know if I could handle a $3.5 billion budget."

20 September 1993, La Grande

Max Peterson, ex-Forest Service chief, called to tell me he was alarmed that I would accept a political appointment. He said that he had no objection to my being chief, but that he considered a political appointment the beginning of the end for professional leadership of the Forest Service. I told him that I had initially refused the appointment as a political appointment and had accepted only after being told (1) there was simply no way I could be appointed as a careerist because I did not hold credentials in the Senior Executive Service; and (2) there were simply no other folks in the Senior Executive Service in the Forest Service that the administration believed to be the kind of person needed to head the "new" Forest Service. So either I accepted the appointment or the administration would go outside the Forest Service, for the first time in the agency's history, for a new chief.

Max was unimpressed by my logic and gave the impression that he would exercise what influence he could to thwart my appointment. He assured me that his opposition was based on principle and was not personal in any way.

Shortly after Max's call, Jim Bradley, a staffer for Congressman Bruce Vento and a Forest Service veteran of some fifteen years, called to tell me of Mr. Vento's concern with the political appointment. He told me that the White House Personnel Office had told him that the entire action was a "plot" using my name recognition and "temporary reputation" to make the appointment a political one (i.e., break the tradition) and then to replace me within a year with "the person they really wanted."

He asked me to withdraw; I told him that I had given my word to Assistant Secretary Lyons and would keep my word. In response, he assured me that my nomination would be withdrawn but that this action was based on principle and there was nothing personal involved.

He seemed somewhat taken aback when I said that I sincerely wished him luck. Nothing would give me greater pleasure than to have this "honor" pass to another, as the job will be brutal, not in my family's best interest, and financially irresponsible.

Bradley asked me one last time to withdraw "for the good of the Forest Service." Once again, I told him that I had given Jim Lyons my word and that he should talk to Lyons. If Lyons released me from my reluctant promise, I would immediately and gratefully withdraw my name from consideration or it could be ignored that the offer was ever made. His voice was cold as we said goodbye.

All of this makes me wonder why anyone would be interested in either running for office or seeking a politically appointed position of power.

8 October 1993, La Grande

Assistant Secretary Jim Lyons called from his mother's home in New Jersey at 12:30 A.M. Eastern Standard Time. His message was simple: President Clinton had signed off on my appointment as the next chief of the Forest Service earlier in the day. The next step is a call from White House attorneys to make certain there is nothing in my background that would preclude my appointment or that might prove an embarrassment to the president of the United States.

The waiting and uncertainty have come to an end. Mr. Lyons was still uncertain as to the exact mechanism of making the appointment public knowledge. I told him that whatever his intention, I was not available for the next ten days because of a long-standing speaking engagement and an elk season that begins next week. He asked if elk season was mandatory so far as my participation was concerned. I told him that I had not missed an elk-hunting season in twenty years and didn't intend to start now.

What was not relayed to him was how badly I needed this hunting season in the high Wallowas, particularly just now. I have a real need to draw strength from the wilderness and the isolation and the majesty and the solitude. What lies just ahead—and now with certainty—is the awesome responsibility of rebuilding the Forest Service and the loss of my life's partner and the light of my life.

If I had my sweetheart with me, there is little doubt that the journey would be exciting and joyful. She would make certain of that, as she always has when I was shy and withdrawn and a little afraid. Just now, the contemplation of that journey without her fills me with trepidation.

27 October 1993, La Grande

Last Friday, I spoke with Assistant Secretary Jim Lyons, and he told me to come to Washington on Tuesday for the announcement of my assignment as chief. When we checked into our hotel on Tuesday night, the desk clerk handed me a note that said, "Dr. and Mrs. Thomas are expected to visit with the President at the White House at 10:00 A.M. tomorrow." President Clinton had told me that he wanted to visit with Meg to tell her how much he appreciated her sacrifice in supporting my ninety-four-day stint of the Forest Ecosystem Management and Assessment Team effort. Now, he (or somebody on his staff, more likely) had remembered. I was impressed.

At 10:00 A.M. the next day, along with Katie McGinty (presidential adviser for environmental affairs) and Jim Lyons, Meg and I were ushered into the Oval Office to be greeted by President Clinton and Vice-President Al Gore. We visited about fifteen minutes, making small talk. But the real purpose of recognition and appreciation for Meg was obvious and achieved. Small enough recognition, perhaps, for the agony of the FEMAT effort and ninety-four days away from our home in La Grande, during what may very well be the last year of Meg's life. But it seems like a lot, and it represents a considerable change from the Bush years. As I look at the tears in Meg's eyes, and in the president's, it seems a special moment.

There is a regional foresters and experiment station directors meeting going on today. Chief Dale Robertson and Associate Chief George Leonard will announce that they are being shifted to other jobs, and that the associate deputy chief for the National Forest System will become associate chief and, temporarily, acting chief. It seems to be common knowledge that I will be named chief at some future date.

Within the past week or so there has been an organized effort among some line officers (three regional foresters and seventy-two forest supervisors) to force the secretary of Agriculture to alter course in appointing me chief through a political appointment. What they have to say is nothing more than I have said. However, using the Forest Service computer network and telephone system and paid working time is quite inappropriate.

President Clinton with Margaret and Jack Ward Thomas, Oval Office, October 1993. Photo courtesy of Jack Ward Thomas.

In the letters they have sent to the president and the secretary, these line officers emphasize how well the Forest Service is respected by political leaders, the Congress, and the American people compared with other natural resources management agencies and attribute this to Forest Service leadership by "career professionals." They are, in my opinion, in deep denial of how much the agency's reputation has declined over the past several years. For example, the *Washington Post* described the Forest Service as "fundamentally one of the government's most deeply troubled and frequently criticized agencies." And it further stated that "Robertson's tenure was marked by charges that he was too willing to accommodate the timber industry, and by widespread internal dissent, charges of reprisals against whistle-blowers, criticism of money-losing timber sales and allegations of large-scale timber theft abetted by agency employees."

Assistant Secretary Lyons considers the critical aspect of my appointment is that it preserves the importance of appointing the chief from the professional ranks of the Forest Service. Further, he has promised that I will have the opportunity to qualify for the Senior Executive Service and then the appointment will be converted back to Civil Service. Or I will have the option of resigning and being returned to La Grande as an ST-17 research scientist, and a new chief selected from the ranks of those qualified in the Senior Executive Service.

It seems likely to me that much of the resistance to my appointment by line officers is that my Forest Service pedigree does not include service in National Forest System as a forester who has been through the chairs of being a district ranger, forest supervisor, Washington Office staff officer, and regional forester. A biologist (as opposed to a forester) and a research scientist (as opposed to a National Forest System administrator) is simply outside the brotherhood considered worthy of appointment as chief.

On the other hand, they have legitimate concerns that a future administration will appoint a chief who resembles former Secretary of the Interior James Watt in outlook—toward relatively unconstrained exploitation of natural resources. I just hope all of this doesn't get in the way of bringing the Forest Service into the twenty-first century and restoring credibility with the Congress and the public and internal pride in the agency and its mission.

Thursday, 4 November 1993, Washington, D.C.

The time since the meeting with the president and vice-president has been filled with frustration. The day after that meeting (last Wednesday), it became obvious that the anticipated announcement on Thursday was a pipe dream. Somebody dropped the ball and did not send me the required paperwork nearly a month ago. There were four hurdles to clear: (1) the president's approval, (2) clearance by White House folks for ethics and "clean living," (3) certification by the Internal Revenue Service that there were no tax problems, and (4) clearance by the Federal Bureau of Investigation as to no criminal record and "loyalty to the United States."

At the time I filled out the paperwork, I was number 152 on the waiting list for White House clearance. The president moved my paperwork to the number 1 position and that clearance took only two days. However, it seems that not even the president can hurry the Internal Revenue Service or the Federal Bureau of Investigation. So, it became a prolonged game of wait and see. The Forest Service is "idling," waiting for clearance to fill critical positions long delayed by the administration's distrust of the agency and inaction on a number of key policy decisions.

Understandably enough, the new associate chief, Dave Unger (also acting chief), wants guidance from the new chief on how to proceed. As a result, I have met with Unger several times in the assistant secretary's office (it is not appropriate for me to show my face in the Forest Service building until my appointment as chief is confirmed) to provide guidance on just exactly what I want ready to present to Assistant Secretary Lyons for approval at the earliest possible date.

It appears likely that several members of the Senior Executive Service (high-ranking Washington Office officials and regional foresters) have been actively engaged in dealing with members of Congress—all Republicans— and the Democratic governor of Idaho, Cecil Andrus, to derail my appointment. Four Republican senators—Mark Hatfield of Oregon, Larry Craig of Idaho, Max Baucus of Montana, and Malcolm Wallop of Wyoming— have sent a letter to the president, the secretary of Agriculture, and the assistant secretary of Agriculture protesting my appointment, or rather, the use of a Schedule C appointment.

Senator Larry Craig spoke on the Senate floor on Wednesday (November 3) to that effect. The script of his text contained information that was obviously provided by sources from within the Forest Service. An anonymous note slipped under Assistant Secretary Lyons's door from "concerned Forest Service employees" identified George Leonard, Tom Mills, Bjorn Dahl, Gray Reynolds, and Dave Jolly as the most active in that group. Lyons has already removed Dahl from his post as legislative liaison. Of course, this merely involves shifting him to another position (responsibility for the Resource Planning Act activities or a deputy regional forester post). Either of these would be a promotion. Such are the consequences of inappropriate action in government service: screw up and move up. But given the agony of disciplinary action and the time required to achieve a conclusion, it is the only feasible course of action.

Bob Smith (my Oregon congressman) stated in the press this week, "Ordinarily, a congressman would be pleased with the appointment of a constituent to such a high office as chief of the Forest Service. I can't say that I am pleased with the appointment of Jack Ward Thomas to that position. He is a preservationist . . . " Of course, Bob is toying with running for governor of Oregon and this is likely a part of that strategy.

16 November 1993, airborne,
en route from Washington, D.C., to La Grande

We are on our way home after twenty days in Washington, D.C., and there is time to reflect on the events of the past three weeks. It has been a most frustrating time. My appointment as chief of the Forest Service was to have been announced nineteen days ago and that announcement is yet to occur, with no definite date in sight. Such delays seem to be the norm regardless of the chaos and frustration resulting from those delays. Evidently, nearly a year after the election of the new president, there are still hundreds of appointed positions that have not yet been filled because of the exhaustive process of ensuring that none of those appointments might prove politically embarrassing to the president. Obviously, the political imperatives are more significant than the efficient functioning of government. Such seems to put the cart before the horse and begs the question of whether government exists to serve the needs of politicians or the needs of the nation. The answer seems obvious.

Our plane had a fifty-minute stop in Chicago. While strolling up and down the concourse, I heard myself being paged over the loudspeaker system. The message was to call Jim Lyons in Washington. When I reached Jim, he told me that the FBI had given final clearances to my appointment. He said that Secretary of Agriculture Mike Espy wanted me to return to Washington immediately. The secretary wanted to make the announcement at a 10:00 A.M. press conference tomorrow.

I told him that I couldn't return as Margaret is worn out and needs to get to Portland, that she has chemotherapy treatment on Thursday in Portland, and that I am scheduled to give the Starker Lecture at Oregon State University on that same day.

He made one more try at persuading me to return for the press conference. Again, I declined. Jim said that as an alternative, they would simply make a press release and that would be that. I agreed. Jim made it clear that he was merely trying to ensure that Margaret was present at our "big moment." I told him that we appreciated his efforts (he had finally gotten Secretary Espy to make a personal plea to the FBI director to get the clearance) but that it was simply not acceptable to me to stress Margaret any further. I didn't say it to him, but our patience was simply exhausted. We had gone to Washington and then were on standby for twenty days. Enough is enough.

Jim was disappointed, but he acquiesced in my decision. So the big moment will occur in the form of a press release. That is just as well, as I don't like press conferences and would just as soon forgo the honor. This whole thing has simply dragged on so long that the final approval and announcement seem an anticlimax as opposed to a high point.

17 December 1993, airborne,
en route from Washington, D.C., to La Grande

I am in the air on my way home after my first seventeen days as chief. The days have been very revealing and educational. A number of very tough decisions have been accumulating over the past several months. It seems that almost every briefing begins with the statement, "If we do not take a certain action, it is likely that we will be sued. And if we do take that action, it is likely we will be sued." I have taken the attitude that since being sued seems inevitable, the Forest Service might as well do the right thing and take the suits one at a time.

Scaling logs at Ketchikan Pulp Mill, 1957. USDA Forest Service photo 485139.

Among the issues was the decision to review the Forest Service timber sale on Prince of Wales Island to supply timber to satisfy the fifty-year contract to the Ketchikan Pulp Company. The simple decision to review the sale set off Senators Ted Stevens and Frank Murkowski. I have not yet heard from Congressman Don Young but that seems certain, given time.

It was also necessary to face up to the fact that given the situation in Alaska, for the first time the Forest Service simply cannot meet the volumes of timber we are contractually obligated to meet for the two long-term contracts. And that does not even consider the demand by the independent mill operators in Alaska, who are not protected by the certainty of contracts and who must, like all the other timber companies in the United States, compete for timber to feed their mills in the open market.

That was not made clear to me by staff, who asked me to sign, or let Regional Forester Mike Barton sign, a letter that tells Alaska Pulp Company at Sitka that its mill closure was in breach of contract and that we have supplied and would continue to supply the contracted timber volumes

Experimental block cutting of timber in the Maybe-so drainage, Prince of Wales Island, Tongass National Forest, Alaska, July 8, 1958. The Forest Service's fifty-year contract with the Ketchikan Pulp and Paper Company, signed in 1951, became an issue for the agency in the 1990s. USDA Forest Service photo 486654, by Leland J. Prater; Forest History Society Photo Collection.

to meet the needs of the mill.[1] Upon detailed questioning, it became obvious that the initial estimate of annual sale quantity projected in the draft forest plans of 400 million board feet was, in reality, likely less than 200 million. That is the volume of timber necessary to meet only one of the contract obligations.

I refused to sign the letter on the grounds that it simply was not true: the Forest Service cannot meet the contractually obligated timber volume over the longer term. I did, however, agree (along with Assistant Secretary of Agriculture Jim Lyons) that we would send a letter over Mike Barton's signature saying that we had met our contract up to this point and therefore the company was the one in breach of contract. If this turns out, in

1. The company was petitioning the Forest Service to allow continued cutting of timber, with sawlogs going to its mill in Wrangell and pulp materials to the Ketchikan Pulp and Paper Company.

whatever fashion, to be a cancellation of the contract, it will be a significant change for the better for the forest resources of Alaska.

Chief and staff also reached a conclusion that the Forest Service would, in the new Resources Planning Act report, announce the reversal of a long-standing policy on how roadless areas are treated in calculating the allowable sale quantity of timber. In the past, the policy under the forest planning process was to designate all of the land areas that would be open to logging. This became known as the timber base on which the ASQ was calculated. Included in this base were extensive acreages of roadless areas that were scheduled for roading and timber cutting. In reality, the vast majority of these areas were never "entered" (i.e., roaded and the timber sold) because of lawsuits or extreme controversy.

As time passed, the ASQ was never reduced and the areas available for logging activities were reentered more frequently than had been anticipated. This led, over time, to overcutting of those areas.

In the new policy, the Forest Service will calculate the ASQ on roaded areas only. The volume to be anticipated from roadless areas will be calculated separately and will not be included in the overall estimate until those areas are actually entered. I suspect that the timber industry will come unhinged over that. I don't care; it seems the logical and right thing to do, and for once, all the deputy chiefs concur.

1994

1 February 1994, Washington, D.C.

Over the past month I have sat through several meetings with the political appointees and their lawyers. They have come to dominate decision making concerning the actions to be taken in the Pacific Northwest in terms of management of the federal lands. These meetings often involved over thirty people, heavily dominated by the folks from the Department of the Interior, although the issue largely centers on the national forests, which are under the jurisdiction of the Department of Agriculture.

The issue has evolved to the point that the "decision makers" (who that includes is not at all clear and seems to change from meeting to meeting) are making continual changes in Option 9 to satisfy the environmentalists and the Fish and Wildlife Service, the Environmental Protection Agency, and the National Marine Fisheries Service field biologists. The "one government" approach to the Forest Ecosystem Management and Assessment Team effort almost instantly dissolved upon completion of the report, as the individual agencies involved reverted to turf battles and power games. In their efforts to mollify the environmentalists, "they" have agreed to double the buffer widths on ephemeral streams as a mitigation for species judged to be at significant risk and to provide additional dispersal habitat for spotted owls. In addition, they will provide permanent reserves of old-growth (no matter if they burn or blow down) and will confer with "appropriate people" as to where those reserves should be.

Since Tom Collier, Secretary of the Interior Bruce Babbitt's chief of staff, pushed this idea, it may be a good assumption that this represents some deal reached by the secretary. Neither he nor the group as a whole seemed impressed by the lack of a scientific rationale for this picking and choos-

ing of "crown jewels" by "appropriate persons." My impression was that
the whole thing was a done deal and it was pointless to argue.

It was obvious that the consuming political desire was to get clearance
for Option 9 (adequately modified, of course) from Judge Dwyer, and mol-
lifying the environmentalists was the way to do that. They don't seem to
get the point that you simply cannot mollify the environmentalists, or the
forest industry gladiators, for that matter. The clear strategy of the environmentalists has for years been to cut a deal to get what they can and then
come back for more. And why not? As a strategy, it works!

A strange thing has happened as the end game approaches: both sides
(the enviros and the industry) smell victory in the same set of circumstances.
Industry strategists seem to believe that their best chance to return to the
good old days is to cause a train wreck—a complete shutdown of the timber harvesting program on federal lands (or at least on the national forests).
This, in theory, would have such an effect on increased wood prices and
economic and social upheaval in the Pacific Northwest as to produce a legislative fix that might even include repeal of the Endangered Species Act
and other environmental legislation.

The environmentalists, on the other hand, smell total victory defined
as a shutdown of tree cutting on national forests. I would guess that they
would camouflage this desire with a lawsuit to demand Option 1 from the
FEMAT report with an attendant cut of 200 million board feet per year,
demonstrating they are not after a total shutdown in the timber program.
Of course, this cut level and attendant safeguards would make, in effect, a
total shutdown. They can't both be right. I do feel that the professional
resource managers are now on the outside looking in as the final scene in
this protracted drama plays out.

The technical aspects of the issue have been reduced to slings and arrows
wielded by political players and their lawyers. Perhaps we technocrats simply misunderstood what was actually going on and believed, at some point
in the past, that there was a technical fix that would transcend politics. The
lesson to be learned is that technical expertise has a role in defining alternatives and evaluating some crude probability of success (or failure) of alternative courses of action. Making the final resource management decision
is a moral issue best served in a democracy through political and legal means,
no matter how inefficient that may prove to be.

What is more frightening to me is that the multitude of "fixes" being agreed to in Option 9 may well interact in such a way that they reduce the already very low timber sale projections even lower. That would be all right, except that when a resolution is achieved, it is likely, very likely, that the politicians who are cutting the deals will have promised more, much more, than we can deliver.

The technical aspects of carrying out the plan will likely be further hampered by obviously inadequate funding and an inexorable diminution of the workforce needed to do the job the political operatives will have finally struck. This could prove very embarrassing to the Forest Service, BLM, and the politicians. Dean Bibles, Oregon state director of BLM; Regional Forester John Lowe; and I have gone over this with the appointees—over and over. They simply do not want to hear what we say. The striking of a deal (which they as former congressional staffers deem achievement in and of itself) is the paramount objective. Though the outcome will be the opposite of what has been advertised, the meeting of the promised timber targets will once more be the measure of success for the Forest Service. The timber industry has played the game beautifully in this regard. At every turn, they express doubt that FS and BLM can produce the timber volumes advertised. The political appointees (ignoring the warnings in FEMAT and the admonitions from FS and BLM analysts and line officers) make stronger and stronger promises to concerned senators and representatives that the targets can and will be met. As a result, success and failure will be judged on the basis of achieving the timber targets. This seems a perfect example of being hoisted on one's own petard.

13 February 1994, airborne,
en route from Washington, D.C., to La Grande

I am on a plane with Meg's ashes in a cardboard box. She wanted memorial services in La Grande, and that will happen next Tuesday. She fought the good fight for fourteen months and never complained a single time that I heard. But in a way, she won her battle in that she lived to see her first grandchild born and her husband become chief of the Forest Service. And, if indeed courage is grace under pressure, she showed her family what courage is.

She simply slipped away over the past three months, losing a little weight and a little strength each day. But I never saw any slippage of will until three days before she died [on February 10], when she simply said, "I want to let go now." And that is what she did. Within a short period she lapsed into a comatose state and then, when her two boys and grandson were all present, at home in Roslyn [Virginia], she roused herself to hug each one and say goodbye—one last exhibition of pure will and grace under pressure. Her husband and sons wept in that most remarkable moment, but she did not. Perhaps she was just too weak, but I don't think so. She was in control, and she was simply letting go after telling us that she loved us and that we would be all right. And we will be all right, as she raised no weaklings—not sons, and not husbands.

She had class and she was a lady. I will miss her sorely, but I will never be without her, either. We had somehow become one over our forty-one years as sweethearts and dear friends, and as long as I live, she will live as well. We asked God only for the courage to face her illness with grace and dignity. We were granted that, and I am grateful.

**17 February 1994, airborne,
en route from La Grande to Washington, D.C.**

Margaret's memorial service was supposed to signal the end of our life together and the beginning of a new day, but somehow, it doesn't seem that way. I have never felt more alone or seen life so bleakly as I do now. I didn't want to leave La Grande—it seems a mistake to have ever left. My heart is where she is and I desperately want to stay. There is simply no heart for what lies ahead in Washington.

But she wouldn't like what I am feeling now and wouldn't tolerate it— not in herself and not in me. There is good work—important work—to be done and a chance to make a difference.

There can be no flinching now. She knew what was to come for her and she never stepped back. And she knew that I wouldn't, either. She was always there to put some steel in my spine when it was needed—and she always will. I won't let you down—I won't let us down.

And so I will write no more of my Margaret, but she is with me always. Goodbye.

24 February 1994, airborne,
en route fromWashington, D.C., to Portland

I am in the air on the way to Portland to give a speech at the tenth annual meeting of the Rocky Mountain Elk Foundation. They have asked me to give a speech in the same time slot I filled at the first such meeting in Spokane in 1984. Unfortunately, I will arrive too late for the speech because of a late start from Washington. That late start resulted from a hearing in front of the Senate Appropriations Committee on the budgets for Interior and Related Agencies, which include the Forest Service. Secretary Babbitt handled testimony for the Department of the Interior and I testified for the Forest Service.

The hearing turned out to be nothing but a roasting of Secretary Babbitt and had very little to do with the budget. Senator Johnston of Louisiana chaired the committee and led the charge, and then turned Babbitt over to the tender mercies of Republican Senators Larry Craig of Idaho, Malcolm Wallop of Wyoming, Pete Domenici of New Mexico, and Mark Hatfield of Oregon. Senator Lighthorse Campbell of Colorado also charged in. Democrats, particularly any with a kind word for Mr. Babbitt, were conspicuous by their absence. Was this a signal? Or were they merely not interested in the budgets? Perhaps they were simply being cowards.

I sense a change in mood in the Congress. There is a weariness with environmental issues, and the conservative western senators suddenly have the opening to attack viciously and skillfully what they have identified as Babbitt's "War on the West." The Democrats (at least on this committee) do not seem eager to spring to the secretary's defense. In a sense the secretary has invited such attacks with his very, very high profile on these issues and the making of promises that could not be kept and now seem empty. He is a consummate politician, fast on his feet, and seems to love to perform. All in all, it was not a good day for Mr. Babbitt or for the good stewardship he espouses.

Very late in a drawn-out, three-and-one-half-hour session, Senator Dale Bumpers of Arkansas, the committee chair, arrived and spoke of what a tough job the secretary has and what a good job he was doing. That seemed to calm things down to some extent. Senator Akaka of Hawaii followed Bumpers and spoke well of the secretary, and the hearing ended.

The only questions directed to me of any substance were from Senator Hatfield, who made a point of asking the same questions that Congressman DeFazio had asked in an earlier hearing on the House side. How much money did the Forest Service ask for and how much was received in the president's budget to carry out the President's Forest Plan for the Pacific Northwest? The Forest Service essentially received one-half of what was requested from the Office of Management and Budget. The senator seemed to think that such was a rather serious difference. So do I and that was obvious.

As I rode back to my office, I had mixed feelings. The first was the lesson that the politicians will turn away from any appointee at any time if they perceive that such is politically expedient. The second was that sharing a witness table with a Cabinet secretary ensures that you will receive little attention from the senators, and that can be good or bad, I guess. The third was that having the Forest Service chief sit at the table with the secretary of the Interior seems to, or has the potential to, emphasize how completely the Department of the Interior is dominating administration action in the area of natural resources. All in all, it was an interesting day.

16 March 1994, airborne, en route from Sitka to Anchorage

It has been a long twelve-hour day of meetings with a variety of groups, first in Ketchikan and then in Sitka. The conversations centered on the fifty-year timber contracts negotiated after World War II to encourage economic development in Alaska. The groups are cleanly and clearly divided with no middle ground that I can see.

Elected officials at national, state, and local levels seem in essentially unanimous support of the continued fifty-year contracts. This was most easy to see in Sitka. Though the town and borough officials speak of jobs and social conditions, they seem most concerned by how they will meet their bonded indebtedness that they encumbered to build the power supply for the Alaska Pulp Company mill, and by the loss of their biggest source of tax revenue. The entire community has been sucked into heavy dependence on the fifty-year subsidy represented by the Alaska Pulp Company contract.

They pled, understandably enough, for the Forest Service to renegotiate a deal with APC to convert their pulp operations to a mill to produce medium-density fiberboard. This will produce fewer jobs but will buy Sitka a reprieve from their bonded indebtedness. The Sitka government had petitioned the state to forgive their multimillion-dollar debt for power facilities. During our meeting, the representative from the state of Alaska gave them the bad news that no relief was on the way.

However, the state appears willing to come up with $44 million to help convert the mill to MDF manufacture in order to produce 125 jobs. That is about $352,000 per job, and so far as I can see, there is inadequate feasibility or market research to give any measure of the probability of success. We also met with a group of some fifty or so folks of the environmentalist persuasion who told a very different story and asked that we cancel the fifty-year contracts.

All the meetings were polite and the presentations cogent and well made. It was obvious that the procontract groups could not conceive that an independent mill could survive and that the "stability" afforded by the contracts was essential to any chance of success. Further, they assume that if the APC contracts were canceled, there would be no significant timber program in Alaska. The city manager of Sitka said that he was convinced that there was "a widespread and well-organized conspiracy" to stop all timber harvest in Alaska.

Late in the day we visited the APC facility and were given a tour and an explanation of how the company was planning to convert the mill to an MDF facility. It was obvious from the briefing that such a conversion is a long way from a done deal. They have consultants looking at feasibility of the conversion as well as the suitability of the raw material for use in MDF manufacture. They vaguely discussed what quality of MDF might be manufactured to suit the available market. There was no clear vision of what and where those markets might be.

I watched the Alaska development fellow closely. He did not seem at all disturbed that the discussion was so vague. Some $44 million seems a rather large amount of money to commit on such a sketchy proposal.

28 March 1994, airborne, en route from Alaska to Washington, D.C.
We (Assistant Secretary Jim Lyons, Regional Forester Barton, and I) met with Governor Walter Hickel and his chief of staff for natural resources. Their pitch was direct and to the point. They want the Alaska Pulp Com-

pany contract modified and extended and the state is willing to put $40–$60 million on the line to see that happen. Further, they will oppose the institution of the PACFISH—Pacific fisheries—strategy[1] in Alaska, contending that there is no problem with fish habitat in Alaska. In fact, they seem to argue that the logging that has taken place has been beneficial to salmon.

The chief of staff for natural resources consistently pressed us for an answer to the question: "Why wouldn't it be a good idea to simply delay our decision for six months until Alaska Pulp Company finishes its feasibility studies for manufacture of medium-density fiberboard, the markets for that product, and the availability of necessary financing? Why wouldn't that be a good idea?"

There are answers to those questions that we cannot make without weakening the Forest Service's case as this issue moves into the courts. Our best case for the termination of the fifty-year contract can be made right now based on the situation of the moment. The APC pulp mill is closed (obviously permanently) and the employees terminated. If, as the Forest Service argues, APC has a contractual obligation to operate a pulp mill, APC is in obvious breach of that agreement. APC says they shut down because of uncertainty of wood supply under the fifty-year contract. That seems an incongruous argument in that they believe that the Forest Service would meet that level of wood supply if the mill were converted to MDF manufacture.

It is all damned complicated. This seems, in context, an incredible focus. In order to provide 125 jobs (down from 400 at the pulp mill) in an MDF plant in Sitka, we are considering continuing a highly controversial fifty-year contract that is straining the capability for sensitive ecosystem management and Forest Service capability to meet contractual levels while obeying environmental law and regulations. In addition, these sales have traditionally been below-cost, sales, which are becoming much less politically acceptable. And the state of Alaska will put an additional $40–$60 million ($320–$480 thousand per job) into the pot.

Compared with the situation in the Pacific Northwest, where the likely management strategy will reduce the annual cut from over 4 billion board

1. U.S. Department of Agriculture and U.S. Department of Interior. 1995. Decision notice/decision record, FONSI. *Environmental assessment for the interim strategies for managing anadromous fish–producing watersheds in eastern Oregon and Washington, Idaho, and portions of California (PACFISH)*. Washington, DC: USDA Forest Service.

feet per year to less than 1 bbf per year, this is extremely minor. And when it is considered that these jobs will not appear for three to five years, the situation becomes even more puzzling as to its viability. Would not political figures in Washington and Oregon question the double standard of governmental concern over 125 jobs in Sitka compared with thousands of timber-related jobs in the Northwest? Is it not inevitable that there will be listings of threatened species or problems with the viability of certain species in the foreseeable future? If so, what is the long-term trend in timber harvest on Alaska's national forests? There is little doubt that this trend will be down over time under current law.

What to do? We will meet in Washington next week to make the promised decision. It was, somehow, easier to be a research scientist able to pontificate freely while not being responsible for decisions or the consequences of those decisions.

29 March 1994, airborne,
en route from Washington, D.C., to New York

This has been an interesting day with two significant issues coming simultaneously to bear. Some months ago, the Friends of Animals threatened to sue the Forest Service over allowing states to proceed with spring hunting of black bears over bait. I made the decision to respond to that threat by publishing a notice in the *Federal Register* to the effect that the policy of the Forest Service was to defer to the state wildlife agencies to make such decisions. In that notice, the Forest Service requested public comment for a period of sixty days on that decision. The emphasis fell on Wyoming. The Forest Service had excluded all bear baiting in categories 1, 2, and 3 of grizzly bear habitat. Deputy Chief James Overbay had earlier agreed, in a letter to the Friends of Animals, to solicit public comment on the proposed policy prior to allowing bear baiting in Wyoming in the spring of 1994.

Because of the change in Forest Service chiefs and the change in deputy chiefs for National Forest System, the decisions on bear baiting got delayed until it was too late to submit the policy for public comment. I didn't consider what we did as a change in policy but rather as a continuation of present policy.

The Forest Service had tried, without success, to get the states to withdraw from baiting of bears as a legitimate hunting technique. We tried, over and over, to get the states to recognize that bear baiting was repugnant, not in keeping with any aspect of fair chase, and was a perfect target for the animal rights groups that wanted to push the Forest Service (and other federal land management agencies) into the business of regulating the hunting (or other treatment of animals) on federal lands.

The accepted practice has been for the federal government to defer to the states in the matter of the regulation of hunting on federal lands. The states maintain that such is a constitutional right. Many lawyers, including many who represent the Forest Service, maintain that the Forest Service has not only the right but the clear responsibility to regulate such matters.

The Friends of Animals, and other such groups, very much want to press that issue. They deem that it would be easier to influence wildlife management on national forests from the federal level than through the states. This would seem likely, since the dedication to hunting and trapping is more powerful in the more rural, less populated states where national forests largely occur than in more densely populated states without national forests.

It seems likely that the Friends of Animals want to use the bear baiting issue (which is very easy to attack) to press a federal agency to prepare an environmental assessment or an environmental impact statement on the agency's "permission" to conduct a particular type of hunt or means of taking. Once that line is crossed, there is, in my opinion, no going back. Federal land management agencies will be in the wildlife management business.

The Department of Justice attorneys declined to defend our position in court. So they will go in and offer to have the Forest Service cancel the bear baiting in Wyoming for this spring hunt and put out the proposed rule for an extended comment period. This is being done in the hope that it will avoid a judgment that will do even more damage to the Forest Service's position that the states have jurisdiction over the hunting of wildlife—that is, that we are not required to perform an environmental assessment of state exercise of that prerogative.

I did not argue with either the conclusions or the strategy essentially forced on me by the Department of Justice attorneys and the general counsel of the Department of Agriculture, James Gilliland. However, they gave me no choice. I would have preferred arguing the case and taking the risk of having the judge granting a temporary restraining order halting the bear season. At least it would have been clear to all that we had been dragged over the state's-rights line against our will and through none of our doing.

The International Association of Game and Fish Commissioners inter-vened on the side of the Forest Service position, and the case was decided in favor of the Forest Service; that is, the states retained the right to set hunting regulations.
—*JWT*, 2003

8 April 1994, Washington, D.C.

We (Assistant Secretary Jim Lyons; Adela Backiel, deputy assistant sec-retary and the acting chief of staff to the secretary; and a lawyer from the Office of the General Counsel) met with Secretary of Agriculture Mike Espy concerning the decision on the status of the Alaska Pulp Company's fifty-year timber contract. He was presented with two clear options: terminate the contract or provide an extension to consider the possibility of the con-struction of a medium-density fiberboard plant as a substitute for the Sitka pulp mill permanently shut down by Alaska Pulp Company.

We discussed the pros and cons of the two options for nearly an hour. The unanimous staff recommendation was to terminate the contract. The secretary agreed and departed.

Lyons instructed me to prepare two decision letters and news releases to match and to continue as if both decisions were viable and still under consideration. It is a tough political decision but the correct legal and pro-fessional decision for the forests of the Tongass National Forest. We were, collectively, proud of Secretary Espy.

12 April 94, Washington, D.C.

Late in the afternoon, I was called to Assistant Secretary Lyons's office to discuss the APC decision. I thought that was somewhat strange in that, so far as I knew, the decision had been made to terminate the contract. My assumption was that the meeting was to discuss the plans for announcing the decision and dealing with the political ramifications.

When I arrived, the message was that Secretary Espy wanted our pres-ence for a meeting between the secretary and Senators Stevens and

Murkowski of Alaska. The meeting was intense. Senator Stevens, in his best bullying manner, made the case for Alaska Pulp Company. He ran the entire gamut of polemic statements, including what a debt we owed to the Japanese investors that had financed APC (though they had just closed an antiquated pulp mill after they had "used it up"); the 400 jobs that would be lost in Sitka (not mentioning that the jobs were already gone and that a medium-density fiberboard plant would employ 125 people three to five years hence); the likelihood that we would be successfully sued by APC (not mentioning that the Forest Service was already being sued by APC and that the Forest Service's chances of losing that suit would be increased by an extension); that he could not believe that the Forest Service—and by implication the chief—would be so stupid and shortsighted (knowing full well that the decision was being made at the political levels in USDA); and implying that the secretary should elevate such an important decision to the secretary himself (clearly understanding that the secretary would never step into such a position to be a clear target for criticism).

Senator Murkowski was more subdued and polite. He eschewed the legal arguments—for whatever reason—and concentrated on the policy issues of potentially discouraging Japanese and other foreign investors in American enterprise, the social and political consequences of "reneging" on a solemn promise of the U.S. government to the citizens of Alaska, and the retention of the good will of the Alaska delegation to the business of the secretary of Agriculture.

The secretary listened politely and nodded understanding or agreement (it could be interpreted either way) at appropriate points during the senators' presentation. I could see something in his eyes and demeanor that I didn't like.

The meeting ended and we each shook hands with the senators, with the secretary being first. While the rest of us were saying goodbye, the secretary without saying a word walked away to his private office, followed in short order by the acting chief of staff, who told us to wait.

She returned after about twenty minutes, obviously somewhat upset, and told us that the secretary, in spite of her arguments, had reversed his decision and we were now instructed to proceed with the decision to give APC four additional months to complete the assessment for an MDF plant.

He did give instructions that the agreement to be negotiated for the extension was "to fully protect the interests of the United States."

We were dumbfounded, to say the very least. Why the sudden switch? Was the secretary persuaded by Senator Stevens's bluster and legal arguments? Was he swayed by Senator Murkowski's appeal and examination of policy implications? Both? Or was he fearful of alienating two senators while he was trying to get the USDA reorganization plans through the Senate?

13 April 1994, Washington, D.C.

Deputy Assistant Secretary Adela Backiel called to tell me that Secretary Espy had changed his decision yet again. Now we have been instructed to terminate the contract and, simultaneously, offer to try to negotiate a new contract for ten years under the authority of the Forest Service. My new instructions were to prepare news releases to cover this new option.

After I hung up, I puzzled over the logic in this new decision. The termination will certainly anger the Alaska delegation, and the pie-in-the-sky of a new negotiated contract will not assuage their anger and feeling of betrayal. Conversely, the termination will please most of the people who care one way or the other. But the promised negotiation will feed their deep distrust of the Forest Service and erode the reputation of the Clinton administration as strong environmental advocates even further.

There is one certain consequence: the Forest Service will get caught in the middle of the nasty fight to come. The industry and the politicians will pound the agency to produce a contract, which is highly unlikely given the developing circumstances of timber yields' likely decreasing in Alaska. The environmentalists will bring down the wrath of their constituencies and their champions in Congress on the Forest Service as the people who are negotiating the contract, which in their minds and mythology was a sneaky act by the chief.

[nd] April 1994, Washington, D.C.

I discussed the final decision concerning Alaska Pulp Company's fifty-year contract with Regional Forester Mike Barton. I told him that if he did not agree with the decision, I would take the decision out of his hands and make it at my level. To his great credit, Mike said, "No, I am the regional forester and the designated contracting officer and I will make the decision. It's my job." He stood tall at that moment.

I told him to make the transition as easy as possible for APC in order to minimize the negative impact on the folks working in the woods and in the Wrangell mill. That will probably cause some consternation with the environmental community and some political pressure. But it is the decent thing to do and, I believe, it is good business to carry out the already sold sales.

So the decision is finally made. It is, without doubt, the right decision for the land. But how will it play in Alaska politics? It seems likely that each such controversial decision nibbles away at the length of tenure of the chief. Maybe that is for the best. Make the tough calls until you are used up and move on. It seems right to recognize that probability and be prepared for the inevitable.

[nd] April 1994, Washington, D.C.

This seems, to me at least, a historic day for the Forest Service and for conservation in the United States. The final record of decision for institution of the President's Forest Plan for the Pacific Northwest has been signed. This plan is simply the biggest, boldest plan for forestland management ever put into action in this or any other country. The plan covers federal lands stretching from the Canadian border to central California. A series of late-successional and old-growth forest reserves are arrayed down the Cascade and Coast Ranges, each stream is protected by buffers varying from 300 to 100 feet in width, and there are provisions for protection in the 22 percent of the federal forestlands open to timber harvest and management. It is a new day and represents the end of a five-year effort by various teams of scientists and managers of which I have been part: the Interagency Scientific Team, the Gang-of-Four, the Scientific Assessment Team, and the Forest Ecosystem Management Assessment Team.

After all the debate, after all the political maneuverings, after all the technical and personal attacks, after all the legal battles, the strategy so carefully put together stands as the president's plan. This produces a strange feeling and some bemusement. The assignments given to me seemed likely to end my career and have instead led to my appointment as chief of the

Forest Service. Now, I and the heads of the Bureau of Land Management, the Fish and Wildlife Service, and the National Marine Fisheries Service must make the intricate plan work. Is that poetic justice or, more likely, a simply bizarre turn of fate?

It was announced yesterday that the Forest Service had cancelled a fifty-year timber sale contract with Alaska Pulp Company based on that company's closing its pulp mill in Sitka. This decision was made in the face of intense political pressure from the entire Alaska congressional delegation (Senators Stevens and Murkowski and Congressman Young), Governor Walter Hickel, and most of the Alaska state legislators. It was a difficult task to persuade Secretary Espy to support the decision, and then to stay with that decision once it was made.

As I was traveling to the airport after briefings in the Department of the Interior with three groups—the interested members of Congress or staff, the press, and key interest groups—Senator Murkowski caught me on the car phone. He seemed more frustrated and pained than angry. At the end of the conversation he essentially challenged me to appear in Wrangell and Sitka to defend the decision to the workers affected by that decision. He seemed taken aback when I said that I would do that. On that, the conversation ended.

So, on this single day, the Forest Service took a momentous step away from the timber industry–dominated policies of the past and took a giant step toward becoming the "Conservation Leaders for the 21st Century"— my dream and goal. While I doubt that the significance of this day will be immediately apparent, it seems likely that historians will see it for the turning point that it is.

2 June 1994, Washington, D.C.

I had a several-hour meeting today with my range management folks and the attorneys who are representing the Forest Service in one of four ongoing suits over grazing. This particular suit is called the California Trout case, or CalTrout, after the litigants. The gist of the case, which concerns the Forest Service in the northern Sierra, is that the agency is allowing the grazing permittee to continue grazing even though there is no present compliance with the National Environmental Policy Act. CalTrout is asking that grazing on the allotment in question—and all the allotments on that

national forest—be shut down until NEPA requirements for an environmental impact statement are met. These four cases are all similar in that
respect.

There is no doubt in anybody's mind that the Forest Service is in violation of the requirements of NEPA in these cases and that the plaintiffs
will prevail if the cases go to trial. The Forest Service attorneys know that,
the grazing permittees know that, and the plaintiffs know that. Yet the
plaintiffs want to settle the case before trial. Why?

The objective of the suit is not to force NEPA compliance except over
the longer term. The plaintiffs do not want to shut down grazing on the
vast majority of allotments that are not in compliance with NEPA. Such
would be extremely risky, given the current political uproar over a long
string of perceived environmentalist victories in the federal courts and the
political backlash that is growing among conservative western members of
Congress, particularly in the Senate. The grazing issue on public lands has
become a cause célèbre following Secretary Babbitt's thrust in "grazing
reform," which has been fairly badly botched. Babbitt was forced to back
up on the issue by the president's need to placate those western senators
and play down the growing political impact of the "War on the West."

What the plaintiffs clearly want is for the Forest Service, and other federal agencies, to make some significant moves to more firmly control livestock grazing in order to produce an improving range condition, particularly
in riparian zones and aquatic habitats. So the strategy becomes using the
handle of noncompliance with NEPA to leverage a settlement whereby the
Forest Service and the permittees will agree to monitored standards of performance that will ensure that negative impacts on riparian habitats are
decreased to the point that recovery of streamside vegetation, including
shrubs that provide shade over the stream, will be enhanced as streams narrow, with decreased bank sloughing caused by livestock trampling.

It becomes immediately evident, to me at least, that by this agreement
the Forest Service (and probably all federal agencies) has set the process of
management for all of our allotments that are not in compliance with the
requirements of NEPA. Very probably this precedent may well determine
our grazing program as each allotment is brought into NEPA compliance
on a five- to ten-year schedule. This seems, again at least to me, a strange
way to set grazing policy and totally outside the planning approaches pre-

scribed by the National Forest Management Act and the National Environmental Policy Act.

23 June 1994, airborne, en route from Houston to Missoula

A meeting of the top leadership of the Forest Service has filled the past week. Chief and staff, Washington office staff directors, forest supervisors, experiment station directors and assistant directors, and a staff of fifty or so that put on the meeting. There were nearly three hundred people in attendance.

Holding this meeting was something of a gamble from several viewpoints. The first risk was one of perception. The Forest Service has just reduced its rolls by 2,300 positions, and holding this meeting now makes the agency vulnerable to the criticism of wasting money on a fancy meeting for the brass while the troops suffer.

The second risk was getting a negative response from those attending the meeting. The two previous meetings of forest supervisors were a disaster of dissension that produced continuing rancor. Given the current state of morale, this was a real risk. Most of the deputy chiefs recommended against such a meeting and have remained negative and somewhat critical as the preparation went on over the past months.

I thought the meeting was essential to take advantage of the change in leadership and the window of opportunity it provided to perhaps get a turnaround in attitude, from one of pessimism and cynicism to one of optimism and faith. That was a tall order. Assistant Secretary Lyons backed me completely, and that was critical to getting the job done. He absorbed the criticism from the Department of Agriculture so that we could go ahead. Obviously, he stood ready to take some of the heat that will come from members of Congress and the press. You can't ask for more than that.

I was most pleased that Lyons was willing to ask, by questionnaire before the meeting, about how the participants felt about relations between the Department of Agriculture and the Forest Service. The troops let him have it, with a barrage of criticism for micromanagement and lack of trust for the Forest Service. He took that criticism to heart and made his peace with a speech that was a masterstroke. It was a pleasure to watch. He wants so badly for the Forest Service to make the transition to the new Forest Service that will be the conservation leaders for the 21st century. He put out his hand and the Forest Service took it. Good.

The theme of the meeting was leadership. I told them—over and over—that I wanted and expected each of them to be a leader. I told them—over and over—that I wanted and expected them to instill, cultivate, and cherish leadership in their people. Each person in the Forest Service is to receive a pin that shows a Forest Service shield superimposed on a map of the world. The words around the outside read "Conservation Leader." That pin will be a part of the Forest Service uniform for the next year.

I made them a promise to remain as chief so long as my presence was desired by the administration in power and so long as I considered my presence more a help than a hindrance to the agency. It seemed that most of them liked that idea. I admit to being caught up in the spirit of the meeting, but it seemed to be the right thing to do. The change we wanted is on the way. The genie is out of the bottle. Give me just two more years and there will be no going back. In two more years, I can make enough right appointments to nail the change in place.

23 June 1994 [later that same day], Washington, D.C.

I was intensely bothered by one set of behaviors so evident at the leadership meeting. People are incredibly uptight about being exactly "politically correct" in language and demeanor. One of our invited speakers—Sherry Sheng, director of the Portland Zoo and member of the National Forest Foundation—gave a great speech. Throughout, she used humor directed at the white males (or merely males) in the audience. When I stepped to the lectern, I said, "I am trying to learn to be a sensitive, caring kind of guy. By the way, Sherry, that was a great speech—for a girl." She, and most of the audience, cracked up. But later in the work groups, some voiced serious upset with my lack of political correctness and insensitivity.

In my final speech, I advised folks to lighten up. I said that we talked constantly about honoring cultural diversity but the fact that the culture of someone like me (southern born, remembering the Depression and World War II, lower middle class, pioneer in conservation in an all-male environment) was invalid aggravated the hell out of me. I said that honoring cultural diversity was O.K. by me, but I demanded respect for my

Chief Thomas receiving "pooper scooper salute" at Wilderness Training Center, Nine Mile Station, Montana, June 1994. Photo courtesy of Jack Ward Thomas.

culture in return. I said that if being chief required me to give up my personality, the price was simply too high. I meant it. Most folks applauded.

I said that the Forest Service was no longer settling personnel cases simply because it was expedient and, superficially, cheaper than carrying the case to conclusion. If we are wrong or if we have a weak case, we will settle. If we have a strong case, we will hang tough.

The situation has simply spun out of control. Settling every case may have made sense in the beginning, although I doubt it. This course of action leads to the belief on the part of some employees that any grievance will be settled. Such is both inefficient and extremely costly in terms of troop morale.

Personnel shops will resist my decision because it puts pressure on them to do very detailed, very good personnel actions. Their lives are complicated by fighting personnel cases through to the end.

I am convinced that this is the correct course of action. From the applause, it was my impression that the audience felt the same. We will see how much resistance comes from the Forest Service personnel shops and how much from the "Civil Rights Mafia" in the Department of Agriculture.

A system that rewards bad performers, troublemakers, and scofflaws because it is cheaper and easier than doing right can produce nothing good over the longer term. Not to stand up against such a circumstance is dereliction of duty. Now, let me see if I can make it happen. I don't think it will be easy.

[nd] July 1994, airborne,
en route from Colorado to Washington, D.C.

I spent four days in the area of the Canyon Creek fire on the outskirts of Glenwood Springs, Colorado. I first got word of the fire when Deputy Chief Lamar Beasley called me at home at about midnight with the message that ten firefighters were confirmed dead and forty were missing on Storm King Mountain in western Colorado. Lamar had already formed a review team headed by Deputy Chief Mark Reimers, and they would depart at 6:00 A.M. He didn't know much more than that, except that the crew was a mix of Forest Service and Bureau of Land Management firefighters. I then called Mike Dombeck, acting director of BLM, and made arrangements for the two of us to travel to Colorado the next afternoon.

We were met in Denver and driven to the Forest Service regional headquarters, where we were given a quick briefing on the situation. Things were obviously very confused, but it was now clear that there were twelve confirmed dead, two missing, and the rest accounted for. Four of the confirmed dead were women, out of the five engaged. The fire on Storm King Mountain, just above Glenwood Springs, had been ignited by dry lightning several days before but because of its location on a ridge and other factors, it had been given a lower priority for attack than other fires in the vicinity.

Mike Dombeck and I spent several hours visiting with the survivors of the incident. They were badly shaken by their experiences and many were crying off and on, others simply stared into space, some were sullen, and some seemed quite calm and able to tell jokes and laugh. They had two messages for us: Get them home and away from the madness, and keep the "stress counselors" in white shirts and ties away from them. They seemed appreciative of our presence and concern. It took some time for them to

Prineville Hotshots, Storm King Mountain fire, Colorado. Photo by David Frey.

recognize us as the chief of the Forest Service and the director of BLM, quite understandable for seasoned firefighters.

I spent much of my time with the crew chief of the Prineville Hotshots, as requested by his crew members. These elite firefighters seemed most concerned about his welfare. They were anxious that there would be a witch hunt and that their crew chief, as the low-ranking person involved, would be the sacrificial lamb. Many of them said, over and over, that they were worried about their leader and how much they owed to his leadership and mentoring. Several of them said they would go back up the mountain if he asked them to.

When I introduced myself to him, he said, "I'm sorry. I'm so sorry. I let my crew down and half of them are dead. It is my fault—but I don't know what I did wrong." I told him he shared responsibility with me and the others. He was not alone in this thing. I told him what his crew had said about him—that they loved and admired him. He put his face in my chest, and when my arms encircled him, he began to sob uncontrollably, and so did I.

Many of the survivors asked the same question over and over: What happened? What went wrong? We told them that the investigation team would figure that out. In my own mind I was much less certain that it would be very clear and certain—ever.

In retrospect, however, it will likely be true that there were a dozen decisions that, if made differently, would have yielded a different result. In the end, I suspect, these top-of-the-line wildland firefighters will emerge as "victims" of some ineptitude or callous decision makers. There will be calls for "independent" investigations. And it will go on for years and years.

The pounding in the press conference made it clear that some will insist that someone be found at fault. The intimations in the questions hinted, rather strongly, that a coverup was under way. There is no truth in that. The joint BLM–Forest Service team is composed of the best of technical experts and headed by senior officials of unquestioned integrity. I suspect that the results will simply be that a freak event occurred and, given that wilderness firefighting is inherently dangerous, fourteen people died.

We made all the arrangements possible to ensure that survivors and the families of those killed received all the attention and assistance that the Forest Service could provide. I believe that this has been and will be done. For example, I instructed that each of the bodies be escorted by a regional forester as it was delivered to the place designated by the family. A Forest Service DC-3 delivered the bodies to Prineville, Oregon, and McCall, Idaho.

The response of the people of Glenwood Springs was most heartwarming. I was wearing a uniform, and people would stop me on the street or walk up to me in hotel lobbies and restaurants to say, "Thanks and God bless you." Restaurants would not accept payment from us for meals. A fund was started by a Glenwood Springs bank for the families of those killed and injured in the line of duty, and money flowed in. Everyone was wearing a piece of purple ribbon in remembrance of those who died. Purple ribbons were tied to phone poles, gates, car antennas, doorknobs, and elsewhere. Somehow, seeing people sharing in loss and, at the same time, expressing appreciation was uplifting at a time of very low spirits.

Though the press behaved well enough overall, there was enough poor behavior to last me for more than a while. Some reporters somehow obtained the fire-resistant yellow shirts and green trousers of firefighters and, after an appropriate "dirtying," passed themselves off as firefighters just in off the firelines. They intruded into stress debriefing sessions for fire crews, sidled up to conversations between people working with the firefighting coordination, tape-recorded conversations, etc. Fire crews, particularly those who had survived the Storm King Mountain incident, complained of being harassed by the press. Then, to top it all, some press people from the *Denver Post* violated the restrictions on entering the burned area, imposed to protect evidence and prevent injury from rolling rocks, to get pictures so that they could scoop the competition. They were arrested by law enforcement officers. Others threatened to take the risk and pay the fines to get the pictures they wanted. We had to agree to take a "pool" cameraman into the burned area to get pictures to be shared by all—an expedient compromise with the devil.

The role of the politicians was something else again. There is a fine line between politicians' playing out their roles as leaders of government and using such circumstances for personal aggrandizement. Governor Roy Romer, running hard for reelection as governor of Colorado, was on the scene almost immediately and took a prominent role in dealing with the media. Secretary of the Interior Bruce Babbitt was not far behind and likewise provided "press opportunities." Secretary Espy did not show up (he was dealing with flood catastrophes in Georgia) until things had calmed down. Bless them all.

At the end, I decided to go back to Washington rather than go on the plane delivering the bodies home. After talking to Secretary Espy, it seemed likely that we can ride the political wave to get some benefits for the families of those killed and injured through an amendment to the Appropriations Bill for Interior and Related Agencies that is pending joint action by the House and Senate. I decided that is a better service to the fallen and injured than the symbolism of escorting the bodies home. I hope that is the right decision.

12 July 1994, Washington, D.C.

Two days after arriving home from the Canyon Creek Fire, Deputy Chief Lamar Beasley called me at home at about 9:00 P.M. with the word that a contract helicopter was down near Silver City, New Mexico, with four Forest Service firefighters on board. There were three dead and two injured. The helicopter was ferrying firefighters between two fires when it crashed while trying to land on a ridge. The helicopter had burst into flames and started yet another fire. Retardant bombers and helicopters dropped retardant and water, respectively, directly onto the crash site. That was all he knew.

He reminded me that we needed to get an investigation team formed and on the way to the scene. Beasley recommended those who should be included in the Forest Service contingent. I agreed.

I felt sick to my stomach; adding these deaths to those of the previous week in the Canyon Creek Fire, we have lost seventeen halfway through the fire season. Heartsick, I called my administrative assistant, Sue Addington, and told her to arrange my travel to Silver City for the next day.

I arrived about 7:00 P.M. and proceeded directly to the parents of one of the dead firefighters. The father was a third-generation Forest Service employee, and his son had been training to be a wildlife biologist and was to be a fourth-generation employee.

The atmosphere was tense and strained at the beginning of the visit. The father had said earlier, "The Forest Service killed my son." Finally, the father began to talk and to cry. I walked over to him, put my arms around him, and began to cry myself. I sat in front of him on the floor and held his hand as words began to tumble out. He talked about his boy, about the Forest Service, about the accident, about his family. He had worked in the firefighter ranks all his life and had been in the spotter plane that arrived at the crash site only minutes after the accident. As he watched the helicopter burn and directed another helicopter with emergency medical technicians to the crash site, he did not know that he was watching his own son burn.

We went to the hospital to visit with one of the young men who survived the crash and who was hospitalized with a wrenched ankle and knee. The hospital had a guard on duty to exclude the press, at the request of the family. He appeared to be in good spirits and happy to be alive. He described the crash in simple terms. The helicopter used a straight-in landing approach, and very near the ground something happened—maybe a wind gust—and the helicopter began to spin to the left and then turned over and crashed and tumbled down the mountain. He did not know how he got out of the helicopter. He repeated several times that he did not want to talk to the press. I promised him and his mother that we would have someone at the hospital the next day to help them with whatever paperwork was required to pay the hospital bill and gain workers' compensation.

We ran into the inspection team at supper and I talked with them for a few minutes. They didn't seem to know much more than we did, having only gotten the team fully organized at 4:00 P.M., just twenty-four hours after the accident. Not too shabby! I was extremely tired and a little jet-lagged. The bed felt great.

We were up and going at 5:30 A.M. for a visit to the ranger district on the Gila National Forest that was home base for the firefighters. When we arrived at the ranger district, a reporter and a photographer were waiting. I told them that we would visit as soon as our family meeting had concluded.

The entire firefighter contingent in attendance was Hispanic males ranging in age from late teens to the midforties. I spoke to them briefly about the sadness that everybody in the Forest Service felt about the three people killed in New Mexico, in addition to the fourteen who had died the previous week in Colorado. I then asked if they had anything to ask and to tell me. The silence was deafening though I detected no animosity. It seemed much more likely that they simply had no idea who the chief of the Forest Service was, nor did they care. Theirs was a much smaller, much more self-contained world, and their grief and healing were contained in that world and not in the larger Forest Service. That is understandable enough.

We drove back to Silver City and boarded our charter plane for a short hop to Las Cruces, the home of the second firefighter who was killed. The mayor of Las Cruces, who was the uncle of the dead firefighter, met us upon arrival. At the home when we arrived were the young man's grandparents, uncles and aunt, and his wife, seven months pregnant with her first child.

Some of the family seemed upset that they were reading answers in the newspapers and intimated that we were withholding information from them. I explained that most of what I had read in the press was either wrong, partially correct, or represented information that we did not have. I gave instructions for the Forest Service to provide updates twice a day on information that we were certain was correct.

I felt good about the visit, perhaps for my sake as well as that of the family. I had done what I considered the right thing to do and felt better for having done so.

These past two weeks seem like a bad dream that simply will not end. I have seen the best and most noble of human behavior and the most despicable—the good, the bad, and the ugly. I pray to God that there are no more casualties this fire season.

17 July 1994

I received word yesterday that Secretary of Agriculture Mike Espy intended to attend the memorial service in Silver City, New Mexico, for the three killed in the helicopter accident on 12 July: Bob Boomer, the pilot, and Sean Gutierrez and Sammy Smith, crew. Espy asked me to attend and speak at the service.

I flew down from Portland to Albuquerque (meeting up with Jim Lyons in Denver) and on to Silver City by charter aircraft. We were escorted from Albuquerque by Chip Cartwright, regional forester.

The memorial service was held at an outdoor auditorium on the campus of Western New Mexico State. The families of Boomer, Gutierrez, and Smith were present, as were the two survivors of the crash, Johnny Lopez and Charlie Sanchez. There was a crowd of some 300—a mixture of Forest Service folks (mostly firefighters), families and friends, and townspeople. Speakers included Senator Bingaman of New Mexico, Secretary Espy, the lieutenant governor, Silver City's fire marshal, Cartwright, and me.

The irony of these terrible losses is that the wildland firefighters have become so adept at their jobs that we tend to forget that the work is inherently dangerous. Flying in small aircraft over rough terrain, jumping out of airplanes, and parachuting down to fight fire—this is dangerous. Flying in helicopters and hovering over rugged terrain and rappelling from helicopters—the entire thing is a very dangerous business, and there is no choice. My responsibility is to minimize the risk and maximize the effectiveness of our firefighting efforts under those safety constraints.

18 July 1994, Portland, Oregon

I left Denver at 5:30 A.M. and arrived in Portland in time for an 11:00 A.M. briefing with BLM Director Mike Dombeck and staffs from Washington and the Northwest (about thirty, including John Lowe, regional forester for Region 6; Elaine Zielinski, state director for BLM; and Charles Philpot, director for the Pacific Northwest Station). The subject was PAC-FISH—the Pacific fisheries standards for habitat protection for aquatic systems on federal lands within the range of anadromous fish. These standards call for interim protection of 300 feet along perennial streams, 100 feet along tributaries, and 50 feet on ephemeral streams until watershed assessments are available to guide the treatment of these areas. Mike and I made the decision four to five months ago to institute this strategy throughout the entire Columbia basin on federal lands until a full assessment becomes available in two to three years.

Of course, this decision has proven controversial as it has potential negative impacts on both timber yields and grazing. Arguments have ranged from "there are no habitat problems for spawners and rearing, the problems are dams, fishing, etc., etc.," to arguments over process and we should do an environmental impact statement instead of an environmental assessment, etc. As in other cases, when one wants to ignore technical questions (which always go to the courts' assumptions of technical credence of the government), it is well to argue the technical aspects of the process. That is not a bad strategy, as it has prevailed in a number of cases. The result of a victory for the plaintiffs is a stay on the activity proposed, bringing management to a standstill. This train wreck strategy produces a cessation of management activity that is assumed to be so politically unacceptable as to produce a legislative fix.

Such an approach has never worked to the satisfaction of either extreme side in the debate over appropriate use of federally owned land. However, it is still used as a last-ditch defense when the execution of the proposed management action is perceived as even worse. Therefore, I would anticipate legal action on the process of the institution of the PACFISH strategy on an interim basis. It is not clear what the federal courts would decide. Our lawyers say that our approach is legally risky but that we have a chance of prevailing on the merits of the case—particularly now that several races of salmon and the bull trout are listed as threatened by the National Marine Fisheries Service.

However, we have a more immediate problem with "consulting," as defined by the Endangered Species Act, with the National Marine Fisheries Service in its regulatory role as enforcers of the law for threatened or endangered anadromous fish, and with the Fish and Wildlife Service, which serves the same role for resident fish. The result? The Forest Service has stream stretches that have both the threatened salmon and bull trout, and it is necessary to consult with two separate federal agencies on any of the activities that may have an influence on that stretch of stream. It seems that one agency could defer to the other in such circumstances. But that time is not yet.

It seems likely, to me at least, that the National Marine Fisheries Service, left to its own devices, would never have had the foresight nor the pure guts to install PACFISH as a management strategy for anadromous fish throughout the range of those fish in the Pacific Northwest. However, since the Forest Service and the Bureau of Land Management took the lead, they have not been at all hesitant to begin to impose additional standards for consultation on proposed activities.

These standards are being imposed by relatively young, inexperienced biologists who seem remarkably naïve as to the social/political/ecological game in which they are engaged. The entire scenario is frighteningly reminiscent of the Fish and Wildlife Service's handling of the issues surrounding the consultations on the northern spotted owl and the marbled murrelet, both old-growth forest–dependent species in the Pacific Northwest. The situation with the Fish and Wildlife Service has been dragging on for nearly five years, and as their young biologists continue to make calls and impose additional standards, they keep the Forest Service and the Bureau of Land Management from any type of methodical approach to management of the forests of the Pacific Northwest.

The answer is, in both cases, for the heads of the National Marine Fisheries Service (Rollie Schmitten) and the Fish and Wildlife Service (Mollie Beattie) to name regional directors who have the guts and smarts and the experience to deal with the issues astutely. Issues of this significance— economically, politically, and ecologically—simply cannot be left to the whims of young, idealistic biologists. What they conceive of as standing at Armageddon and battling for the Lord is, in my opinion, ill advised and an overreaction. They go too far and, in doing so, risk throwing away one of the greatest victories for good natural resources management ever achieved, or ever likely to attained again.

This is simply too important to be left to any but those in the highest positions. I have asked for an immediate meeting of the directors of the Bureau of Land Management, Fish and Wildlife Service, National Marine Fisheries Service, and the Forest Service to lay this out on the table and see if we can inject some sanity into the situation. Otherwise, I think the Forest Service and the Bureau of Land Management will be unable to produce even the much-reduced timber levels expected under the President's Forest Plan for the Pacific Northwest, and I predict a stagnation of all forest

Chief Thomas with John Twiss, director of Wilderness Management, and Deputy Chief Grey Reynolds, Bridger Wilderness Area, Wyoming, July 1994. Photo courtesy of Jack Ward Thomas.

operations in the Columbia basin. Those who care about good, cautious, environmentally sound management of renewable natural resources are, again in my opinion, well on the way to snatching defeat from the jaws of victory.

At 2:00 P.M., the meeting ended and I went on annual leave, called a cab, and was off to the Portland airport for a flight to Pendleton. Kay Pennell, who was my administrative officer when I was at the La Grande laboratory, met me at the Pendleton airport and delivered me to La Grande. Tomorrow Bill Brown, my old hunting partner, and I are off for the high Wallowas in the Eagle Cap Wilderness.

I need the wilderness now, as a parched man needs water. I have had more of death and anguish in the past two weeks than I ever want to see again. These two weeks have been the worst of my life save the week that Meg died. There is need to "lift up mine eyes unto the hills from which cometh my strength." Things will come to order when we are packed, mounted, and on the trail. I hope that in the solitude will come peace and the will to try yet again.

29 July 1994, Washington, D.C.

Office of General Counsel lawyers just gave me the word that the environmental interest group Pacific Rivers had prevailed in its case in the Ninth Circuit. The court is requiring consultation with the National Marine Fisheries Service on forest plans for forests within the range of any species listed as threatened or endangered. This was quickly followed by a court injunction to cease all operations that are deemed in the environmental impact assessments both "likely to affect" and "not likely to affect" listed salmon species on the Wallowa-Whitman and Umatilla national forests. The injunction specifically listed ongoing timber sales, road construction, and grazing. This decision, particularly the immediate shutdown of grazing, will have a profound effect on local economies.

The political fallout—given the fires that are raging across the West, the irritation building among western representatives in Congress, the pending reconsideration of the Endangered Species Act, and the recent history of the western livestock industry's manhandling of the secretary of the Interior over his foray into grazing reform—is sure to be severe and construed by many as a continuation of a "War on the West" by the administration. This victory by the environmentalists could well be Pyrrhic.

It seems that such has dawned on the environmentalists, as they quickly offered to negotiate a settlement whereby they and the Forest Service would agree to a set of conditions that would satisfy the injunction. This seemed attractive on the surface but would leave the impression that the Forest Service was in collusion with the environmentalists. In fact, my conversations with politicians in eastern Oregon—primarily State Representative Ray Baum—made it clear that rumors are rife that this entire matter was a setup ordered by the administration. Such is not true, of course, but perceptions are what count at the moment.

I made the decision not to negotiate with Pacific Rivers and the others involved in the lawsuit. Rather, we will go to court with our proposed course of action to comply with the injunction, which is not too far removed from what was suggested by the plaintiffs. At that point, the plaintiffs can support that course of action or object or suggest another course of action.

The environmentalists issued a detailed press release—which they obviously had ready—when the court made its ruling. This release, which was a carefully crafted set of half-truths, seems to have had the inverse effect of what they had hoped. It was either ignored or deemed a set of half-truths.

I believe that the power players have decided that the environmentalists, and the Ninth Circuit Court, have simply gone too far this time. The Forest Service had already moved to stop all ongoing activities that were deemed by our biologists to be likely to adversely affect salmon habitat in the streams in which the listed fish spawn. Arrangements had been made to have all livestock out of pastures with spawning streams by the time the fish arrive to spawn. Activities that were deemed by our biologists as not likely to affect salmon habitat were proceeding and had been submitted for consultation. If the National Marine Fisheries Service did not agree, we were prepared to respond. The Forest Service and the Bureau of Land Management are in the process of installing the most broad-ranging anadromous fish spawning habitat strategy yet contemplated throughout the entire Columbia basin. This PACFISH strategy is scheduled to be put into action on an interim basis on October 1.

Two years ago, the Pacific Northwest regional director of the National Marine Fisheries Service (Rollie Schmitten, now director of the agency) sent the Forest Service a letter saying that it was inappropriate, even technically stupid, to consult on the forest plans. Rather, he said, it was more appropriate to consult on individual activities, and that was what we were doing.

It was pointed out to me that the Eighth Circuit Court, in a similar case, had given an exactly opposite decision on the requirement, need, and appropriateness of consulting on forest plans when a new listing occurs. This seems likely to be a good reason to appeal the decision to the Supreme Court. I have recommended that course of action but it is up to the Department of Justice to make that decision.

The ramifications of this decision are potentially quite far-reaching. If consultation on the forest plans within the range of the species in question is required each time a new species is listed, it is likely, for some forests at least, that the consultations would be almost perpetual. And if all the activities that could adversely affect the species in question must cease pending that consultation, chaos would result.

For example, listing of the bull trout throughout the Columbia basin seems likely within several months. If that were to occur, some thirty-four forest plans would be subject to consultation with the Fish and Wildlife Service, which deals with nonmigratory fish species (the National Marine Fisheries Service deals with species that migrate from fresh water to the ocean). Would all activities cease that could adversely affect the habitat of the bull trout in the meantime? Such seems likely.

Even simple consultations on individual projects can take from several months to a year or more. That is at the present time. It seems likely that continuing budget cuts, personnel reductions, and an increasing workload will make consultation periods longer and longer.

Mark Rutzick, attorney for the timber industry in the Pacific Northwest, tried in a series of court cases (some of which are ongoing) to produce a train wreck that would force Congress to deal with the issue. This could involve a legislated plan not subject to challenge, a change in the environmental laws, or both. That, so far, has not been a good strategy, as people have become conditioned to a reduced timber supply from the Pacific Northwest over a five- to six-year period.

Now, the environmentalists may have produced what the timber industry strategy was unable to achieve—a West-wide train wreck that brings in another set of players that have more power than the timber industry because all the western states are involved. The livestock graziers typify the mythical western lifestyle that is still so idolized in our national culture. This decision could have much more influence in coalescing the political forces in the West than anything that has occurred previously, including the president's plan and Secretary Babbitt's grazing reform activities. It will be most interesting to watch this play out.

13 August 1994, Washington, D.C.

I visited by telephone with Bob Doppelt, the executive director of the Pacific Rivers Council, concerning its case. While I told him that I was not in a position to negotiate with him over the case, I did discuss the situation with him.

I suggested that Pacific Rivers would be well advised to take their victory and run and paraphrased Congressman George Miller's admonition to them: "You have pulled off the great stagecoach robbery. But if you don't leave the watches and rings alone, the sheriff will surely get you."

I suggested that the better part of discretion, given the extreme backlash that is developing, is to go to the judge and ask that no injunction on activities be imposed except that no activities that may adversely affect salmon habitat in the Snake River drainages be started in 1995 unless cleared by the consultation between the Forest Service and the National Marine Fisheries Service.

I pointed out that any chance of a significant adverse effect due to carrying out ongoing activities in 1994 was infinitesimally small. Why should the people who are grazing cattle and carrying out timber sales that are judged not likely to adversely affect the habitat of the listed salmon runs be punished by the clash between Pacific Rivers and the Forest Service? Those people are not to blame for the situation but can be damaged significantly.

The Pacific Rivers Council has won their point in court. The Forest Service has already moved to comply with the court's order to begin consultation with National Marine Fisheries Service on the plans for the Umatilla and Wallowa-Whitman national forests.

They have won! There is no need for an injunction that does nothing for the listed fish and can do serious damage to graziers and loggers affected. This injunction can only be carried out at a huge cost to the reputations of the involved environmentalists and add fuel to the fire in Congress against the Endangered Species Act. He seemed to listen and appeared to be open-minded. I hope so.

15 August 1994, Washington, D.C.

I worked extensively this week with the team that is revising the regulations issued pursuant to the National Forest Management Act that deal with planning. These revisions are directed at streamlining the regulations and making the process more efficient.

One of the most confusing and stringent portions of the existing regulations that has caused the most problems in application and in subsequent court actions is the viability regulations. These regulations require the Forest Service, in its planning, to maintain all native and nonnative vertebrates in "viable" status. This is to be accomplished by maintaining habitat for those species in the size, amount, and distribution that will maintain the numbers and distribution necessary to ensure viability "within the planning area."

This regulation is even more stringent than the requirements of the Endangered Species Act, in that all vertebrate species must be considered and that viability must be maintained on each planning area. This has come to be interpreted as the national forest covered by the plan in question. The Endangered Species Act, in contrast, applies only to those species that are officially judged to be threatened or endangered, and the recovery plan applies across the range of that species, and not on a piece-by-piece basis.

It seems likely that the Committee of Scientists, established by the National Forest Management Act to give advice on the regulations to be issued pursuant to that act, based the viability regulation on the instruction in the act that the "diversity of plant and animal communities be preserved." They meant the viability regulation to be a statement of policy as opposed to a requirement for a rigorous assessment of the viability of every vertebrate species within a planning area.

Since the development of the present planning regulations nearly twenty years ago, there has been the development of rigorous technical processes for evaluating viability of species (a risk assessment) under some array of projected conditions. Unfortunately, there are very few species for which adequate data exist to make such assessments. Even in the case of those species, such as the northern spotted owl, well-credentialed and experienced biological scientists debate vigorously over the validity of those assessments.

Environmentalists have come to love the viability regulation because it makes demands of the Forest Service that are, pragmatically, impossible to meet—certainly, impossible to achieve in a way that cannot be challenged on technical grounds. As a result, the viability regulation and the adequacy of the plans to meet the requirement have been increasingly successfully employed in numerous appeals and lawsuits.

Lawyers for the timber industry have argued—though as yet, unsuccessfully—that the regulation goes far beyond the intent of legislation. They have further argued that courts have interpreted the regulation to have meaning and requirements of performance far beyond that envisioned by the Committee of Scientists that proposed the regulation. I concur with that argument.

The team that is revising the regulations struggled mightily with the problem of the viability regulation and finally came back and recommended that the regulation be left intact. They know that the environmentalists will protest vehemently against any weakening of the viability regulation for two reasons. First, they believe the underlying intent of the regulation is in keeping with the Leopoldian tenet, that "to save every cog and wheel is the first rule of intelligent tinkering." They are correct. Second, they know from experience that the regulation provides an increasingly viable mechanism to place a burden of proof—or at least assessment—on Forest Service planners and biologists that is impossible to meet at any agreed-upon level of technical and legal achievement. They are correct about that as well. Ergo, to change that regulation at all will certainly be anathema to the environmentalists and will form the basis for a real debate.

I did not accept the revision team's recommendation and insisted that they attempt one more time to produce a substitute for the current viability regulation that is in keeping with what I consider the prime directive in this area, the Endangered Species Act. I asked that the regulation being prepared provide some standard of judgment as to when a species is to be considered "sensitive." The objective of the new regulation should be to meet the intent of the National Forest Management Act that the diversity of plant and animal communities be preserved through the intent and purpose of the Endangered Species Act, stated as "the preservation of the ecosystems upon which threatened and endangered species depend." The intent of the new regulation, then, is to prevent any species from being declared threatened or endangered, and thereby preserving the ecosystems in question. This is to be further enhanced through the mechanism of "sensitive species" that might well slide to threatened status if care is not taken to ensure that their required habitat does not decline to the extent that consideration for listing is deemed necessary. A crosscheck on the validity of declaring a species as sensitive is that the species must be so certified on two separate lists prepared by recognized experts.

The work on this alternative rule is almost complete and in my opinion is a great technical improvement over the existing rule and will be much easier to satisfy, both legally and technically. Yet therein lies the rub. Anything that achieves *that* lessens the usefulness of the existing rule to cause confusion, fuel legal debate, and serve as a weapon in the continuing debate over land-use planning and the execution of those plans one action at a time.

When we discussed this alternative with Assistant Secretary Jim Lyons and Deputy Assistant Secretary Adela Backiel, they immediately saw the situation as political as opposed to technical. It was clear that political expediency might well carry the day even though Forest Service activities could well continue to be stymied as a result. Live and learn is the story of my tenure as chief.

16 August 1994, Washington, D.C.

Dealing with the biopolitics of the issue of spawning and rearing habitat of the anadromous fish of the Pacific Northwest becomes more bizarre by the day. This week, the National Marine Fisheries Service upped the ante by changing the status of the Snake River salmon stocks from threatened to endangered. In theory, at least, this means that extirpation or extinction of these runs is imminent.

On the stretch of the river that runs through the Sawtooth National Recreation Area in Idaho, there are eight miles that essentially contain all of the spawning area for the perhaps twenty spawners that will make it past the Indian nets in the Columbia and past the dams. If these twenty fish fail to reproduce, one entire year class is missing from the subpopulation and the end of the run of salmon in the drainage is at hand.

Because of prolonged drought, the river stands at 28 percent of the average flow level for this time of year. As a result, the river is narrower and shallower than the long-term average. This simply means that the spawning fish and redds that hold the eggs are much more susceptible to disturbance than usual.

This river receives very heavy use from recreational rafting, and most of this rafting is by commercial operators. Commercial operators, of any kind, have more at stake than others. After all, as an economics professor at Texas A&M said, the most sensitive nerve in the human body is the nerve that runs between a person's pocketbook and the heart.

Dale Bosworth, regional forester for the area, had been in consultation with the National Marine Fisheries Service on what to do about the floaters, as required by the Endangered Species Act. Given the circumstances of the change in status of the fish from threatened to endangered, the very low population in this run, the low water levels, and the probability that heavy rafting traffic would repeatedly flush salmon off the redds, it was jointly determined that continued floating would likely cause a "take" and that such would be both biologically and legally indefensible. The two agencies concurred, and the decision was made to shut down floating for the remainder of the season.

Bosworth called me to give me the news. We discussed the economic and political ramifications of such a decision, as the biology and the regulations were clear enough. There seemed to be no other decision possible.

Senator Larry Craig of Idaho already had a call in to me, which I put off answering until after I had discussed things with Bosworth. I called the senator to give him the word. He was not happy but listened patiently to the rationale and understood that if National Marine Fisheries would not give the Forest Service some leeway on take, there was no choice.

We had no more finished our conversation when Rollie Schmitten, director of the National Marine Fisheries Service, called to discuss the negotiations between our two agencies over court-ordered consultation on the approved forest plans for the Umatilla and Wallowa-Whitman national forests. At the end of the conversation, he remarked that they had decided to give the Forest Service a take permit on the Sawtooth and the rafting on the river could continue.

Puzzled, I called Bosworth back to confirm. He was totally baffled but said he would check things out and be back to me the next day. I called Senator Craig and told him that my previous statement to him was "inoperative" and said I'd let him know the score as soon as I knew.

Bosworth called about 2:00 P.M. today and gave me the story. The Idaho Fish and Game Department director, Jerry Conley, evidently received pressure from Governor Andrus and announced that his department saw no reason to stop the rafting. Conley emphasized that there was no science (how about common sense?) to support the idea that scaring spawning fish off their redds would cause any harm, stated that there was no problem with spawning habitat in Idaho, put the entire blame on the dams, and said his department would openly oppose the decision through press releases and other means. The words were exactly those that I had heard from Governor Andrus several times. So Conley folded to pressure. So the National Marine Fisheries Service, without consulting the Forest Service, folded to the pressure and announced it was giving the Forest Service a take permit and rafting could continue.

22 August 1994, en route from Ontario, California, to Denver

I've just finished a three-day visit to the Angeles, San Bernardino, and Cleveland national forests in southern California. The occasion was a gathering of forest supervisors from the twelve "urban forests"—those forests that have over one million people living within a one-hour drive of the forest boundary. These supervisors wanted me to have some appreciation of the very different problems they face in the management of these urban forests, compared with the vast majority of the national forests.

I was familiar with their statistics on the incredible number of "user days" of recreation. But the impact is difficult to imagine until it is seen up close and personal. The cities of Los Angeles, Pasadena, San Bernardino, etc. are built on the flatlands that extend from the ocean to the mountain ranges that define the boundaries of the national forests. This is the urban-forest interface at its most dramatic. The slopes are covered predominantly by manzanita, shrub oak, etc. that become highly volatile fuels in the late summer and fall. Fires that begin in the national forests sometimes spread into the houses built in the foothills, with stunning results—hundreds of homes destroyed.

Then, houses are rebuilt right back on the original locations, often with no improvement in the circumstances that led directly to their previous loss to fire. This has led to the establishment of a very large state/local/federal force of firefighters with expensive and increasingly sophisticated equipment. This has led, in turn, to a segment of the Forest Service workforce who more and more identify themselves with the firefighter contingent than with the Forest Service per se.

The rugged topography along the front that runs parallel to the ocean produces another phenomenon in the recreational use of the forests. The use is dramatically concentrated in the canyons that come out of the national forests. So not only is the use heavy by the city dwellers trying to escape the congestion of the cities, that use is highly concentrated, with dramatic effects on the recreational experience, the environment, and the Forest Service management style and expenditure of resources.

We visited the main canyon on the Angeles National Forest on what

Secretary of Agriculture Mike Espy presenting to Assistant Secretary Jim Lyons
a memento related to the fiftieth anniversary of Smokey Bear, with Chief
Thomas, August 1994. Photo courtesy of Jack Ward Thomas.

was described as a light-use day. It was an eye opener for me. The recreational use is concentrated in the canyon bottom with devastating impacts
on the riparian zone, which has been reduced to bare ground except for
mature trees. There seemed to be at least fifty people per one hundred yards
of stream in the stream itself and thousands more occupying every designated campground and every other place flat enough to accommodate a
blanket or some folding chairs.

The vast majority of the users were Hispanics, and most, at least the ones
we visited with, spoke only Spanish. Graffiti was everywhere, particularly
on any buildings or signs. The managers said that the restrooms are
repainted on a weekly basis to cover the graffiti. The litter was appalling:
fecal material, disposable diapers, and the standard cans and bottles were
everywhere.

The managers told me, however, that the situation was much improved over the situation of several years ago through the advent of a cooperative fee assessment program of the county and the Forest Service. The county collects a $3 per vehicle fee and after charging for overhead returns the balance to the Forest Service for improvements of the area. This has provided about $100,000 to $200,000 per year for improvements, which is most welcome, given the miserable state of Forest Service funding for these forests.

In addition, the Forest Service has tapped into some funding for "eco-corps." These are young people—predominantly Hispanic, bilingual youth—who walk through the heavy-use areas to visit with the users about appropriate behavior and to distribute plastic garbage bags. This effort is said to have yielded good results. The people with whom we visited seemed enthused about their work and were certainly bright, attractive young people.

I am amazed that EPA or state environmental agencies have not chastised the Forest Service for the conditions. I can't believe, for example, that water samples would pass muster for fecal coliform, and there are thousands of people (mostly children) playing in that water.

But the pressure is so heavy in the city and the use so heavy and the need so great that it is probably politically expedient to look the other way, for the Forest Service and other agencies as well. As the city—with all of its problems—moves into the woods, those problems become Forest Service problems. For example, three murders took place on the Angeles National Forest in the week before my visit, and that, according to my briefing, is not unusual. The body count from all causes on that forest is three to five per week. Dealing with these fatalities plus the wounded is said to take a heavy toll on Forest Service people.

The law enforcement people are more numerous on this forest compared with others but are woefully understaffed at that. They are most concerned about the decision to phase out Level 4 law enforcement officers (those who perform part-time enforcement duty and do not carry weapons) over the next three years. The law enforcement brass in Washington have issued this edict in an attempt to reduce Forest Service liability that results from inappropriate action by poorly or inadequately trained Level 4 enforcement folks.

The officers I talked to believe this to be a serious mistake for several reasons. First, they have no idea how the activities carried out on an as-needed basis can be continued. Second, they think the probability of increased liability is unfounded, based on accumulated experience. Third, if there is a problem, it should be addressed through increased or enhanced training and not through elimination of the Level 4 enforcement officers.

I promised to specify that a workshop be conducted of law enforcement people and line officers to discuss the problems and future of the law enforcement operations. This should be organized using an approach similar to that which was used to put on the Houston meeting—i.e., appoint a committee of the rank and file and let them put it together. I suspect that our Washington-level law enforcement staff won't like that idea much as they resent anything that smacks to them of diminishing their authority or power. They have, to say the very least, a command-and-control mentality and management style. We will do it anyway.

The southern California forest supervisors, recognizing that increased Forest Service funding is unlikely, have taken on a new way of handling their programs. It is a combination of volunteers, partnerships (with state and local governments and private business), grantsmanship, donations, support groups, etc. They are "breaking new ground" and that is making the traditional Forest Service—particularly the bean counters—a little nervous. The forest supervisors admit to walking a tight line with the rules but continue, through constant consultation with our legal people, to assure that the entrepreneurial activities are legal and ethical. They make it clear that the primary roadblock they face is that of traditional Forest Service folks with staff ranks at both regional and national levels. They asked me not for money but for support of innovative and entrepreneurial efforts to get the job done.

I promised to give that support and encouraged them to use the opportunity presented by the "reinvention" effort to move ahead. Just maybe, the Forest Service could be the bright spot that Vice-President Gore is seeking in his efforts to develop more efficient government programs.

22 August 1994 [later that same day],
en route from Denver to Washington, D.C.

BLM Director Mike Dombeck, Lester Rosenkrance (state director, Arizona BLM), Mark Reimers (deputy chief, FS), and I held a press conference today to release the report "South Canyon Fire Investigation." Dombeck and I started with brief opening statements, each describing the sympathy we felt for the survivors of the dead firefighters and the firefighters who survived the incident.

Reimers and Rosenkrance did the briefing and answered the questions from the press. The conclusion of the report was that a series of circumstances and decisions interacted to bring about the tragedy. Any break in the chain of events might well have produced a different outcome. There were decisions made to not attack the fire for two days as attention was diverted to other, higher-priority fires. There was confusion in dispatching the crews. There was no request for a briefing—and no effort to brief the crew—on an expected cold front with high winds, although some did know the weather forecast. They were building downhill in high winds, a dangerous situation. The escape routes were too steep and too long. There were inadequate lookouts. The fuel danger was dramatically understated, and so on. Just one decision in a string of decisions may have been the difference. Just one circumstance different in a long string of interacting consequences may have made all the difference.

But the press vultures could not accept that. Who exactly was at fault? Who would be punished? Who screwed up? Who, who? Who is at fault? Blame someone! Now all the second-guessing will begin and the disagreements will come hot and fast—from the press, from the survivors, from the families of those who died, from the politicians, and from all the "experts" who were not there. All will be given their day in the press, which makes no effort to differentiate between fact and unsubstantiated opinion.

Soon the lawsuits will begin and drag on for months and years—that is certain. And millions will be settled on the survivors of those who died, for how could any jury not grant millions from the government to those brave young innocent firefighters? It seems strange that we will not even grant some of those firefighters benefits for their year-in and year-out journeys into harm's way, but we pay their survivors millions for their deaths.

28 August 1994, airborne between Washington, D.C., and Boise, Idaho

I am on my way to the fires burning in Idaho and Montana. My original intent was to testify for the administration in field hearings being held in Boise on 28 August by Senator Thomas Daschle of South Dakota and

Senator Larry Craig of Idaho, representing the Senate Committee on Agriculture, Nutrition, and Forestry. The subject of the hearings is "Forest Health Conditions in the West."

Since this hearing was scheduled, the forest fire situation in Idaho and Montana has moved into the realm of crisis. The fire readiness situation has been at Level 5 for nearly two months. This simply means that all available resources are committed, the military and National Guard have been called on for resources, and air tanker support has been requested from the Canadians. At Level 5, those in command are unable to assign firefighters and equipment to all those incident commanders making requests. Priorities are being set and adhered to by those charged with allocating resources. Firefighters and equipment are fully committed from federal agencies, seven battalions of military personnel have been assigned, National Guard aircraft capable of delivering retardant and helicopters are in action, all—or nearly all—states with wildland firefighting capability have provided firefighters and/or equipment, and additional crews are being recruited and trained.

The situation in Montana is such that the order was given ten days ago to act on a new priority: concentrate on the protection of communities. The tenseness of the situation was evidenced by calls to me from Senators Max Baucus and Conrad Burns of Montana, urgently requesting me to visit the firelines in Montana.

I explained to Senator Baucus that we had already committed about 20–25 percent of our available firefighters and equipment to Montana and that we had already obtained the required funding from the Federal Emergency Management Agency to carry out all the activities we were capable of exercising. I further explained that ordinarily Forest Service brass do not show up on a fire unless invited for a special reason. This is so for three reasons. First, the presence of brass diverts people and resources from the mission at hand. Second, a visit from the chief (or an equivalent) can create a false impression that there is lack of confidence in those in command or that the chief has arrived "to take charge." Third, it can lead to resentment among the troops about a desk jockey showboating for press and TV. The senators know all of that but still want me to visit so that they can say the visit occurred in response to their influence and request, or demand, depending on how they want to play it with the media. Fair enough. There will come a time when they may be able to help me help forward the Forest Service mission. It is through such small actions that good working relationships are formed. Besides, it is time for me to visit the firelines in Idaho and Montana.

29 August 1994, Boise, Idaho

We returned to the hotel to clean up before our command appearance at a field hearing of the Senate Committee on Agriculture, Nutrition, and Forestry. I removed the soiled Nomex, a shower and soap removed the ash and dirt from my face and body, and the donning of a clean, pressed Forest Service uniform completed the transition in both circumstances and role.

A Forest Service security detail picked us up at the hotel and warned us that demonstrators from Earth First! and other extreme environmental organizations were present at the front door of City Hall, where the hearings are to be held. The security detail had arranged for us to go through a side entrance in order to avoid the demonstrators. They seemed a bit surprised when I told them that unless they had some reason to expect a dangerous situation, my custom was to walk straight up to demonstrators and offer to talk.

We went in through the front door, and most of the demonstrators were too busy performing for TV cameras and giving interviews to the men and women of the media to pay any attention to our arrival. One tall, pudgy fellow saluted, presumably in disdain for our uniforms and association with the Forest Service. I returned a crisp salute and said, "At ease. Those shoes need a shine." The reporters laughed as the demonstrator, unthinkingly, looked down at his shower shoes. He struggled to come up with a witty response, then yelled "F— you!" at our backs as we strode past. I turned and said with a smile, "Good thinking on your feet, Trooper." The reporters laughed even harder. The demonstrator made no further comment.

The hearing began exactly on time. Senator Thomas Daschle (South Dakota) was in the chair and Senator Larry Craig (Idaho) was the ranking minority member. Senator Kempthorne (Idaho), though not a member of the committee, was invited to sit on the dais and participate as a guest of the committee.

Each of the senators made a statement. Senator Daschle's comments were relatively neutral. Senators Craig and Kempthorne made statements that made it clear that their sympathies rested with prompt attention to the problems of "forest health," including salvage of dead trees, particularly from burned areas; thinning—both precommercial and commercial—of densely stocked stands; and reductions of fuel levels in selected stands. It is an interesting phenomenon that senators and congressmen make statements at the start of a hearing, clearly stating their positions before the evidence to be adduced at the hearing is even heard. Ostensibly, the purpose of a hearing is to gather evidence for the use of the committee in making decisions. If minds are already made up, what is the purpose of the hearing? It seems likely that there are several more important reasons than gathering evidence.

The hearings give the members of the committee a chance to focus attention on themselves, to pay off debts to friends and financial supporters by saying the right things, to afford the appropriate witnesses a chance to say their piece, and to develop a record that might be useful to any member of the House or Senate wishing to study the situation in some detail.

The invited witnesses included me, a representative of the timber industry, a labor union spokesman, the Idaho state forester, two college professors—one economist-sociologist and one forest economist—and two representatives of the environmentalists. The presenters, while seemingly very diverse, were in reality singing the same tune. The timber industry lobbyist, who made the pitch that much salvage and thinning are needed to prevent catastrophic fires and to restore forest health, posited one extreme. The other extreme was voiced by the environmental spokesmen, who declared that the Forest Service, under the influence of industry—which is only interested in profits—would treat every forested acre in a prescription that would be applied "across the board." The others, including me, took the middle ground on the issue. Some salvage, thinning, and fuel reduction should be carried out. In each case, the selection of areas for treatment should be based on an assessment of the ecological condition and capability of the site in question, efficiency of the investment, desired ecosystem condition, location, and other appropriate factors.

In turn, the prescription would be appropriately tailored to the situation. Assuming that the spokesmen for industry and the environmentalists are posturing for their respective clientele—which seems highly likely to me—there is some agreement that salvage and thinning will be applied. The questions are where and how much.

The test will be if the Forest Service professionals can be politically astute enough to screen opportunities and handle public participation in such a way as to fend off appeals and lawsuits. Somehow, I doubt that it is possible to appease the hardcore environmental groups. They will always object to any proposed manipulation of vegetation, particularly cutting trees, on public lands. They may agree in this forum to some proposed manipulation to avoid appearing unreasonable, but I think it likely that the opposition will shift to an action-by-action opposition. They have experience in how to string out all decisions through administrative appeals, appeals of decisions on administrative appeals, and lawsuits.

The war for public opinion is being waged in full-page newspaper advertisements. In today's Boise paper was an ad from the timber industry, which left the impression that full-scale forestry could provide jobs and alleviate the curse of declining forest health. The environmentalists' full-page ad did not even pretend to deal with technical questions. Their ad identified Steve Mealey, supervisor of the Boise National Forest, as the "Butcher of Boise" for his past actions in salvage and rehabilitation efforts after the Foothills Fire, and for which the Forest Service personnel involved received the Chief's Award for Salvage Activities. Salvage, thinning, and fuel reduction to address forest health problems were identified as "voodoo forestry." The timber industry was accused of promoting such forestry activities for the gross purposes of profit.

1 September 1994, Boise, Idaho

I have just finished a tour of the fire situation in Montana. I arrived in Missoula in midmorning and was met by Regional Forester Dave Jolly and Dick Bacon, the staff director for Fire and Aviation for Region 1. We departed almost immediately for Libby, Montana, which is the headquarters for the Kootenai National Forest. The air was filled with smoke as far as the eye could see in every direction, although I was assured that conditions were much improved over the past few days as the strong winds had abated.

We went to the parish hall of the local Catholic church, which had been made available for meetings and briefings, for lunch and a briefing on the

Chief Thomas being briefed on Libby, Montana, forest fires by Operations Section Chief Jim Payne and Incident Commander Doug Porter, September 1994. USDA Forest Service photo by Mike Ferris.

current status of fires in northern Montana. All available crews and equipment had been committed, and there were at least six fires in wilderness areas that were not manned. The decision had been made not to actively fight those fires given their location, the danger associated with the rough terrain, lack of access save by air or a long hike, and the absence of immediate danger to life and property, the highest priorities for protection. There simply are not enough resources available to man all of the fires, even after the commitment of one Marine and three Army battalions. This will be a good decision *if* the weather holds. If there are high winds, these fires have the potential to spread down the east slopes of the Rockies. If that happens, the criticism will be severe. They were telling me the situation. They realized the decision was theirs and not mine. They are in charge and that's the way they want it and the way it should be.

As we drive through town after the briefing, I notice the signs that all say—one way or another—"Thank you, firefighters." There is nothing like real danger to unite people. Two weeks ago the Forest Service was catching criticism from all sides, and two weeks after the fire danger is over the controversies will return, focused primarily on what to do about forest management at that moment. But just now, it is all appreciation.

A Rudyard Kipling poem I memorized in high school suddenly pops into my head—a poem about the British soldier, "Tommy," and how in peacetime there is disdain, but in time of war nothing is too good for Tommy. This is really the same thing over and over again. Human nature does not change. For those in public service to expect appreciation on a continuing basis is to expect too much. But there are moments like this one that can be cherished when things return to normal.

15 September 1994, Washington, D.C.

Today I attended an all-day session of the assistant secretaries and some of the agency heads who deal with natural resources and environmental affairs. The meeting was held at the National Arboretum and included folks from USDA (Forest Service and Soil Conservation Service), State Department, Office of Budget and Management, Commerce (National Marine Fisheries Service), Interior (Fish and Wildlife Service, Park Service, Bureau of Land Management), Justice, Environmental Protection Agency, and the Office of Executive Policy.

It seemed a pessimistic meeting filled with the recognition that we were getting bogged down in executing new thrusts in environmental and natural resources management. There was recognition that many of the environmental laws need to be adjusted, but there was no stomach for even trying, as the "political climate isn't right."

There was fear that moving to correct problems would not sit well with "our friends," which after some puzzlement I interpreted as the environmentalists. There was considerable reference to the idea that we had to do a better job of explaining situations and circumstances to "our friends" and to be cautious not to offend "our friends." Finally, late in the day, I could contain myself no longer and explained that as chief of the Forest Service—a multiple-use agency by heritage and by law—I did not consider the timber industry, or graziers, or miners as any lesser friends than those of environmental bent. Further, we tried to think of all these people as customers, stakeholders, and friends. They listened respectfully enough, but the tenor of the conversation did not change.

I do not profess to be an astute political analyst, and most of the people in the room were from political backgrounds as former House or Senate staffers, former activists (lobbyists) from various environmental groups, or former elected officials. However, I thought that they were dramatically misreading the political situation.

It is obvious that a backlash is under way, particularly in the West, against the tightening screws of environmental regulations. It is likewise obvious that there will be dramatic gains for the Republicans in the upcoming midterm elections. Further, the environmentalists are insatiable in their demands and are trashing the administration on the environmental front even more frequently and viciously than the extractive industries and conservative politicians. Besides, are the enviros going to abandon the administration in favor of the increasingly conservative Republicans? Not likely.

To me, the smartest political strategy would be to move now to modify laws to assure a more efficient and rational mechanism to carry out the spirit of the environmental laws, or to at least study such a modification. Why? First, it needs to be done. Second, such a move would demonstrate concern for those affected by the consequences of such laws. Third, it would steal a significant issue from the opposition, just before the November elections.

I believe that the more mainstream environmental organizations could be brought to the support of such action—at least in principle—as they are being increasingly criticized as irresponsible and reckless and uncaring toward people adversely affected by the consequences and cumulative effect of carrying out these myriad laws. I propose what is captured in the old slogan, "Trying to protect environmental values while returning management to the professionals," instead of turning over that management to the federal courts or to regulatory agencies.

There may be some small window of opportunity that the Forest Service can surface as symbolic of the problem. That is the salvage/forest health/sustainability issue that we will focus attention on in the very near future. Though not inclusive, it will be symptomatic of the larger problem, which is the tangle of detailed line-item budgeting, the Endangered Species Act requirements for consultation, the requirements of the National Forest Management Act (the law and the pursuant regulations), the analyses required by the National Environmental Policy Act, the necessity of achieving public participation and still following the mandates of the Federal Advisory Committee Act, and the restrictions imposed by the Environmental Protection Agency on clean air and clean water. That seems to be enough confounding to clearly demonstrate the problem of dealing with a single, intertwined issue.

It became clear in the presentation by Peter Coppelman of the Department of Justice that it is not at all clear that taking on the enlarged concept of ecosystem management is legally permissible or possible.

So we are snared in a tangled web of confounding laws, rapidly evolving case law, myriad appeals, interagency interaction, turf protection, and agency disruption caused by downsizing and reinvention. That produces a morass into which we sink more deeply with each passing day, and our entanglement worsens by our poorly coordinated struggles to extract ourselves.

On balance, I believe the environmentalists have a much better chance to lose from the looming train wreck than does industry. But they (the environmentalists) are so used to "victory" and are so certain of their righteousness and so consumed by hubris that they seem likely to take the bait of the impending "total victory."

The powers in government should be aware of these scenarios and devise policies and strategies to produce the outcome they desire and, by virtue of empowerment by the citizenry, are charged with developing and pursuing. But such does not seem likely for there is no vision, no focused will, and no leadership. The strategy is simply to muddle through. Such is made mandatory by the acceptance of the statement, "The political circumstances are right for us to suggest that we have serious problems, and the laws need significant adjustments to facilitate ecosystem management of at least the public lands."

30 September 1994, airborne,
en route from Edmonton, Alberta, to Washington, D.C.

I am en route home after a five-day meeting of the North American Forestry Commission at the lodge in Jasper National Park. It was my honor to be the leader of the delegation from the United States to this eighteenth biennial meeting of the commission. Yvan Hardy was leader of the Canadian delegation and Victor Sosa was head of the delegation from Mexico.

Dr. Jerry Sesco, deputy chief for Research, and Dr. David Harcharik, acting deputy chief for International Forestry, accompanied me. The aim of this commission is the maintenance of communication and the institution of joint ventures that serve the purpose of forest husbandry in North America.

Although there were a number of matters discussed, two seemed to me to be of paramount importance. The first involved the commitment of the three countries to forestry practices that would ensure "sustainable forests." President Clinton pledged that the United States would achieve this goal by the year 2000. Attaining this lofty objective will require the development of and international agreement upon the measures and standards of assessment that define and measure the attainment of sustainable forestry activities. Teams are now at work to develop, test, and promulgate these measures.

Related to this subject are the ongoing efforts in the nongovernmental sector to certify products from the forest as having been produced in an "environmentally acceptable" fashion. This is also called green labeling, which seems to be well under way in Europe. The Canadians seem most aware of this trend as they export significant amounts of wood products to Europe and hope to increase that trade over time.

As there is no means by which governments can control the development and voluntary institution of such standards by nongovernment organizations, the best bet seems to be to cooperate to the extent possible by suggesting the attributes of forest practices that should be of concern and the standards of measure of those attributes. It is deemed essential that one set of standards be developed and universally applied, or chaos seems likely. Developing such measures for boreal, temperate, and tropical forests will be a real technical and political challenge. It does not seem pragmatic, however, to ignore what is happening and is almost certain to continue in Canada, the United States, and Europe.

I was impressed by the emphasis that the Canadian delegation put on this developing situation; I have heard very little about it in the United States. We will hear more, I am certain.

Some ten years ago, I was the sole speaker from the United States at a forestry congress held in Canada and attended by everybody who was anybody in the forest products industry in Canada. In my speech I emphasized that they could expect an increasing environmental concern in Canada as had been experienced in the United States. I remember the body language of the audience, having never seen so many crossed arms and legs and jutted jaws in one place at one time. They did not respond favorably, nor did they believe me. Now that time has arrived, but to my surprise, via green labeling. Developing environmental consciousness in Canada seems to be much less of a causative factor. But by whatever route, the Canadian foresters have become environmentally aware.

On the last day of the meeting of the North American Forestry Commission there was one last bit of business. Yvan Hardy and I had discussed several times during the meeting that the meaning and practice of forestry in North America were rapidly evolving, at long last, into a much broader context than the efficient production and harvesting of wood fiber. And that the legitimate business of the commission extended beyond "forestry" to the business of dealing with forests in whatever context. We agreed that a symbolic change in the commission's name to the North American Forest Commission was in order. Yvan said he would discuss the matter with our colleagues from Mexico and obtain their acceptance.

This change was made but not without debate from some of my staff. After they had their say, Yvan looked at me a bit quizzically, and I smiled and nodded. The motion passed unanimously. And again, the walls come a-tumbling down.

26 October 1994, Washington, D.C.

I met today at the Old Executive Office Building with representatives of the departments of Agriculture, Interior, Commerce, State, and Justice and the White House. The subjects of discussion included the treatment of two species—broadleaf mahogany and African elephants—under the Convention on International Trade in Endangered Species of Wild Flora and Fauna.

The Fish and Wildlife Service is the lead agency in CITES matters. The Forest Service was involved only because broadleaf mahogany, which ranges from southern Mexico to northern South America, is a tree of commercial value. This situation, through convention, requires the Fish and Wildlife Service to ask for advice from the Forest Service. Such advice is to be based on four criteria provided by the Fish and Wildlife Service, reflecting on the probability that the species of concern is likely to become extinct or is likely to approach extinction.

Several months ago, my staff dealing with threatened and endangered species presented this question to chief and staff for decision, with the recommendation that the Forest Service support the listing of broadleaf mahogany on Appendix I of CITES. Upon examination, it was clear that the species met none of the four evaluation criteria for such action. The staff still pushed for listing on the basis that the ecosystem dominated by mahogany needed protection, an interesting extension of CITES to cover the recent ecosystem and sustainable forestry movement in the United States. Obviously, the staff's agenda was to see the species listed.

I rejected their recommendation on the basis that the Forest Service was expected to provide technical advice based on evaluation of the available data or informed scientific opinion. The data were clearly inadequate to permit such a decision and those data which did exist did not, in my opinion, satisfy the listing criteria. Further, the opinions of the identified experts were split.

There was a lesson in this for me: do not casually accept staff advice. The closer to the top of the power pyramid one gets, the more limited the time that can be devoted to any one subject. The more limited the time, the less likely the person making decisions will probe the staff recommendation and the more likely the staff's agenda can be achieved. And it is always well to assume, in this Byzantine world, that there is an agenda.

The staff was not pleased with my decision. I made a mental note to carefully examine any additional documents prepared in this case to detect any subtle attempts to weaken that decision.

As the issue gained momentum over the next few months, it became obvious that this decision or recommendation by the chief of the Forest Service was likely to derail the effort to list mahogany under CITES as an approach to the problem of tropical forest exploitation. The leaks that are almost immediate in such circumstances, to the timber industry by one staff and to the environmentalists by another staff, made my decision a focal point. The timber industry strongly questioned how the United States could support listing when the "foremost authority in the government" (the chief of the Forest Service) said the evidence was too weak to support such action. The environmentalists concerned about the issue were appalled that a Forest Service chief thought to be an environmentalist could oppose their agenda, the protection of the tropical forest.

For two weeks before this meeting, the calls flew between agency heads and, I suspect, others in government and between lobbyists and officials. I stood fast on the theme that the Forest Service was to provide technical assessment and that our (my) opinion remained unchanged. The next ploy was to organize today's meeting and leave the Forest Service chief out of attendance and to elevate the matter to the assistant deputy secretary level. That ploy failed when Deputy Assistant Secretary Backiel insisted that I accompany her to the meeting.

This meeting was one of the most educational experiences of my life. The interplay between technical assessment, law, domestic politics, international politics, and commerce was fascinating—biopolitics to the max.

The technical issues were quickly handled: the data were inadequate and expert opinion so divided that listing mahogany could not be supported on those grounds. The political discussion was devoted to pros and cons.

At this point, deadlock ensued. The chairwoman, in the best tradition, decided to move on to the next question on the agenda. African elephants have long been on the list; CITES prohibits all trade in elephant parts, with particular emphasis on ivory. South Africa has a stable elephant population and culls over 200 elephants a year to maintain populations at an acceptable level. The South Africans have excellent data on the population, and their request seems well justified on that basis. They want to market in international trade only hides and products made from hides.

Now, interestingly enough, the postures flipped 180 degrees. The Fish and Wildlife Service does not want to comply with the request of the South Africans, as the decision will not be popular with our domestic environmentalists and that portion of the populace that cares about such things. The representative for the Commerce Department, whose job it is to be concerned about marine species, opposes changing the elephants' CITES listing given the effect such may have on the position of the United States on marine megafauna, especially whales. The representative of the Environmental Protection Agency supports the position of the Fish and Wildlife Service—again without saying why. The representative for the National Marine Fisheries Service does likewise.

So once again, the Forest Service and I are isolated in the position of being inadequately "green." Now I am beginning to understand the meaning of being politically correct.

The issues seem clear to me and the outcome is predictable. The United States simply can't support the listing of broadleaf mahogany. So the position will be to try to get the Dutch to withdraw the petition. If that fails, the U.S. will offer to back the petition if the range states, or a majority thereof, do likewise. That won't happen. So the next position will be to abstain from the vote. That will probably kill the issue, for this go-around at least. In the case of South Africa's request, the position will be to back the change in CITES listing if the ban on trade in ivory is retained as offered by the South Africans and "key African states" (i.e., Kenya) support that position.

In typical fashion, the decision will be delayed until another meeting has occurred. In the next meeting all of this will be rehashed and the decision will be as described above; I'll bet on it. But infinite discussion is the hallmark of consensual decision making, and that's simply the way of things.

As we leave the Old Executive Office Building, there are twenty-five or so demonstrators at the street corner, chanting "Hey, hey, how many elephants did you kill today?" Each has a placard that says "Stop the Slaughter of Elephants" in one form or another. How did they know about the meeting? Leaking to fulfill personal agendas is a fine art in Washington. This has been a most educational day.

9 November 1994, Washington, D.C.

Yesterday was the day of midterm elections. The results represent a sea change in American politics. The Republicans took over control of both bodies of Congress and control the majority of governorships. A new day has dawned. For those interested in conservation, it should be clear that for the foreseeable future, the environmentalists have reached their zenith of influence in American politics. Now the rearguard actions begin to hold on to the gains of the past three decades.

The window of opportunity for significant change in enhancing concern and protection and sustainable use of natural resources has closed, at least for the rest of this decade. There is still some opportunity for "reinvention" of the Forest Service to occur, and the real change protector will be the appointments made to the positions of regional forester, forest supervisor, and district ranger over the next two years, assuming that President Clinton is defeated in the next election. Even if he survives, it seems unlikely he will attempt any bold moves in the area of conservation or environmental protection.

Environmentalists have gone from a questionable political asset to an outright liability. The Republicans will find them most useful as a foil and the Democrats will find environmentalists' support—which has been weak and fitful at best—a liability instead of an asset. The environmentalists, convinced of their own righteousness, have in their hubris and arrogance thrown away their political power to influence events.

In the short term, the environmentalists will continue their assaults through the courts, where they have been most successful, and other victories lie ahead. But each new victory will bring them ever closer to the ultimate defeat: alteration or repeal of the legislation upon which their lawsuits are based.

The first test will be the contest in the next Congress—probably next year—over the reauthorization of the Endangered Species Act. It seems likely that the ultimate outcome will be severe weakening of the act. It also seems probable that the National Environmental Policy Act will be significantly modified to reduce the costs and complexities of compliance.

18 November 1995, airborne,
en route from Albuquerque to Washington, D.C.
The wilderness conference was a roaring success. Assistant Secretaries of Interior Bob Armstrong (who handles the Bureau of Land Management) and George Frampton (who handles the Park Service and the Fish and Wildlife Service) and I were the banquet speakers. While we were eating, John Twiss (my staffer who deals with wilderness affairs) told me that I needed to work on George Frampton "to get him to assign someone from F&WS, Park Service, and BLM to the Interagency Arthur Carhart Wilderness Training Center" and to seek assignment of a scientist from the National Biological Survey to the Aldo Leopold Wilderness Research Center.

I made the pitch to Frampton, pointing out the benefits and the appropriateness of making such an announcement during his after-dinner speech. He listened, smiled, and said, "That's a cheap date. I'll do it." And he did.

Twiss and the Forest Service directors of the Carhart and Leopold centers nearly fell out of their chairs. A goal they had worked for within the appropriate channels had been suddenly achieved in a five-minute conversation over dinner. It was an example of the dictum *carpe diem* at its best. The stage was set, the time was ripe, and the deed was done.

It seems likely that wilderness will be under attack for the next few years as the conservative Congress flexes its muscles. But they who choose to kill that spirit had better bring a lunch, as the fight will be prolonged and it won't be easy.

The lesson I have learned after nearly forty years in the conservation business is this: Make all the gains you can when the circumstances allow and make those gains in yards. When the tide turns, play tough defense and give only in inches. The trick is to have a long-term view and not be deceived by either temporary victories or temporary defeats. After all, the earth abides.

22 November 1994, airborne,
en route from Washington, D.C., to Denver

There were several meetings this week with the Office of General Counsel attorneys assigned to the Forest Service. At issue was the decision concerning whether to appeal the ruling of the Ninth Circuit Court of Appeals that any time a species is listed by the Fish and Wildlife Service or the National Marine Fisheries Service, the forest plans for those forests harboring such species must be submitted to the appropriate regulatory agency for consultation.

The Forest Service's contention has been that a forest plan is not the appropriate mechanism on which consultations should occur. Rather, consultation should occur at the project level where actual activities are projected and analyzed. Two federal district courts have upheld that contention. As a result, it is possible to appeal the Ninth Circuit decision to the Supreme Court for resolution. I have recommended that such an appeal be filed.

The National Marine Fisheries Service, in a complete reversal by the director of his position taken while a regional director, opposes the appeal. The current Ninth Circuit ruling dramatically increases the power of the regulatory agencies to influence land-use allocations by federal land management agencies without any of the requirements of the National Environmental Policy Act to assess the ecological, economic, and social impacts of a decision influencing land use.

Since species are being listed as threatened or endangered at a steady rate, it is likely that it would be increasingly difficult to carry out any forest plan. Operations would be sporadically shut down pending consultation, and plans would be subject to constant revision to attain appropriate go-aheads from the regulatory agencies. To make things even worse, the opinions of the regulatory agencies are based on the opinions and "feelings" of a cadre of relatively young biologists. It is unclear whether this backing represents the administrators' confidence in their field staff or whether they are simply unwilling to take the criticism of the environmental community, the press, and some politicians by overruling these defenders of the environment.

One of the more stunning facts that I have learned over the past year is that, in its ability to independently determine whether or not to proceed with any legal activity, the Department of Justice wields the greatest capacity to set policy of any agency of the government. I naively assumed that the chief of the Forest Service made the decision as to whether to pursue a court action. Not even the undersecretary or the secretary makes those decisions. Such can merely request and suggest. The Department of Justice decides—the agency can propose and the Department of Justice disposes. That power is not well understood even by students of the internal workings of government. If the policy-setting power of the lawyers in the Department of Justice were well understood, I don't think anybody—Congress, the persons affected, or politically appointed agency administrators—would appreciate that fact.

28 November 1994, Washington, D.C.

I met today with Senators Byron Dorgan and Kent Conrad of North Dakota and Congressman Earl Pomeroy. The subject of discussion was the management of the national grasslands. The meeting started with a bluffing maneuver by the three politicians. They expressed their opinion that the Forest Service was well equipped to manage forests but knew nothing about grasslands and, therefore, was applying the standards for the management of the national forests to the national grasslands. And these standards did not fit. Therefore, if I did not immediately see things that way and make the changes they required, they would explore transfer to the Natural Resources Conservation Service or the Bureau of Land Management.

I replied that such a transfer would require legislation and such was certainly within their purview as legislators. I further asked them what I might to do to facilitate such a transfer and indicated that if such was to occur, this was a most opportune time, given the ongoing reinvention of the government in general and the Forest Service in particular.

That response was not what they expected, and the tenor of the meeting began to change. Senator Dorgan quickly discerned that their initial "threat" would buy nothing and quickly changed tactics to an argument that the Forest Service was misinterpreting its mandate to manage the grasslands. This was based on a contention that the Bankhead-Jones Act was the operative enabling legislation, and the Forest Service was inappropriately applying the mandates of the National Forest Management Act. The insinuation was that their delegation would so instruct through legislation if we did not change our interpretation of applicable law.

I replied that I was convinced by consultation with my attorneys that our interpretation was the correct one, and we would continue management on that basis. In doing so, I very politely encouraged the delegation to clarify the situation through legislation and offered the help of the Forest Service in drafting such legislation. And as an aside, I asked if they had consulted with the delegations from the other states that contained national grasslands within their borders. Of course, such consultation had not taken place, and it was crystal clear that the delegation was putting on a show for their constituent sitting in on the conversation, ostensibly representing the permittees on the grasslands. I had taken care before the meeting to check out the degree to which he was representative and was informed that he was a "wannabe" power broker and actually carried little weight with the permittees' associations.

As I had already been considering the potential advantages of placing all of the grasslands under a single grasslands supervisor, it seemed a good maneuver to offer the study of such action as a cure for their desire to placate their constituents who utilize the grasslands. Seeing that their initial ploys were not going to be successful, they quickly seized on this "concession" by the Forest Service chief as a satisfactory conclusion to the meeting.

This was a political win for all concerned in their own minds, when in fact, nothing substantive took place. One can be entertained by this sort of thing. It is a game of bluffs made and called, power and fealty expressed, debts paid, favors earned and dispensed, and trade-offs consummated.

The only danger that I can see in such games is that some involved may not understand that there is a game under way and take the results too seriously.

11 December 1994, Washington, D.C.

The reinvention plan for the Department of Agriculture, including the Forest Service, was released to Congress and to the public. The bottom line for the Forest Service is that personnel numbers, in full-time equivalents, have declined nearly 4,000 from a level of some 45,000 in 1992, and are expected to decline another 3,000 by 1999. And that is the projection under a Democratic president who anticipated a Democrat-dominated Congress. The real decline is now likely to be even more severe. It is essential that the Forest Service shift resources, people, and money to the levels in the organization where the agency mission is concentrated. That means a shift from headquarters staffs in Washington and regional offices to national forests, ranger districts, and research work units.

One way to accomplish this is to cut down the number of regional offices and experiment stations by combining them with other units. So the reinvention plan put forward called for seven regions instead of nine (regional offices in Missoula and Juneau are to be closed), and for seven experiment stations instead of eight. Tuesday was taken up with briefings at the regional foresters and directors meeting in Washington, Senate staffers, House staffers, and selected congressional personnel.

The first such briefing was with the staff of Senator Max Baucus of Montana. The meeting was not pleasant. Curt Rich, Baucus's chief of staff, went ballistic and shouted obscenities at Undersecretary Lyons. I have never heard such an obscene, bullying, and totally unprofessional performance by a congressional aide. The man's attempts to be intimidating were petulant and, in a sense, amusing or pathetic, depending on the moment.

We did not visit with the Alaska delegation, as they canceled the scheduled meeting. However, they blasted the proposal immediately as to the proposed closure of the regional office in Juneau. All in Congress want to cut government spending and reduce bureaucracy until it comes to closing government offices in their states.

Forest Service personnel who were directly affected were likewise upset. That is natural enough, I suppose. However, even the leaders (or at least some) can't seem to come to grips with the reality—that a Forest Service that is reduced about 16 percent in personnel must make significant adjustments in its organization and ways of doing business.

Undersecretary Lyons had his staff prepare a list of senators and representatives for selected staff to call. When I received my list, it was obvious that I had been assigned every member who was likely to snarl and bite. The gist of the phone calls was to inform the members that we—the Forest Service and Department of Agriculture—understood clearly the requirement in the appropriations language that states that no appropriated funds can be expended to alter any regional boundaries or to close any regional offices. Our attorneys have advised me that this is not constitutional and therefore there is no such requirement. That seems a bit of an esoteric point as those committees (Appropriations and Natural Resources in the Senate, and Natural Resources and Agriculture in the House) have us by the throat in terms of budget and authorizing legislation.

Therefore, the plan that we have put forward is a proposal subject to approval of the named committees. Upset by not being consulted ahead of time and because Oregon and Washington would be divided between two ecoregions, Senator Mark Hatfield, chairman of Senate Appropriations, issued a press release the next day declaring the proposal "dead on arrival." All members of the Alaska delegation issued individual news releases blasting the proposal, as did the delegation from Montana.

Regional foresters do not like the proposal, either, and even for regions left intact, they see the separation of operations and removal of State and Private Forestry from the regional foresters' purview as a significant loss of prestige and power. It will be interesting to see how many support the decision (which they should feel bound to do) and how many will resist overtly, or covertly; I suspect there will be some of each and I think I can guess who is likely to do what.

Many of the old-line Forest Service brass are upset at what they believe to be interference by the political appointees who direct the Forest Service operations. They suggest that I must resist the political appointees' directions and influence to the point of getting fired. One of them told me, "The Forest Service has a duty to the American people that transcends our responsibility to carry out the wishes of the 'politicians' appointed to positions of authority in the Department of Agriculture." Now there is an interesting concept. I wonder if he would concur that the same principle applies to the American military.

I told them that I considered it my duty to recognize the authority of those in legitimate power over the agency, to forcefully argue the agency's view in areas of disagreement, to offer courses of action and advantages and disadvantages of each course, and to make decisions on issues within the agency's purview. Then, if we are given orders, we should obey them if they are legal. Each of the professionals must then judge their actions in keeping with their professional and personal ethics and respectfully resign if the conflict with conscience is too great.

I do not think it in keeping with honor and ethics to take these matters of disagreement to others. Particularly, I think it is inappropriate, disloyal, and unethical to play politics on the Hill to undercut those in legitimate power in the executive branch of the government. To my mind, such is the antithesis of ethical behavior.

13 December 1994, Washington, D.C.

Bob Nelson, Forest Service director of Fish and Wildlife, and I ate breakfast with two representatives of the Wilderness Society to discuss the present state of affairs. They are participants in the Pacific Rivers lawsuit that has forced a consultation with the Fish and Wildlife Service and the National Marine Fisheries Service each time another species is listed as threatened or endangered. All action that it is determined may adversely affect a listed species must be stopped in the meantime.

The Wilderness Society representatives understand that the political situation has changed dramatically with the Republican triumph in the last election, and they are testing the waters. I told them frankly that they simply had pushed too far and it was time to back off. Even if they do back off now, the consequences of their pushing too far were likely to occur regardless. This will take the form of significant weakening of the Endangered Species Act. If the decision goes against the government in the Seattle case before Judge Dwyer, it seems likely that Congress will quickly legislate a timber cut level with the appended words, "all other laws not withstanding." Senator Mark Hatfield has already told me that he will put forward such legislation and that it will pass. Congressman Norm Dicks has told me that he and the other three surviving Democrats in the Northwest delegation will support that action. My sense is that such a move would easily succeed.

I told them that I was prepared to lead the charge on the salvage of burned timber and forest health initiatives. Such salvage will provide timber to a short market, help with the balance of trade, provide employment, provide profits for industry, and help reduce fire danger. That adds up to a good thing and we will proceed with vigor.

Senator Larry Craig of Idaho, who in the new Congress will be head of the Energy and Natural Resources subcommittee that deals with forestry issues, has told me that Congress will quickly provide the Forest Service an exemption from appeals of salvage sales if that is requested. In the alternative, an exemption may be provided with an instruction to use that exemption even if we don't ask for it.

I told our environmental friends that it was time to get behind the Forest Service in terms of some rational management actions. The tide of public opinion and certainly the mood in Congress are shifting dramatically and the gains in conservation over the past several years could erode quickly. They got the point and did not disagree. The question is whether or not they can move to a proactive position and still maintain the support of their membership, which sends in dues to fight "environmental destruction," not to promote collaboration with government land management agencies.

That necessary move to some level of collaboration with good multiple-use management may well be the test of the long-term viability of the environmental movement. The mainstream groups may be able to accomplish that shift. The more extreme groups will not, either internally or in terms of their memberships. They are the true believers who stand, in their minds and souls, at Armageddon and battle for the Lord. For them, there can be no compromise without loss of faith and the certainty of righteousness that goes with true faith. But they serve their purpose of staking out one edge of the decision space, which will contain the ultimate consensus that will last for a while.

21 December 1994, Washington, D.C.

I was at a meeting in the Interior Department when I was handed a message that read, "Judge William Dwyer has ruled in the case in Seattle in which the President's Plan was challenged on numerous grounds. The judge ruled for the government on every single issue in question. Total sweep for the government. A 'slam dunk.' Congratulations."

I nearly jumped out of my seat. We did it. We actually did it, despite every prediction that we would never be able to get a plan past the federal court. And we did the job in record time. The president gambled on us and we came through. The plaintiffs included sixteen environmental organizations, and the timber industry lawyers represented some nine cointervenors. Quite a lineup from both ends of the spectrum of those interested in the management of Northwest forests.

Judge Dwyer concluded that "All claims and arguments asserted . . . have been considered, and none would justify invalidation of the plan or a remand to the agencies." There could have been no better result for the government. Every single issue was resolved in favor of the government's position.

So after nearly five years of effort by a core group of scientists with a supporting cast that finally reached well over five hundred persons, it has finally ended. These efforts included the Interagency Scientific Committee to Address the Conservation of the Northern Spotted Owl, the Gang-of-Four, the Scientific Assessment Team, and the Forest Ecosystem Management and Assessment Team. It was my honor to have served as the team leader for ISC, SAT, and FEMAT efforts. The efforts of all those teams have now been vindicated by today's ruling by Judge Dwyer.

I think that historians may write of these efforts and this five-year period as a turning point in forest management and conservation in the United States at least, and perhaps in the world. The issue quickly evolved from a concern over the welfare of a single species labeled as threatened under the Endangered Species Act to the first broad-scale consideration of ecosystem management.

In the first flush of victory after nearly five years of slashing attacks by

Chief's Christmas reception, 1994. Photo courtesy of Jack Ward Thomas.

elements of both environmental and timber industry hired guns, it is easy to be elated. After all, the professionals in conservation who worked so long and so hard against the political grain have been, at long last, vindicated. But that elation is tempered by the certainty that the environmentalists' advocates have lost decisively. Such is also true of the timber industry. But that matters little as the timber industry squandered their credibility some time ago. Now the environmentalists have again squandered both scarce resources and rapidly declining credibility in a battle over an issue that they had, essentially, already won. Coupled with the potentially disastrous results of the recent midterm elections to their positions, this is another significant setback.

22 December 1994, Washington, D.C.
I received a letter today from Mollie Beattie, director of the Fish and Wildlife Service. She informed me that she, as head of the delegation from the United States at the Convention on International Trade in Endangered

Species, had voted in favor of listing broadleaf mahogany in CITES Appendix I as a potentially threatened species. This was contrary to the agreement we had reached in earlier deliberative meetings.

This vote was cast on the direct order of Vice-President Gore, who had evidently, based on conversations with Director Beattie, been told by representatives of the Department of the Interior that USDA approved this action.

As I had not been contacted, I assumed that such approval had originated with either Undersecretary Lyons or, more likely, Deputy Undersecretary Adela Backiel, who had been present with me in the multiagency discussions on the matter. I was a bit distressed at the possibility of their changing position without consulting with me.

Backiel had also been informed of the situation, and as neither she nor Lyons had been consulted by Interior, she was upset. She told me that she has so told the president's counsel on environmental affairs, Katie McGinty.

By afternoon, it became apparent that Secretary of the Interior Bruce Babbitt had made a dramatic change in policy over the proposed range reform package that addressed grazing by permittees on BLM and Forest Service lands. Though the Forest Service was supposedly a full partner in this effort, there was no consultation prior to the decision or afterward. We were informed by reading about Mr. Babbitt's reversal in the *Washington Post*. After a year of turmoil caused by packing grazing activity considerations with the issue of grazing fees, the secretary now reads the political tea leaves as requiring splitting the issue and turning over grazing fees to the Congress. Oddly, this was exactly the approach that USDA and the Forest Service had advocated a year earlier.

We also read in the papers that Secretary Babbitt had decided to back off his decision to turn over 9,000 acres of Forest Service lands to an Indian tribe in New Mexico. USDA and the Forest Service did not even know he had made a decision to transfer the lands. In fact, in an earlier consultation, I had been told the circumstances of the case and had been encouraged to acquiesce in such a decision. My Office of General Counsel attorneys advised otherwise, and I declined to approve. The representatives of the Department of the Interior accepted that and assured me that they would not proceed without my approval. They then went over my head to Undersecretary Lyons, who also refused and was given the same promise.

In that same week, we read in the *Post* that Secretary Babbitt had criticized the Forest Service for not "debarring" Kaibab Industries from any additional purchases of timber from National Forest System lands. The Forest Service had determined that it was likely that Kaibab Industries had illegally cut and milled trees from national forest land. Kaibab Industries was indicted on the charge but accepted a plea bargain from the Department of Justice that involved restitution and a fine and no debarment. That decision was not in the hands of the Forest Service. Rather, it lay with the Department of Justice. Further, this issue was not in Secretary Babbitt's department and he did not know the details.

All in all, this was not a good week for enhanced working relationships between the departments of Interior and Agriculture (primarily, the Forest Service). These problems are to be laid directly at the feet of Secretary Babbitt.

Things looked even worse when Babbitt announced that he was overriding the decision of the Park Service on the number of cruise ships per week to be allowed to visit Glacier Bay in Alaska. While not particularly significant—in my opinion, anyway—in the environmental sense, the decision further emphasized an apparent retreat from proenvironmental positions. That may have been the intent as the politicians all over Washington begin to scramble away from environmentalist positions.

27 December 1994, airborne, en route from Louisville, Kentucky, to Washington, D.C.

"Ecosystem management" is much discussed in conservation circles these days. Under Chief Robertson, the Forest Service defined it as "the use of an ecological approach to achieve multiple-use management . . . by blending the needs of people and environmental values in such a way that national forests and grasslands represent diverse, healthy, productive, and sustainable ecosystems." The term draws comments ranging from "an idea whose time has come" to "ecosystem management—whatever that is" to "a plot to destroy the constitutional guarantees of the ownership of private property." The Clinton administration has essentially declared ecosystem management, with its analog of "sustainable" agriculture and forestry, as our national policy. As chief of the Forest Service, I have declared the institution of ecosystem management to be a primary building block of the agency's *Course to the Future.*[2]

To my mind, while the problems of bringing the concept into practice can be formidable, given its complexity, it can conversely be made acceptably simple enough to have meaning. Ecosystem management requires us to think in three unusual ways: at landscape scale and across artificial boundaries, at expanded timeframes, and with an underlying objective of maintaining biodiversity, which conceptually evolves into keeping every cog and wheel, which leads to the maintenance of ecosystem function. Therein lies the assurance of the continuation of the ability to extract the raw materials to support human life.

These concepts become more meaningful when an expanded concept of the *eco* in ecosystem management is pondered. The meaning of the Greek word *oikos,* from which *eco-,* is derived can be interpreted simply as "house," even more appropriately as "home." *Homo sapiens* is, without doubt, the most ecologically significant vertebrate on the planet. There can be no landscape-scale, long-term vision of ecosystems without the consideration of the role of humans. It is not by accident that the word economics contains the same root word as ecology.

Now the time has come, by fortuitous confluence in the evolutionary streams of biological sciences, data-handling technology, philosophy, religion, transportation, trade, and politics. Those primarily concerned with guiding their drift boats down the separate feeder streams have arrived—more or less simultaneously—at the mainstream river. Now, guiding their drift boats past bigger shoals, rocks, and rapids is complicated by having to work in concert with the other boats.

At this stage in the journey, perhaps the best that can be hoped for is a recognition that there are indeed other craft moving along the same river in the same direction. It may be better over the short term to lash the boats together to lessen the chances of an upset. Perhaps sometime when the boats are farther down the river and the dangers increase, it will seem expedient

---

2. Jack Ward Thomas, 1994. *A course to the future.* Washington, DC: USDA Forest Service.

to use the materials in each boat and the skills and experience of those at the oars to build a bigger, safer boat to face the vicissitudes of the growing river.

There is an adage in ecology—and in economics: Everything is connected to everything else and there is no such thing as a free lunch. Growing concerns about—or symptoms of—the failures in addressing emerging interests piecemeal have brought us to the point where the Gulliver of natural resource management is increasingly bound by a plethora of threads, each held by a specialist (forester, soils scientist, hydrologist, fisheries biologist, et al.) with a particular interest to serve and likely a particular ax to grind.

The resulting creeping paralysis by analysis of the single-minded approaches of technical specialists and the overlapping responsibilities of agencies within the federal government and within state governments and between state and federal governments—and sometimes between nations— has led to a system that is simply bogging down. That complexity is exacerbated by the constant bombardment of legal actions, and the courts constantly issue new decisions that clarify (or confuse) the appropriate execution of the laws that apply to natural resource management—particularly the management of public lands.

There seem to be only two ways out. The first is the abandonment of the applicable laws, either by outright revocation or modification, or by congressional targets in the appropriations to be achieved by the land management agencies "all other laws notwithstanding." Such would, in my opinion, be a tragedy of national and world significance. I know of few natural resource management professionals who do not believe in the objectives and purposes of environmental laws. But more and more of these professionals are becoming more and more frustrated with the constantly escalating costs— social, economic, political, and ecological—of the deepening morass.

The second course to the future of natural resource management is ecosystem management as a guiding concept. That is becoming obvious. For example, Judge William Dwyer, in his recent decision on the President's Forest Plan for the Pacific Northwest, clearly answered the question whether ecosystem management and interagency land-use planning were in keeping with the law. He made a dramatic decision by saying, "Given the current state of the forests, there is no way that agencies can comply with the environmental laws *without* planning on an ecosystem basis."

But there are obstacles in this path that has now been taken—and from which, I believe, there can be no turning back. The very laws that Judge Dwyer considered in his decision were not written with the attributes and possibilities of ecosystem management approaches in mind. These laws are "functional" in nature, in that each was written to emphasize a single or several values and name a champion agency to pursue the Holy Grail of that particular law. The agencies then staffed their ranks with the most appropriate functional specialists, who quickly adapted to the task of the agency with dedication and zeal—as they should.

Turf battles between agencies and the zealousness of the troops produced the chaos of functional collision. And though there has been some marked improvement in the coordination between agencies at higher levels, the attitude of the lower-echelon folks is as single-minded as ever. This is particularly obvious in the regulatory agencies—Fish and Wildlife Service, National Marine Fisheries Service, and the Environmental Protection Agency. To make matters worse, there seems to be a great reluctance in these agencies to alter the positions taken by field personnel at the journeyman level.

In my heart, I don't believe that we can make things work over the long term without modification in the environmental laws and regulations, including some reduction in the absolute power of the regulatory agencies. Perhaps that can be achieved with the new Congress. The trick will be to hold on to the purpose of those laws in the process.

1995

13 January 1995, Washington, D.C.

Today, Undersecretary Jim Lyons and I met with Senator Conrad Burns of Montana regarding Forest Service reinvention and other matters. His only interest is, I believe, protecting the Forest Service presence—i.e., the regional office—in Missoula that is proposed for closure in the reinvention strategy.

The senator fired the opening shot across my bow by stating, "Maybe we should turn the Forest Service lands over to the state." The statement was designed to throw us on the defensive. My reply was based on the Republican rhetoric about "unfunded mandates" to the states. I replied, "Senator, that would be the ultimate unfunded mandate. You might want to check on how much money was spent fighting fire in Montana last summer."

His next ploy was to say, "I don't think we ought to be fighting all these forest fires." My response was to ask if he was the same Senator Conrad Burns who called me late at night last August demanding that I go to Montana to personally show concern and to provide additional firefighting capability to Montana. He smiled and allowed that I was not making things easy for him.

His next shot was to demand a detailed cost-benefit analysis of closing out the regional office in Missoula and transferring responsibility to the regional forester in Denver. Of course, the intimation was that such an analysis would lead to a different conclusion as to the proper location of the regional office. I responded that such a detailed analysis had not been done as part of this decision, but such analyses have been done in the past and there is little reason to believe that anything would change. This—

coupled with considerations of cooperation with other agencies, ease and cost of transportation, ecosystem management considerations, etc.— would be likely to reinforce the proposed closure, not diminish it. I asked, pointedly, if such an analysis did not make the point he desired, would he change his position? He allowed that he probably would not change his position on that basis.

He then tried to make the point that the Forest Service has disproportionately located personnel in the "big city" of Denver and slighted the "rural area" of Missoula. He was a bit taken aback when I told him that there were more Forest Service personnel in Missoula (830 or so) than in any other location in the country, including the Washington office.

Senator Burns's next tack was to stress the importance of having a regional forester where "he could at least see a tree," rather than in the big city of Denver. I made the point that it was our desire to move such decision making downward in the organization to the forest supervisors. When he insisted on his point, I suggested that then he would not object if the decision was to maintain a regional forester in Missoula, and that except for technical support staff, support functions would be transferred to Denver or elsewhere. This would free up several hundred slots for moves to field locations. He didn't like that either and admitted that he would be distressed by anything other than the status quo.

I then went through the necessity to downsize the Forest Service by 7,000 positions and the crisis that was developing at the forest and district levels due to lack of personnel and scarcity of appropriately skilled people in the right places in the organization. The senator was becoming more uncomfortable as he realized that we were coming face to face with the old adage, "There ain't no such thing as a free lunch."

It was time to put the ball in his court by making the point that if he did not like the proposal put forward by the Forest Service, it was his responsibility to suggest alternative solutions. This obviously made him uncomfortable as he was caught between the rhetoric of "cutting government" and "reducing the number of bureaucrats," and protecting the dollars and people located in his state. This is, obviously, disproportionately difficult for an outspoken conservative.

The conversation was becoming a bit tense at this point, so I changed the subject to the safe area of people and places known in common and times gone by. On that note of amiability, the interview came to an end.

25 January 1995, Washington, D.C.

Late in the afternoon, Undersecretary Jim Lyons, Deputy Under-secretary Adela Backiel, Alaska Regional Forester Phil Janik, and I met with Senators Stevens and Murkowski and Congressman Young of Alaska. This is a powerful trio in that Young heads the Resources Committee in the House and Murkowski heads a similar committee in the Senate. Stevens is on the subcommittee of Appropriations that deals with the Forest Service.

This was the most bizarre meeting I have yet attended with congressional people. The meeting was an hour-and-a-half tirade by the three Alaskans about the Forest Service "plot" to hold down the timber available for market in Alaska. They insisted that this failure was due to a unilateral decision by the Forest Service in response to the petitions to list the Queen Charlotte goshawk and the Alexander Archipelago wolf as threatened species. Of course, the accusation was not true. The decision was made in concert with the Fish and Wildlife Service to prevent a hurry-up listing of either of those species as threatened under the Endangered Species Act.

Murkowski and Young claimed that the Alaska regional director for the Fish and Wildlife Service had told them such listings were totally unjustified and that the Forest Service made its decision totally on its own and without Fish and Wildlife Service participation. Janik didn't believe that, and this was certainly not in keeping with my conversation with Fish and Wildlife Service Director Mollie Beattie a week ago.

Janik tried to respond to the Alaska delegation's questions but was shouted down time after time. The Alaskans said that they did not wish to discuss the matter. They simply demanded enough timber to satisfy "the demands" of industry. Phil was very calm throughout, and I said nothing but looked directly at each one in turn as they talked, with no expression on my face and no body language.

Threats escalated as the harassment continued. First, Young said unless he got what he wanted, he would transfer the Forest Service lands to another agency. Later, he upped the ante by saying that he would turn management over to contractors. To emphasize his perceived power, he told us of that power and the consequences of not heeding his wishes and then strode from the room.

Stevens then took over as the chief scolder. He told us that the Forest Service had no credibility with him, that we were involved in some plot to destroy the timber industry in Alaska, and worse. He made his threats, starting with the threat to reduce Forest Service personnel in Alaska. That soon escalated to the threat to insert a rider in the Appropriations Bill for the Forest Service to specify that none of the funds be used to pay my salary, and by implication, Janik's salary as well.

Finally, Murkowski asked me a question that I tried to answer, but he abruptly cut me off. I tried again. He cut me off again.

I told him if he wanted me to respond he needed to afford me the chance to talk. I said that I had listened to him berate the others and me for the better part of an hour. If that were the purpose of the meeting, then I would again return to silence. They agreed that all they wanted to hear from us was that we would acquiesce to their demands, and according to Stevens, the meeting was over.

Stevens then said he didn't believe a word we said and that he would take the unusual route of placing us under oath when we testified before his subcommittee. By this, he simply called us liars. I told him that I would be pleased to be placed under oath as I considered myself under oath in any such dealings.

The Alaskans promised legislation that would "fix the problem" if we didn't give them what they wanted. By this time, I was beginning to wonder if they did not protest too much.

They are obviously frustrated and feeling their oats as the Republican majority and are impressed with their power as committee and subcommittee chairmen. That makes for a dangerous combination. However, I suspect that their bullying behavior might backfire on them. Use of brute force and bullying are not becoming in those exercising power. It is much more tolerable, even amusing, from those without power.

31 January 1995, Washington, D.C.

I met today with Congressman Ron Wyden and Congresswoman Elizabeth Furse, both Democrats from Oregon, and Katie McGinty, director of the Council on Environmental Quality. Wyden and Furse are interested in presenting a bill to set aside the Bull Run Watershed, which supplies Portland's water, from any additional logging. This bill would significantly expand the protection offered in the President's Forest Plan for the old-growth reserve and the tier 2 watershed. Wyden and Furse seek to add the Little Sandy drainage—which is well roaded and has been significantly logged—to the Bull Run Watershed, along with several significant buffers around the edges of Bull Run.

I had examined the Forest Service maps and remembered the discussions that covered a number of hours by the Forest Ecosystem Management and Assessment Team about how to deal with the Bull Run Watershed issue. The Little Sandy addition is not well justified as a pristine watershed because of previous logging and roading, the necessity to use water treatment, and the fact that the water is allocated to a power company.

The buffers around the Bull Run contain lands that drain into watersheds other than Bull Run. This is the typical environmentalist game of "take what you can get and then come back for just a little more."

Wyden and Furse asked for two things. The first was a letter from the Forest Service declaring that there would be a moratorium on any land-disturbing activities in the area of concern until after the bill to be prepared and introduced by Wyden and Furse has been acted upon. The second request was for the administration to support such legislation.

During the ensuing discussion, I said I would not write a letter declaring a moratorium, as it would be political and a bad precedent that would lead to numerous other such requests across the Pacific Northwest. Besides, such a moratorium is not necessary, as the forest supervisor has already written to Wyden and Furse saying that no assessments are planned for the area until 1997 and no activities could occur before 1999.

I also said that the Forest Service could not, on technical grounds, support the addition of the Little Sandy and the suggested buffers. However, obviously the administration could support whatever bill they chose.

Wyden and several of the aides present said they would be satisfied if the Forest Service would work with them to produce a bill that the Forest Service would find acceptable. He said that Forest Service people had refused to discuss the matter with his and Furse's aides.

I replied that Forest Service people had met with them several times and pointed out the problems they saw with the proposal that led the Forest

Service to testify against the bill introduced in the last session of Congress. This, of course, was de facto the administration's position.

That led to a discussion of the politics of the matter. I spoke up to say that the Forest Service could and would provide technical assessment and advice but that it was inappropriate for us to be involved in politics. That stiffened things up quite a bit, and such was obviously not a politic thing to say.

In closing the meeting, I promised, if requested, to provide a letter from the chief to Wyden and Furse to the effect that no actions would take place in the area of concern prior to 1997, and we would provide technical assistance in evaluating various forms of the proposed bill. The meeting ended on a friendly note.

8 February 1995, Washington, D.C.

The Occupational Safety and Health Administration delivered its report today concerning the deaths of the fourteen firefighters on Storm King Mountain during last summer's fire. This independent report (OSHA had refused to participate in the Forest Service and Bureau of Land Management investigation) was identical in all significant respects to the report of the agencies involved. OSHA also found that the corrective actions already taken by the agencies were appropriate and should remedy the failures they detected.

OSHA then went on to cite the Forest Service and the Bureau of Land Management for "willful negligence" in the failure of "management at all levels" that contributed to the tragedy. As OSHA is the judge, jury, and executioner in such matters, it is difficult if not impossible to dispute its findings. OSHA uses the technique of delivering its report to the press at the same time its decision is announced, which assures that the action is fixed in the public's mind. The system allows for the condemned to appeal to OSHA for a change in findings and interpretation of findings. Of course, if OSHA changes its mind as a result of the appeal, the agency remains culpable in the public's mind because of the attention given the original report. Nice system!

I vehemently disagreed with the statement of "willful negligence," which implies not only was management negligent (which is untrue in my opinion), we were willfully so. That is ridiculous. No one in the chain of command above the level of fire boss on the fireline at Storm King Mountain on that fateful day willfully made decisions that placed the firefighters in jeopardy. At least one of the line bosses made a decision, despite knowing that the crews were in violation of the ten standing firefighter orders, to stay on the line. Perhaps that was willful, but I was not there and do not know the circumstances. Building the fireline, which was also the designated escape route, downhill at a very steep angle, was willful, and certainly it was a mistake. But again, I was not there on the line.

If there was a significant failure in upper-level line officers, it was that somehow the intensive training that these top-of-the-line firefighters received did not impart an adequate appreciation of safety first, last, and always. They were tough and brave and aggressive, and those fine qualities contributed to their deaths when the fire simply blew up. That blowup was extreme and unexpected, but OSHA says we should have known. Maybe.

My response to the press calls was to emphasize that the OSHA report corroborated the agencies' report, that OSHA agreed with the corrective actions taken, and that—as the head of the agency—I accepted responsibility for the safety of our people. I made no excuses and no arguments with the findings. To do so would have come across as evading responsibility and would have demeaned the agency and the memory of those who died on Storm King Mountain.

Of all the press calls, one question-and-answer stands out in memory.

Question: "Chief Thomas, is this devastating report a blow to your agency?"

Answer: "Madam, the blow to the Forest Service was losing fourteen of our people to a firestorm. All else is trivial. Good day."

Sometimes still, in the night, a vision comes to me of charred bodies along the escape route on Storm King Mountain. Those were my people. I did not need a report from OSHA to remind me of my responsibility.

10 February 1995, Washington, D.C.

Today there was a long hearing on the subject of forest health before the Senate's Joint Committees of Agriculture and Natural Resources. Wayne Allard of Colorado chaired the hearing. It was a political sideshow designed by the new Republican majority to showcase the consequences of decades of mismanagement under Democratic rule in Congress on the forests and timber workers of the West. As with most such hearings, there was more shedding of heat than light.

The Democrats were angry with the Republicans about the format, organization, and conduct of the hearing and were noticeable largely by their absence. There were a few embarrassing, childish exchanges between congressmen to keep people awake—at least, the exchanges were embarrassing to me.

The hearing was clumsily structured and conducted, which was likely a reflection of the inexperience of the new Republican majority in the exercise of the prerogatives of power afforded the majority.

The old Republican power players made routine appearances but then failed to reappear after one or other of the numerous interruptions for votes. Nearly all the "questioning" (speechifying) was done by freshman Republicans, who on average are very young, very doctrinaire, and amazingly ignorant of the subject. The Republicans who carried the action were freshmen who looked preppy, gave lectures (which, given their age and inexperience, seemed pompous), and exuded what they perceived as power. I could not help but think that they represented a wonderful argument against term limits.

Even adversarial hearings can be stimulating for witnesses, at least for me. This hearing was deadly dull and drawn out. The hearing started at 9:30 A.M. and was to include seven panels of witnesses. In the normal circumstance, administration witnesses go first in order to spare demands on their time. In this case, the Forest Service and Bureau of Land Management witnesses were third, after testimony for Senators Slade Gorton of Washington and Conrad Burns of Montana and Representative Wally Herger of California. Testimony by members of Congress is always a bit amusing, in the lack of understanding of the subject displayed and the "courtesy" afforded by the committee members to the witnesses' statements. The reason for the courtesy is obvious: What goes around comes around. If there is hard questioning or dispute, the perpetrator will get the same treatment when he or she chooses to pontificate as a witness in front of the committee.

The second panel was made up of spokesmen for labor unions whose members were losing jobs. Of course, they claimed that all these job losses were caused by the lack of timber from Forest Service lands. There was no mention of the very high levels of timber processed in Oregon and Washington and the jobs lost to improved technology, etc. The purpose was to establish the contrast between suffering workers and the uncaring bureaucrats on the third panel. The strategy was good but the execution was abysmal.

The point was made repeatedly that "radical environmentalists" were misusing the "environmental laws" and the laws were being misapplied by "liberal federal judges" and they, the new Republicans, were going to change all that. But it was obvious that they do not know what to change or how to change it. As a result, they keep asking the Forest Service to tell them what the problems are and how to correct them. They don't seem to know that such a response from the Forest Service would be a policy call that must be made by the administration. Or maybe they know that and just want to paint the Forest Service as intransigent and a tool of the environmentalists. Who knows?

We were kept before the committee until after 3:00 P.M. Six hours had gone by and only three of seven panels had been called. At this rate, the hearing was likely to go on until 10–11 P.M. I was glad to get away. Today's experience was not a high point in the democratic process.

13–14 February 1995, the Chief's Office, Washington, D.C.

Last Friday, I received word from Undersecretary Lyons that Leon Panetta, chief of staff to the president, had instructed me to convene the heads of the Fish and Wildlife Service, the Bureau of Land Management, the National Marine Fisheries Service, and a representative from the Environmental Protection Agency. Our mission was clearly defined as "removing barriers to the speedy preparation and sale of timber salvage from Forest Service and Bureau of Land Management lands."

Rollie Schmitten (director, National Marine Fisheries Service), Mike Dombeck (director, Bureau of Land Management), and I were present. Mollie Beattie (director, Fish and Wildlife Service) never showed up but was represented by her staff director for endangered species.

After two days, we were able to identify several factors under our control that could be adjusted to speed things along. Among those were (1) teams of biologists—FS, NMFS, and F&WS—will work together to prepare sale plans and biological opinions so that the final document will be simultaneously approved as passing consultation; (2) NMFS will accept fisheries biologists from FS to handle consultations; (3) FS and BLM will prepare activities for consultation in batches, and NMFS and F&WS will accept consultation in that form; (4) NMFS and F&WS will do simultaneous consultations on proposed actions; (5) NMFS and FS will seek a waiver from the Government Printing Office to print environmental assessments or environmental impact statements for public review; and (6) Director Schmitten will approve helicopter logging of salvage in roadless areas.

Schmitten's decision concerning helicopter logging in roadless areas is very significant in that a hot issue is settled in one decision. I don't know if Rollie really understood the political firestorm that will erupt from that decision. The hardcore environmentalists—particularly Pacific Rivers Council—have been playing the roadless area issue as if the question were road construction. In fact, their intention has been, I think, to maintain those areas as de facto wilderness with an eye to adding them to the Wilderness System at some propitious time in the future. This decision, if he stays with it, will blow up that issue to a cause célèbre. It is almost certain to engender a lawsuit with a request to the court for an injunction on salvage logging—even by helicopter—until the court can hear the case and make a decision.

If that is successful, the Republicans in the Senate (Murkowski, Burns, Hatfield, Craig, Kempthorne, and Packwood) will pounce with "sufficiency language" that will, in my opinion, pass easily. This is no longer a chess game in flux. It is moving to the end game and there are only a few moves left. Those moves seem now preordained and perfectly predictable, the embodiment of a Greek tragedy.

The next move after that will be predicated by the recognition of the Idaho delegation that there is suddenly a reason to push an Idaho wilderness bill. This is because roadless areas identified as wilderness study areas can be excluded from wilderness and through "release language," coupled with sufficiency language, made available for, at least, salvage logging by helicopter.

So on paper, we shorten the time frame for getting salvage ready for sale by some 20 percent if we can actually get things as well organized as described. It will work if NMFS does not change position again, which it has done numerous times in the past. NMFS is without discipline, understaffed, newly staffed, and unable to make decisions that stick. Agency brass continue to let themselves be rolled by journeyman-level biologists. These biologists believe in their mission and that they stand at Armageddon and battle for the Lord as the last hope for salmon. They simply do not see the bigger picture. They do not trust the Forest Service or the Bureau of Land Management to carry out promises. They have good reason, based on past experience, to be distrustful. I know how they feel and how they think, for I have walked a mile in their moccasins.

So it is the job of the graybeards to see a larger picture and to grasp reality. We do not need a final heroic stand and a glorious defeat. Now is the time to tack into the wind.

14 February 1995, Old Executive Office Building, Washington, D.C.

Early in the day, I received a note from my assistant, Sue Addington, that I was to attend a meeting with Katie McGinty, the president's adviser on the environment and head of the Council on Environmental Quality; Undersecretary Lyons; Deputy Undersecretary Backiel; Interior Undersecretaries Bob Armstrong and George Frampton; BLM Director Dombeck; Fish and Wildlife Service Director Beattie; and Deputy Assistant Secretary of Commerce Kate Kimball (National Marine Fisheries). The entire purpose of the meeting was to relate to us conversations between the president and Governor Kitzhaber of Oregon about the political situation in Oregon related to timber supply. The president wants to visit Oregon but has been advised by Governor Kitzhaber not to do so until the administration had "kept its promises" related to timber supply. The White House wants us to understand how important it is "to get the cut out." It is described as essential to the political fortunes of the president to place significant volumes of the timber on the market early this year.

I carefully explained several times that it takes nine months to have a baby, and that the Forest Service is moving as rapidly as possible to prepare sales under all the processes put into place in the record of decision on the President's Forest Plan. It seems to have been conveniently forgotten that the White House was cautioned over and over that the bells and whistles being added to the original Option 9 in attempts to satisfy the environmentalists (who can never be satisfied) would make for a slow, tedious process. So now, as the pressure begins to build, those in power once again begin to press the Forest Service to get the cut out. The more things change, the more they remain the same.

I told McGinty that the Forest Service was moving as fast as it can, and that if it were not for the unpaid overtime being contributed, things would slow even more. I told her bluntly that no matter what the political necessity, the bulk of the timber sales being prepared will not be ready for sale until the last quarter of 1995. She did not like that answer and told me, "The White House thinks the Forest Service is dragging its feet," then turning to Beattie, "and the Fish and Wildlife Service is a pain in the ass."

I again took issue and restated that we were on schedule, moving as rapidly as possible, and the targeted sales cannot be significantly accelerated. She smiled and told me that she knew that but the view from the White House differed. She encouraged us to do the best we can, then made it clear that the administration's position was that no legislation was necessary for us to keep the timber flowing as predicted. My last statement of the meeting was that we were proceeding as planned but that our every action would be monitored, and it was likely that continuing appeals and lawsuits would shut us down. And then I asked, "Why doesn't the administration recognize that some change in environmental laws and regulations is inevitable and set out to lead to make those changes in the most rational, thought-out fashion?" The question was left hanging as the meeting closed.

Frustrations are mounting and pressures are building on us to get out the cut or the salvage or whatever at the same time that the Forest Service is being debilitated by downsizing and the agonizingly slow reinvention process. Something is going to have to give.

23 February 1995, Washington, D.C.

I met today with the Wyoming delegation. Senators Al Simpson and Craig Thomas were present for the entire meeting. The first item of business was the consideration of the case of a permittee who had been given two years of nonuse on a grazing permit and a 50 percent reduction in permitted numbers. These penalties resulted from the permittee's subleasing his grazing rights, illegally modifying the brands of the cattle to his "open 4" brand, and then providing fabricated documentation trying to clear himself of the charges.

The permittee, a former Democratic legislator in Wyoming, aided by his son-in-law, a current member of the Wyoming legislature, exhausted the appeals process and did not choose to go to the courts.

It seems clear that the permittee and his son-in-law did not go the court route because there is very little room to dispute the charges. Instead, they have chosen to go the route of using political influence and pressure to change the agency's decision.

Shortly after coming into the chief's job, I was approached by White House staff and informed that it would be a boon to Wyoming Governor Mike Sullivan (who was running as a Democrat for the Senate) if I were to find a way to reverse the Forest Service's position. After checking the facts, I declined to alter the Forest Service position.

Several months later, Undersecretary of Agriculture James Lyons received some pleas for assistance in this matter by Governor Sullivan and Senator Malcolm Wallop. Lyons asked me about the circumstances and wanted to be certain that the case was sound and that the punishment fit the crime. After a second review, I certified that the case was sound and that, if anything, the penalty for a deliberate violation and then lying to cover the tracks was too slight. I would have simply canceled the permit if I had been the responsible line officer. That ended that round.

Then, last week, Undersecretary Lyons asked me to call Elizabeth Estill, the regional forester for Wyoming, and assure myself that the case was sound and the penalties fair. There was a question of whether we should reconsider the findings, or the penalties, or both, because the permittee was elderly, and with failing eyesight. After checking one more time, I said that the case was solid and the penalties fair. I made it very clear that if the government wanted to back off the case, it would require action above my level and such would occur over my protest.

Now, today, I am talking to the entire Wyoming delegation about the same case. They were, collectively, very polite as I explained the situation. I then offered to provide a detailed briefing and a review of the documentation of the case for either the delegation or for an aide or two. They accepted that offer.

Dealing with such permit violations is extremely difficult. First, the proof of a violation must be ironclad. Second, there is every reason for the line officer to look the other way, for to hold the permittee accountable starts a long chain of events that is just trouble after trouble. In such disputes, the Forest Service employee is an outsider and a "fed" and any actions are against a local (particularly if the permittee is a stalwart scion of pioneer stock, etc.). This makes the situation difficult to begin with.

Then, the range conservationist is aware that the political pressure will be intense and incessant. It will begin with local social pressure and then quickly expand to county officials and state politicians. The next pressure will be from the governor or folks in Congress. During this process, the range con or line officer will find that there are friends to be made— potentially powerful friends—by "reconsidering," and enemies to be made—potentially powerful enemies—by standing firm.

So why take such action in the first place and why stand fast with the decision as the pressure mounts? Many simply look the other way or use a slap on the wrist and a wink to address the problem or flinch when pressure comes. The others who move forcefully and stay the course simply believe that they are doing the right thing as required by their obedience to the law and service to their professional ethic.

The permittee has one card left to play now, assuming he knows going to court is a loser. That card is "War on the West," or "Pioneer Rancher Brutalized by Unfeeling Feds." Given today's atmosphere, that approach could be very effective in stepping up the pressure. We will see.

23 February 1995 [later the same day], Washington, D.C.

Today is the last day we can deliver the decision statement to the printer of the *Federal Register* in order to satisfy the Forest Service and National Marine Fisheries Service pledge to have fully responded to the federal judge's requirements to lift the injunction of all actions that may adversely affect listed salmon species in the Columbia River Basin. This is the completed consultation between the Forest Service and the National Marine Fisheries Service on forest plans for three national forests in Oregon and six in Idaho. The notice also contains the statement of adoption of the Pacific fisheries strategy as the interim management frame for aquatic habitats pending new land-use plans. This PACFISH strategy is essential as the mechanism for consultation on individual management action.

Sometime within the past few days, probably in response to the absolute political paranoia that Secretary of Agriculture Designate Glickman may be challenged in the Senate confirmation hearings over some action by an Agriculture Department agency, all actions that some senator may not like have been placed on hold. As PACFISH was deemed to have potential in this regard and might be judged part of the supposed War on the West, it somehow got bucked up to the president's chief of staff, Leon Panetta.

In late afternoon I was told to be at the White House to meet with White House adviser Katie McGinty to explain the situation. The meeting was scheduled for 6:00 P.M. and then changed to 7:00 P.M. I showed up at 7:00 P.M. and waited until 8:00 P.M.

The White House's extreme nervousness about congressional response was obvious. McGinty carefully questioned me in detail as to the potential impact on grazing, mining, and timber harvest. She specifically wanted to know the potential differences before and after PACFISH. It took some time for me to get the message across that the "before" situation had been dramatically altered by the listing of the Columbia basin runs of salmon as threatened under the Endangered Species Act. This new situation was encapsulated in the injunctions issued by the federal court in the Pacific Rivers v. Thomas case.

That situation, if not responded to by a mechanism for evaluation such as PACFISH, means a dramatic slowdown. That is the new "before PAC-FISH." It is obvious that it is far too late in the game for the White House to suddenly become concerned about the situation. There is no option to approving PACFISH if the Forest Service and National Marine Fisheries Service are to meet the obligation to see the federal court injunctions lifted by the deadline of March 15.

McGinty became convinced of the necessity to sign off on PACFISH and left to go see Panetta to get approval. When I got home, I realized that my hands were shaking from exhaustion and from not having eaten all day. It was 11:30 P.M. and a long time since my arrival at the office at 7:00 A.M. This does become frustrating.

Postscript: We received word of White House approval at 11:00 A.M. on 25 February. But we are prohibited from any news releases or any statements concerning PACFISH. I thought this an interesting instruction, given that I have been briefing, as instructed by Undersecretary Lyons, key and affected members of Congress for the past ten days.

1 March 1995, Washington, D.C.

When I arrived at work, I was informed that the Interior subcommittee of the House Appropriations Committee was working on a rider to the Rescission Bill that would mandate the Forest Service to put up for sale 4 billion board feet of salvage for each of the next two years. I was to visit in the afternoon with Congressman Taylor of North Carolina, who is the designated point man for the subcommittee on the issue.

I arrived at the congressman's office at the appointed time, accompanied by Steve Satterfield from the budget shop. Steve had the numbers from the Forest Service's timber shop. The congressman was tied up on some legislative business and we began our conversation with two staffers. I told them bluntly that there wasn't a snowball's chance in hell that the Forest Service could accomplish the mandate. When asked why, I responded that (1) the time frames were impossible to meet because of the processes required to put sales on the market; (2) there was not adequate staff to perform the chore of more than doubling the anticipated salvage volume; (3) it would not be possible to salvage that volume, including road construction, without violating existing environmental laws and regulations; and (4) it seemed likely to me that we would cross ethical limits for many of our employees, and we could expect problems there.

About that time, Taylor arrived in the room and there was a rerun of the conversation to that moment. We went over three possible scenarios of salvage volume. The first assumed the current workforce, adequate dollars from the salvage fund, and compliance with all current laws and regulations. This option would allow about a half million additional board feet of salvage, to about 1.7 bbf. But the increases would be spread over a three-year period because of timelines, assuming no appeals or injunctions.

The second scenario assumed that we would be able to replenish the workforce with reemployed assistants to handle the increased workloads. Also, there would be some streamlining of processes, such as acceptance of environmental assessments (as opposed to environmental impact statements); no consultation with National Marine Fisheries or Fish and Wildlife Service beyond the overall plan; some additional funding; and "sufficiency language." The first cut at timber yield numbers under this scenario would be about 1 bbf of additional salvage, to about 2.2 bbf.

The third scenario was essentially the no-holds-barred alternative, in which it was assumed that all necessary personnel would be available and only basic requirements of soil and water protection would be considered. This scenario would yield about 1.5 bbf of additional salvage, to a level of 2.7 bbf.

I pointed out that the most salvage volume ever put up by the Forest Service in a single year was less than 3 bbf. That included salvage from the aftermath of Hurricane Hugo in 1990, which involved massive salvage of blown-down timber from flat ground in the Southeast.

I was very careful to point out to Taylor that the Forest Service could not support and could not produce 4 bbf of salvage per year over the next two years. In response to his questions, I told him that my answers were strictly technical and did not imply any position on my part or on the part of the administration.

In response to his questions about the separation of powers, I acknowledged that it was the prerogative of Congress to set national policy through law, and it was the civil servant's job to abide by that law. I also said that it was the professional's job to obey the law up to the point of conflict with personal ethics. Then it was time to obey the law or to resign in protest.

Later that evening, I received calls from Debbie Weatherly, Appropriations subcommittee staff, with an inquiry, "What is going on?" When I gave her the blow-by-blow account of the meeting with Mr. Taylor, she informed me that Taylor had told the subcommittee chair, Mr. Regula of Ohio, that "[Taylor] and the chief had come to a meeting of the minds." Taylor was also reported to have said I was O.K. with the proposed 4 bbf salvage levels for the next two years. I clearly stated that I had not made such an agreement and did not believe that such a level was possible without serious environmental risk.

Later yet, I received calls from Norm Dicks, congressman from Washington, and from Mr. Regula. Similar discussions to that described above took place. It seemed to me that both were sympathetic to my position but neither wanted to stand in the way of the juggernaut that was sweeping the day in the House. They both preferred to hope for a more rational treatment in the Senate.

This seems to be an evolving pattern: don't try to influence matters in the House. In fact, the more radical the House action, the better. The thought is that the Senate will be more deliberative and the delay between the House action and the Senate action will allow some time for public response.

It is now apparent that the salvage mandate and a significant excuse from current laws and regulations will be passed as an amendment to the Rescission Bill (which, paradoxically, cuts the Forest Service budget by about $300 million). This will occur with no hearings. This will be, potentially, the most significant natural resources legislation since 1950, and such is taking place without hearings, without discussion, and without the knowledge of the American people at large. Disgraceful!

3 March 1995, Washington, D.C.

The Rescission Bill (with yet another $3 million reduction in the Forest Service budget) passed today with another surprise. We have been ordered to proceed with the old "318 sales"[1] made as part of the Hatfield-Adams compromise in 1990 in their original form, and court challenges have been precluded. Those sales that have not yet been cut have been significantly modified—some several times—to comply with various plans and strategies, with the latest being Option 9 (the President's Forest Plan). It seems certain that without those modifications, Judge Dwyer would have flunked Option 9 as insufficient to comply with the environmental laws. This has been some week.

7 March 1995, Denver

I met today with some 100 persons (law enforcement, special agents, and line officers) about the management of law enforcement operations in the Forest Service. There has been a tension in the Forest Service concerning how law enforcement personnel in the Forest Service should be organized and how their business should be conducted.

Over two years ago, in response to allegations by some Forest Service law enforcement employees and an assistant U.S. attorney (Justice Department) before Congress that Forest Service line officers were interfering with

---

1. In Section 318 of the Appropriations Act of 1990 for Interior and Related Agencies, each national forest and BLM district was required to put up two old-growth timber sales to provide "stability" while the ISC team was developing a long-range management plan. Some of these sales were not cut and were included in the protected area in the President's Forest Plan for the Pacific Northwest. —*JWT*

timber theft investigations, Congress directed that Forest Service investigators would answer directly to the chief. Chief Dale Robertson then directed
that all law enforcement personnel, including both investigations and
enforcement personnel, would report in a "stovepipe" organization to the
chief.

This decision has never been fully accepted by many in the Forest Service. Foremost among these resisters were a number of line officers in the
National Forest System (district rangers, forest supervisors, and regional
foresters). This resistance was widespread and seemed to be based on a loss
of power and control by line officers.

Many law enforcement officers, particularly those with a natural
resources background, also opposed the change. Others, particularly those
with a straight law enforcement history, were strongly in favor of the
stovepipe organization so that they would be supervised by law enforcement types and not by natural resources people.

This meeting was an attempt to get across, with clarity, some decisions
and to get some ideas of how to address some real and perceived problems.
The meeting was scheduled to last only one day. This was a source of criticism before the meeting: "How can we possibly make any progress on these
problems in only one day?" I was convinced, after sitting through endless
"group gropes," that the primary problems and potential cures for those
problems could be surfaced—at least in crude form—in one day of intense
effort. I felt that we would certainly identify as many problems and cures
to problems as we could handle in a year. Experience tells me that any work
group can get 80 percent or more of the way to a final solution with the
first burst of effort. An additional unit of effort will reach 90 percent, and
yet another unit of effort will get the group to the 95 percent level.

The attendees were divided into work groups to discuss problems and
devise solutions, particularly those that can be addressed within ninety days
at little or no monetary cost. When the ten teams reported at the end of
the day, some rather significant findings began to emerge. Most teams
reported that they did not find as many or as serious problems as was indicated by the questionnaires filled out and returned before the conference.

The teams produced a number of action items that could be addressed
by administrative action—soon, and at little cost in dollars. And there was
a remarkable similarity in the recommendations.

I was satisfied by the end of the day that we had made progress. I instructed the Washington office staff to formulate a report, put forward the ten most important action items, describe how we will execute those action items, and set a time schedule. I instructed staff to reconvene this same group in one year to check on progress.

After the meeting closed, I met in my room with Law Enforcement Director Martinez and the acting director of the Timber Theft Investigation Branch, Al Marion. The purpose of the meeting was for Martinez and me to gain a full understanding of the status of the major cases being carried out by TTIB. Marion briefed us in detail on three ongoing cases. I was surprised to find that all three of the cases were relatively pedestrian in nature and all were likely to be wrapped up in eighteen months and settled as civil as opposed to criminal charges.

I gave Martinez instructions to work with Marion on a high-priority basis to draw final plans for wrapping up these cases and shifting TTIB (with reduced staff) to a national training and consulting role. I made it clear that I wish to wrap this up within a month. I have made the decision to phase out TTIB as it currently exists. The original idea was to sunset the operation in a three-year time frame as the regions picked up the slack. However, the capability to carry out the current investigations will be retained. And as those investigations are completed, the staff will be converted to a training and consulting group.

This will, undoubtedly, bring out a cry that TTIB is being inappropriately disbanded to prevent vigorous pursuit of timber theft. Some members of TTIB are particularly adept at the role of whistleblower and have been building a case for some years in the press and on television as to their "persecution" by Forest Service management.

I am ready to move ahead with the decision described. But the necessary action has been delayed as we wait for the secretary of Agriculture designate to undergo his confirmation hearing and be approved by the Senate and installed by the president. All decisions or actions that are deemed remotely likely to produce hostile or probing questions at the nomination hearing in the Senate have been put on hold. Putting actions on hold has been a principal means of management at USDA for this entire administration. Such delays have produced a situation wherein nearly all the agencies in the USDA have disillusioned employees and deteriorating effectiveness.

10 March 1995, airborne, en route from Washington, D.C., to Rome

The past two days have been occupied with a meeting of regional foresters and directors called to confer on the structure of the Forest Service's Washington office. This "reinvention" effort for the Forest Service has been going on for nearly two years, and all are totally exhausted. The entire thing has taken too long, cost too much, and eaten enormous amounts of institutional energy and morale. Much of this can be laid at the feet of the undersecretary of Agriculture and associates who insisted on ever more extensive public and employee involvement. I believe that they somehow thought that enough involvement would ultimately produce a solution that would be universally acclaimed.

My experience is exactly the opposite. It seems to me that the more drawn out and inclusive the process, the more likely it is to become increasingly contentious and divisive. I believe that the lessons to be drawn from this— and from the land-use planning process—is that efforts at involvement should be intense and short-lived with decisions made forthwith and decisively. Otherwise, the focus devolves quickly to process, and process becomes the issue. In other words, process becomes the focus of the effort instead of a means of reaching a conclusion.

When we polled the public as part of the reinvention process, we were told that while the public expected to be consulted, they expected the Forest Service to lead and to make forceful decisions. Consultation does not diminish the need or responsibility of those in authority to make good decisions based on all available information for guidance, of which the opinions and desires of constituency groups are only a part.

Unfortunately, much of the reinvention effort has been driven by the desire of the undersecretary and associates for change in the Forest Service, with restructuring as an outwardly visible sign. I, like most of the senior Forest Service leadership, was committed to the old admonition, "If it ain't broke, don't fix it." This was interpreted by some, including the undersecretary and associates, as turf protection and unwillingness to change.

In reality, the need for restructuring was based on some distinct realities. There were the orders from the White House, including (1) staff reduction of 25 percent; (2) disproportionate reduction in staff above the level of grade 13; (3) budget reductions; and (4) requirements to double supervisor-to-staff rates.

At any rate, the final decisions were made. I think that the changes are an acceptable compromise between holding on to the best of the old and making enough changes to satisfy the imposed realities of budget, personnel ceilings, and supervisor-employee ratios, while looking different enough to satisfy the undersecretary and associates. In my opinion, it has not and will not be worth the turmoil, costs, and lost productivity. But it is done. Now comes the execution of the decision. I am determined that this be carried out quickly, heeding Machiavelli's admonition to carry out such matters in one stroke to preclude agony and negate the opportunity to be outmaneuvered by those who feel their jobs or power at stake.

21 March 1995, Washington, D.C.

I met with Senator Patty Murray of Washington at her request. Also present were Katie McGinty from the White House and Tom Collier, Secretary of Interior Bruce Babbitt's right-hand man. Senator Murray, a Democrat and supporter of the administration on timber issues in the Northwest, is obviously feeling the heat and the pressure to get out the cut in her state.

She wanted to know if the Forest Service could reasonably increase salvage if we had more money and more people. I told her that more money was not the problem, as there were adequate dollars in the salvage fund. The problem, particularly if there is no relief on the need for environmental assessments or environmental impact statements, was the lack of skilled biologists, hydrologists, and soils scientists to do those assessments. And that is a problem that cannot be solved in the very short term because it takes time to hire and train qualified people to do the work.

The consequences of using the meat ax of mandating personnel reductions without any consideration of what the agency was supposed to do—confusion, demoralization, and failure—are now apparent. There is some aspect of the self-fulfilling prophecy in the situation. Making the assumption that the organization is bloated, inefficient, and inept leads to the further assumption that those faults can be overcome by cutting budget and staff and "reinventing" the structure and ways of doing business. This is carried out on faith that good things will result. However, if the initial, untested assumption proves false, the actions instituted will produce the circumstances of the original assumption: the workforce will become demoralized and confused and there will be a significant period of institutional malaise until new ways of doing business become institutionalized. And so the original erroneous view of reality actually becomes reality. Of course, the longer the reinvention process is drawn out in the guise of public and employee "participation," the more contentious the issue becomes. Contentiousness mixed with time leads to both organized and *ad hoc* resistance to change from both internal and external sources. Individual employees, instead of feeling valued, begin to feel they are being manipulated or are pawns in a game they cannot control. Energy, corporate and individual, is more and more diverted into the seemingly endless process, and gradually, the process becomes the primary work of the agency, to the detriment of its mission.

Therefore it is essential to bring this to an end. But this time I am, and others are, ready to settle on a decision even if that decision, in our opinions, is flawed. Any—well, almost any—rational decision is better than continued controlled chaos. There is a lesson here. Prolonged indecision prepares the mind rather wonderfully for compromise or capitulation. So whoever has the most tolerance for stress and never ceases to press their point of view is apt to prevail—at least partially.

The politicians are faced with dividing up an ever-shrinking pie among hungry constituencies. So Senator Murray asks the questions and I give the answers. This is a zero-sum game and we both know it. If the Forest Service is spared, another agency must absorb the reductions originally assigned to the Forest Service. She and I have different roles. Policy decisions about public land management have come squarely to rest in the lap of Congress and there are no easy answers. The game has traditionally been a win-win game in which all players were satisfied through multiple-use, made possible by ever-increasing levels of intensity of management. Forest planning produced the blueprint. The only problem was that the only portions of the budget that were ever fully funded were those dealing with timber. But the assumption was always, "We will catch up on those things later." Later never came. The game was sustained for a long time because of the existence of virgin forests, whose capital could be drawn upon to make up the difference.

Then the bills began to come due. This was precipitated sooner, by several decades than would have happened anyway, by the controversies and responses necessary to protect threatened and endangered species.

23 March 1995, Washington, D.C.

I met for about two hours late in the afternoon with Senator Frank Murkowski of Alaska (chair of the Senate Committee on Energy and Natural Resources) and his aides, Gregg Renkes and Mark Rey (former head of the American Forest Resource Alliance). David Hessel, director of Timber Management for the Forest Service, accompanied me.

Hessel had provided a report to Senator Murkowski yesterday with detailed answers to a series of questions posed by Mark Rey. These questions were designed to elicit a detailed description of the decline in timber sold from the national forests since the peak year for such sales; a description of increasing costs for sale preparation; and the relationship of increasing stumpage prices related to timber available from national forests.

The atmosphere in this meeting was a marked change from the bullying and antagonistic behavior the senator exhibited in our recent appearance before the Natural Resources Committee. The meeting was relaxed and even friendly. They seemed to be looking for information, and we were able to answer questions to their satisfaction.

Rey told me in a telephone conversation several weeks ago that he wanted to help foster a better relationship between Senator Murkowski and me. That struck me as odd at the time because it was evident in the hearing that Rey had developed the barbs and jabs that the senator used in his attempt at flensing the Forest Service witnesses, primarily me.

As the meeting went on, I was trying to detect whether the senator and his staff sincerely wanted to improve relations or whether I was getting the stick-and-carrot approach. They had, in two previous meetings, demonstrated how unpleasant they could be and intimated that they could and would use naked power. Now, they were demonstrating how reasonable, and even friendly, they could be when I (and the Forest Service) responded in a manner that was more to their liking.

Rey is a master tactician and manipulator and very bright. I can't tell if

he is a mercenary or if he really believes in the cause of the timber industry he serves. He is quite capable of such manipulation and I am quite capable of recognizing his potential.

Senator Murkowski has claimed a joint role for the Natural Resources
Committee and the Agriculture Committee in hearings on Secretary Designate Glickman's suitability for the post at Agriculture. Murkowski bases
this on the fact that nearly half of the Department of Agriculture is made
up of the Forest Service. This is a first and a very clever move on Mark Rey's
part. He knows well that the secretary of Agriculture traditionally knows
and cares little about the Forest Service. This traditionally has shielded the
Forest Service from much direction by the secretary of Agriculture. It seems
likely to me that the first Forest Service chief thought that out and calculated that he would be much more likely to have a free hand in the Department of Agriculture than in the Department of the Interior.

Murkowski (really Mark Rey) has just been handed the information
he needs to show Glickman that the Forest Service can no longer meet its
"true purpose" of making timber available to industry in the right place
at the right price. It will be interesting to see how Mr. Glickman's hearings
work out.

Clearly, the groundwork is being laid for Mark Rey to be undersecretary of Agriculture for natural resources in a Republican administration.
Rey already has very close contacts in the Forest Service with whom he con-
verses regularly, and which contacts, through Rey, seek the reinstitution of
the "old" Forest Service.

24 March 1995, Washington, D.C.

I met today with Senator Larry Pressler, Congressman Tim Johnson, and
the top aide for Senator Daschle (Senate minority leader) about some of
their concerns with Forest Service activities. Their first concern was my decision to perform National Environmental Policy Act–required analysis on
the renewal of grazing permits. I explained that the decision was based on
a court ruling that renewal of existing permits requires a NEPA assessment.

If such assessment is not performed, the permittees are in danger of being
closed off their allotments until NEPA assessments are complete. Pressler,
Johnson, and Daschle's aide were concerned that the Forest Service was
incapable of performing so many assessments so fast.

I suggested to them that the administration had no position on the matter, but if there were a rider to some bill, perhaps appropriations, that no grazing permits would be canceled because of the inability of the government to perform NEPA analysis, it would make everyone involved somewhat more comfortable. That, of course, would include the Forest Service.

I also explained to them that giving priority to the NEPA analyses for renewal of grazing permits would divert attention from NEPA assessments for timber sale preparation and that forest plan revisions were at the end of the queue for attention. They wanted to know how much more money we needed to do all of these activities simultaneously. They seemed shocked when I explained that it was not so much a problem with money as it was inadequate staff of appropriate skills to prepare NEPA assessments, particularly soils scientists, hydrologists, archaeologists, economists, biologists, and range conservationists.

It seems to be dawning slowly on some elected officials that downsizing has serious ramifications for carrying out even traditional tasks. And that all these areas of activity require NEPA assessments and the same people do them all. This requires setting priorities, and that means somebody or something must wait. That, in turn, means that constituents raise hell with both bureaucrats and politicians.

The reactions of the politicians become more and more stressed as they come to grips with the consequences of keeping promises to reduce government without understanding the probable effects. They asked me what I intended to do. I replied that the grazing permits would take first priority, then timber sales, then forest plan revisions. Then, when I told them the consequences if there was another fire season in 1995 that came anywhere close to that of 1994, they got even more sober.

Just to close the conversation on a positive note, I told them about 500 more people leaving the Forest Service with the latest buyout this month, and that by 1999, 2,500 more slots would be lost. It was quiet as we shook hands and took our leave.

29 March 1005, airborne between Minneapolis and Tucson

I spent the last few days attending the North American Wildlife and Natural Resources Conference in Minneapolis. It was a welcome respite from

Washington, and it was particularly good to be among the "wildlifers" I have worked with and known over the years. During the plenary session I gave a paper, "Forest Health: What Is It and What Are We Doing about It?"

The thesis of the paper was that a crisis of forest health is upon us. My discussion centered on the forest health problems in the Intermountain West, on those forests where—because of fire control, livestock grazing, insect control, and logging—formerly green ponderosa pine stands had evolved into fir encroachment or densely stocked pine regeneration. This pattern of stand development set the stage for outbreaks of defoliating insects, such as spruce budworm and Douglas-fir tussock moth, which coupled with drought have produced significant mortality in fir and spruce stands. With the extensive tree mortality in lodgepole pine and ponderosa pine stands caused by an earlier outbreak of mountain pine beetle, the situation has vastly increased fuel loadings that burn easily. And these burns are apt to be more extensive and hotter than the long-term norm for the frequent (every three to ten years) low-intensity burns.

The fires of 1994 were indicative of a situation where many of the ecosystems affected are, as some ecologists put it, outside the historic range of variability. And it seems likely that the fires of 1994 will be repeated.

30 March 1995, airborne between Tucson and Washington, D.C.

I spent yesterday afternoon and this morning at the Marana Training Center at an annual course on firefighting for line officers. This is an interagency effort whose purpose is to give line officers (they call them agency administrators) training in fulfilling their responsibilities in dealing with fire situations.

The fire season of 1994 and the loss of thirty-four firefighters' lives have been a powerful incentive to tighten up our training. The report from the Occupational Safety and Health Administration, which found the Forest Service and the Bureau of Land Management "willfully negligent" in the Storm King Mountain incident, emphasized the responsibility that line officers have for firefighter safety and reinforced the need for such training. Mike Dombeck and I have accepted responsibility for any lack of training and inappropriate attitude. My presence at this session was intended to reinforce my personal commitment to fire readiness and crew safety.

The longer I am in the chief's job, the more impressed I am with what that *position* means to employees in the Forest Service and our sister agencies. It is obviously not so much the person who counts but the position itself. The presence of the chief says more than words in expressing caring or emphasizing a commitment or a concern.

Knowing that, it tears at me that I cannot spend more time in the field with the people who are the guts of the organization. I resent the ever-increasing demand for me to be in Washington to deal with day-to-day crises that seem, on reflection, less significant than being in the field. I am trying my best to extend my presence via videos, teleconferencing, and phone calls. Although I can reach only a few, I think they know I am trying and that I do care about them and the land.

31 March 1995, the Brouha Farm, Alabama

Upon arrival at the cabin, I found a note on the door that read, "Somebody is bombing a national forest. Chief Thomas should call Deputy Chief Gray Reynolds as soon as possible." Thank goodness for mobile phones.

Reynolds explained the situation. There had been a bomb planted on a window ledge at the district ranger's office at Carson, Nevada, on the Toiyabe National Forest. Another bomb had destroyed a comfort station at a campground on the Toiyabe. The explosion at the ranger district occurred at 7:30 P.M. on 30 March (last night). No one was injured and damage was limited to one room. An office of the Bureau of Land Management in the same area had been bombed last year. The "War on the West" now takes a new turn.

FBI agents are on the scene and Forest Service law enforcement officers are assisting and have the district ranger and the forest supervisor protected. Strangely enough, this may be a significant turning point in the situation. The "wise use" coalition, "the militia," and the "War on the West" folks have been continually raising the stakes in the game by issuing more and more outrageous statements intended to provoke a response by the "feds" that could be portrayed as an oppression by "federal occupiers."

Instead, the feds took the case to court by suing Nye County, Nevada, to clarify the questions of landownership and federal limits of authority. Spokesmen for the wise-use folks immediately declared that the feds have taken the bait and "we will carry the day in court."

James E. Kennamer of the National Wild Turkey Foundation and
Chief Thomas during turkey hunt, Brouha Farm, Alabama, April
1995. Photo courtesy of Jack Ward Thomas.

As tensions increased, it became increasingly clear that the side that fired
the first shot would lose in the court of public opinion and sympathy. Their
side screwed up by not realizing that calls for defiance and rebellion, even
if only a clever ploy, will sooner or later entice someone at the fringe of
either sanity or political position to perform a violent act. Maybe that has
happened. We will have to wait for the results of the investigation by the
FBI to know for certain.

Reynolds has authorized a reward of $10,000 for information leading to the arrest and indictment of the perpetrator of the bombing. I will do what I can to stimulate additions to the reward money by the Department of Justice, the National Forest Foundation, the livestock raisers, the timber industry, and some environmental groups. This is an opportunity for the more middle-of-the-road groups to weigh in for rational resolutions of disputes within the law by demonstrating intolerance of violence.

I asked Reynolds if it was advisable for me to go immediately to Carson City. He had discussed that possibility with the Toiyabe forest supervisor. Their advice was for me not to come immediately. Rather, I should stick with the plan to spend a week on the Toiyabe later in the summer. I'll take that advice.

5 April 1995, Washington, D.C.

Today there was a hearing in front of Senator Larry Craig's subcommittee concerning assessments and land-use planning regulations. Mark Rey, who is serving as professional staff to the committee, arranged the hearing. He asked for a large cast, including Jeff Blackwood (director of the Interior Columbia Basin Assessment), Steve Mealey (director of the Upper Columbia Basin Environmental Impact Statement Team), James Space (director of the Sierras Assessment Team), Forrest Carpenter (director of the Southern Appalachian Assessment Team), Christopher Risbrudt (director of Ecosystem Management), Gray Reynolds (deputy chief, National Forest System), and me.

Three Republican senators showed up for the hearings: Craig of Idaho, Conrad Burns of Montana, and Craig Thomas of Wyoming. There were no Democrats. Before the hearing started, Senator Burns made a joking but pointed remark about the size of "the posse" that I had brought along and wondered if there was anyone left to mind the store. Senator Craig began with a statement disparaging the administration's efforts in national forest management, particularly in the realm of producing timber. He did, however, accept some responsibility for the role of Congress in producing a plethora of overlapping and somewhat contradictory laws.

He stated that he and the Republicans newly in power intended to correct those problems and called on the Forest Service to assist in that regard. He finished his statement with a list of the hearings yet to be conducted concerning Forest Service activities, including hearings on whether and what Forest Service lands might best be transferred to the states or into private ownership.

Senators Burns and Thomas chipped in with brief statements questioning, as I interpreted it, why the Forest Service was paying so much attention to trying to comply with the laws or, put another way, trying to avoid federal court actions.

Then it was my turn for an opening statement. I submitted my formal written statement for the record and spoke only briefly, to the effect that assessments on a broader scale than individual national forests and Bureau of Land Management districts were required to deal with issues that transcend the boundaries of administrative units. I used the words of Judge William Dwyer in his decision in the Seattle Audubon case, that not only was ecosystem management (which requires, by definition, broad-scale assessment) legal, such was required to simultaneously obey the myriad laws impinging on land management planning and actions.

Then came the give-and-take between the senators and the panel of witnesses. That lasted for several hours. Most of the senators' questions were, as usual, more short speeches laying out their opinions on one point or another. Each of them, in his own way, said that the Forest Service was not producing enough timber and, to a lesser extent, not keeping livestock permittees happy and content. It was obvious that they cared little for anything but getting the cut out and any other activity that benefits industry of one kind or another.

They came again and again to the point that the Forest Service seemed obsessed with avoiding going to court. I tried over and over to make the point that we are striving to obey the law and that what the law meant was continually being altered or clarified by the courts in case after case. They disagreed, it seemed, that the courts were correct in their decisions, and the senators thought that we paid too much attention to those decisions. I was left wondering if these senators of the United States clearly understood how the government works, in that Congress passes laws that are signed into law by the president or become law over his signature; the executive branch carries out the law; and if its execution is challenged, the federal courts decide if the executive branch is complying with the law; and that decision then further defines what compliance requires.

I explained several times that the Forest Service had learned a lesson in the Pacific Northwest of the consequences of skirting the law. We learned that broad-scale assessment was required and that multiple federal land management units must be coordinated to some extent in their management.

Senator Craig made it clear to me, several times, that he and his colleagues did not think highly of the process or results of the President's Forest Plan for the Pacific Northwest. I made one last response and then addressed the point no further. I explained that the plan had been approved by the federal court, after three previous failures. But Judge Dwyer made it very clear that upholding the plan was a very close call. Senator Craig did not seem impressed.

As our panel was dismissed, I felt that we had done well. I don't know what to make of these hearings in the Senate. Few Republicans show up and no Democrats. There is very little press interest. I don't believe that there is much support for these attacks on the Forest Service, or demand for alterations in approach, outside the Intermountain West.

21 April 1995, Washington, D.C.

We have just finished a three-day meeting of the regional foresters and directors. We discussed the likelihood that Congress will dictate an accelerated timber salvage effort by the Forest Service. The mechanism for the salvage mandate is the 1995 Rescission Bill, which has passed both houses of Congress and is now in conference. Each version contains a mandate for salvage with some modifications in the applicable environmental laws to facilitate the achievement of stated salvage objectives.

I told the group that such a mandated salvage effort, with weakening of the environmental laws, was inevitable. This will bring the Forest Service— and particularly the chief—to a moment of truth. The Forest Service may be placed in a position where we are hard aground between my admonition to obey the law and tell the truth, and the ethics of natural resources professionals and the recently stated agency ethics. If push comes to shove and obedience to the law means violations of personal and agency ethics, what do we do? In response to this question that was hanging in the air unasked, I made the following statement:

It is now obvious that the Congress will pass the Rescission Bill. That bill will contain, in one form or another, an instruction in law for the Forest Service to produce some significant and designated level of salvage timber volume over the next two years. That law, in the words of its proponents on the Hill, will remove any excuses for the Forest Service not to meet those designated salvage targets by some level of modification in the environmental laws to facilitate salvage operations.

I choose to consider this mandate an opportunity to rebuild the spirit and reputation of the Forest Service as the "can do" agency. This mandate will be unequivocal as to the policy of the government of the United States so far as the Congress is concerned. If approved by the president, it will be without question the policy of the United States. Let's be very clear: given such a clear mandate, the Forest Service—and other governmental agencies we work with—are relieved of any doubt as to the legitimacy of such marching orders.

It is the people of the Forest Service that must carry out these mandates, if they do come about. For the Forest Service to succeed in any endeavor, these people must perform well. They do best when they feel comfortable with what they are asked to do. I have told the Forest Service, as a cardinal principle, to obey the law and to tell the truth. Yet, as a foundation to our very being as natural resources management professionals and as a premier agency concerned with such matters, we are pledged to protect and restore the ecosystems within our care so that they can provide for our customers— including those alive now and the many, many more yet to be born.

There are times and circumstances that require an unequivocal stand. For the Forest Service, this is such a time of testing. We will obey the law. We are working with Congress to assure that the salvage levels are within our capabilities. But in the process of obedience, we will not violate the principles and the standards and guidelines in the Forest Service plans put in place to protect the basic resources on which all depends—soil and water.

In the unlikely event that we cannot meet the requirements of the law and conform to the ethics that give us any value as natural resources management professionals or as an agency, we have no real choice but to stand on first principles. We will do just that.

Those are my instructions to you. I will state the same to our fellows that are not here today. This is where we stand. We can do no other.

When I finished and sat down, the attendees applauded. I don't think I ever heard of such a thing before. But obviously, there are core values that we all share.

25 April, 1995, Washington, D.C.

We have been preparing for hearings before Senator Craig's subcommittee. This particular hearing will concern the effect of the environmental laws on the management of the national forests. Mark Rey, chief of staff to the Committee on Natural Resources, has been feeding us questions all week. Today I received a letter from Rey—ostensibly from Senator Craig—that held forth at length on perceived discrepancies in ongoing assessments, planning regulation proposals, application of the National Environmental Policy Act, reorganization of the Forest Service, and several other perceived shortcomings.

I worked with staff to prepare responses to Rey's questions, which had to be coordinated with the Department of Interior, the Council on Environmental Quality, and the Office of Management and Budget. The Forest Service chief had been designated by the committee to give the opening statement and prepare the first draft of answers to the questions submitted by the staff, and the other agencies were obviously very nervous about our being in the lead.

As I played the game and realized that I was enjoying the machinations, I was suddenly filled with a tad of disgust for the corruption in my spirit that was taking place. We were not concerned with truth or solutions; we were concerned with our own power and the contest of outwitting the subcommittee. Such is not becoming in those who came to this town to serve the people and make a better government and a better world.

This condition sneaks up on you on tiny, silent feet. Where once I answered all questions truthfully and without guile, I now catch myself thinking two questions ahead, as if thinking two moves ahead in a game of chess against a closely matched opponent. Beating the adversary becomes a goal in and of itself.

The very atmosphere is corrupting, and anyone with any sensitivity must face that. But at the same time, the game and the illusions of power and influence are seductive.

The compromise with conscience is to rationalize that effectiveness is somewhat dependent on guile—particularly when a new Republican majority outnumbers you and the Democrats do not even attend the hearings. In this compromise, one assumes that guile is justified by the fact that the cause is good, and losing would be intolerable for the agency and for the land.

I think this happens to all the moths drawn to the flame of power and influence—some, of course, more than others. I don't think that most of them even realize the erosion is taking place. Perhaps not knowing is best. Or maybe knowing and resisting the erosion are better—I hope that is true.

26 April 1995, Washington, D.C.

There was a hearing today before Senator Larry Craig's Subcommittee on Forests and Public Land Management of the Committee on Energy and Natural Resources. The subject of the hearing was the interaction of environmental laws and regulations on the management of national forests. Witnesses were Dinah Bear, general counsel to the Council on Environmental Quality; George Frampton, undersecretary of the Interior; Rollie Schmitten, director, National Marine Fisheries Service; a representative of the Environmental Protection Agency, and me. By direction of the committee, I presented the testimony.

After the testimony had been massaged to mollify all the other agencies, the Department of Agriculture, and the Office of Management and Budget, it was too bland to be taken seriously. Mark Rey, chief of staff for the committee, called the testimony a "real batch of Wonder Bread"—all air and no substance—and he wasn't far off. The only reason that it wasn't even worse is that I finally balked and refused to give the testimony if it was weakened any further. No wonder congressional committees become so frustrated in trying to come to grips with complex issues. The committees, collectively at least, know little of the complexities of the issues. But they can't get straight talk out of the agencies, which, even if inclined to straight talk, must be careful to hum the administration's tune. That tune becomes more discordant as more agencies are involved in the ensemble.

The hearings started with the standard three Republicans: Craig Thomas of Wyoming, Conrad Burns of Montana, and the chairman, Larry Craig of Idaho. Mark Rey seemed modestly successful in getting Craig ready for the hearing. Thomas and Burns are obviously not well prepared and sim-ply repeat themes and questions from previous hearings. Craig is some-what better prepared and rather haltingly read a pointed letter (written by Mark Rey) that he had sent me earlier in the week.

Senator Frank Murkowski, chairman of the full committee, came in and read a statement largely aimed at the bureaucracy—particularly the For-est Service—and the overlap of agency jurisdictions in dealing with pub-lic lands. His statement was also obviously prepared by Mark Rey, who whispered the questions into the senator's ear.

The theme of all the opening statements was that the system was bro-ken, and they, by God, meant to fix it. They said they wanted the agencies to help them fix the system or they would do it without us. After a while, it began to dawn on me that perhaps they doth protest too much. It seems that either they don't have the punch to get the changes they want with-out support from the administration (i.e., the agencies), or they don't know what to do to fix the situation, or perhaps some of both. If they had the power and the know-how to get things fixed, as the majority party in the Senate and House they would just ram their solution through and dare the presi-dent to veto it.

The Democrats are boycotting the hearings; only the same three Republi-cans are showing up, and they are not well prepared and seem to be getting bored with the issues. The hearings don't seem to be drawing much of a crowd and are not generating much excitement. Rey is doing good staff work for them from the standpoint of the timber and grazing interests, but the senators are simply not really boring in. They strive to come across as real bad-asses, but they just can't stay in character for the entire hearing. More and more I believe that they don't have the power or the momen-tum to dramatically change the environmental laws.

Senator Bill Bradley of New Jersey, the ranking minority member of the subcommittee, made a brief appearance. He gave me the opening to dis-cuss the problems facing our employees in the execution of their duties. I closed my statement by making it clear the we were doing all we could to carry out our mission without provoking confrontations and moving the issues into the courts and Congress. I did say, emphatically, that we would carry out our duties.

Bradley took the opportunity to roundly chastise those who commit acts of violence and intimidation against federal employees and made it clear that it was the duty of senators to speak out against such acts. The three Republicans rather sheepishly agreed and looked somewhat embarrassed.

Senator Bradley also gave me an opening to clearly state my position that no matter what we were ordered to do by Congress in terms of timber salvage, we would not exceed the limits of professional and agency ethics by failing to protect the basic resources of soil and water. Protection of those resources would be achieved by adhering to the standards and guidelines in the forest plans, with some justified modifications on a case-by-case basis.

I invited the committee to tell me that such a statement was unwarranted. They made no such statement, and in fact, nodded in agreement as I looked each one of them in the eye. Bradley did the Forest Service a good turn.

The Rey strategy of driving a wedge between the agencies did not work, and the hearing was really quite tame. The three Republicans were relatively kind to the Forest Service—even a bit complimentary in several cases—and very hard on EPA, CEQ, F&WS, NMFS, and Justice. They are obviously frustrated and really do want to ease the barriers to efficient management actions by the Forest Service.

They seized upon the promise made by Secretary of Agriculture Dan Glickman to have the Forest Service review the conflicts in environmental laws and regulations. The review is to be ready by June 1.[2] The secretary evidently made that promise without thinking too much about it, and certainly without consulting Secretary of the Interior Bruce Babbitt.

Our Forest Service policy analysis staff have been working for ten days and have less than three weeks to complete their work. This is simply not enough time to do an appropriate job, and the other affected agencies will demand to "cooperate" in the Forest Service effort. That cooperation will certainly make things more difficult and will ensure that whatever comes out will be so toned down, after addressing "concerns" by each agency involved with protecting its power and mission, that boldness will be missing. Be that as it may, the Forest Service will deliver on Mr. Glickman's promise.

---

2. The report was ready on time but remained bottled up in Secretary Glickman's office through the end of the Clinton administration.—*JWT*

Secretary of Agriculture Dan Glickman and Chief Thomas, April 1995. Photo courtesy of Jack Ward Thomas.

2 May 1995, Washington, D.C.

Today was the day for the consideration of the Forest Service budget by the Interior and Related Agencies Subcommittee of Appropriations. Senator Slade Gorton of Washington State is the chairman. As usual, the hearing had very little to do with the Forest Service budget in general and was focused on letting the various senators make speeches about their pet topics or seek funding attention for their favorite items of interest.

Gorton was in the chair at the beginning and was primarily interested in making the point that he wanted more timber produced off Forest Service lands. He ragged on the Forest Service for letting timber cutting rates drop. He was very polite and "senatorial," as opposed to the rather pointed, abrupt manner he displayed in thumping on me during our last private visit.

Gorton then turned us over to Senator Conrad Burns. Burns had visited in a jocular manner, one good ol' boy to another, with Deputy Chief Gray Reynolds and me at the witness table before the session began. Now he was performing for the folks in Montana. First, he stated that he didn't see much point in International Forestry when we didn't have enough money to "get out the cut" and "get the permittees on the range" here at home.

He seemed a bit taken aback when I responded with a rather vigorous defense of the International Forestry budget. My points were aimed at his Republican soul as I described how work by that group had prevented listing of broadleaf mahogany—a prime wood for U.S. furniture manufacturers—on the Committee on International Trade in Endangered Species list of species to be tracked in international trade. I followed that with a description of the recent meeting in Rome of the Committee on Forestry within the Food and Agriculture Organization of the United Nations. At that meeting, FAO considered international certification for forest products in international trade that were produced in a sustainable fashion, and we needed to be at the table.

Senator Burns backed off the International Forestry issue by stating he had an open mind on the subject. He then began to jerk my chain about the lack of timber production from the national forests by questioning why we needed to spend any money on the activities of the State and Private Forestry Division. His point was that the private lands were doing better than the Forest Service, so why should we spread our tentacles into the private lands? Why should we pretend to be able to teach state and private entities anything when we had such a lousy record? It was quite a performance, carried out in good ol' boy, down-home style. I sort of liked the way he played to his audience in Montana—just a good ol' Montana boy giving the hoorah to bureaucrats inside the Beltway.

The only trouble was that he was peddling horse manure by the load, and he knew it. I suspect he will have his state forester and a number of constituents on his back tomorrow. And he will say, "Ah, I was just funnin' those bureaucrats. Ol' Conrad will see to it that you boys get your money."

I considered letting the performance stand on its own merits, but I really couldn't do that. So I delivered the regular commonsense spiel on ecosystem management along with the virtues of what had come about from the program and the importance of increasing good forestry on private lands, both to produce timber and to stay ahead of the certification of products game.

When he came back once more with the point of what a good job the private forest landholders were doing, I gave him the clincher. I said, "Maybe, senator, they are doing so well as the payoff of what the State and Private Forestry programs have helped put in place over the past decades." He didn't seem to want to touch that one, so he passed us on to the next senator.

Senator Bumpers of Arkansas had complimentary things to say about the Forest Service. Then he pressed home his point about wanting to get on with the land exchange in Arkansas between the Forest Service and Weyerhaeuser. This exchange would give Weyerhaeuser some highly productive Forest Service lands in exchange for a larger amount of land, of lower site quality for forestry, around a very popular lake.[3]

I was happy to announce that we were in the final throes of negotiation, with only the hurdle of the mineral rights to be worked out. He said he would handle the exchange by legislation when we had all the details worked out. He left the impression that he would prefer that this be concluded and soon.

At that point, Senator Ted Stevens of Alaska arrived, scowling and grumbling as he took his seat. When called upon, he launched into one of his better tirades, directed at Undersecretary Jim Lyons. The staging was good. This was designed to come across for the cameras and the audience in Alaska as an attack on the administration. This sort of performance, which would be an embarrassment in most states, seems to play well in Alaska, as the entire delegation uses the same tactic.

However, the senator is getting too old to do the job very effectively anymore. He lost his place in his tirade at several points and appeared confused as he fumbled around. His main point was that the Forest Service is an outlaw agency that refuses to obey the law. The law in point that he quoted was the Organic Act of 1897. The quote was to the effect that the sole purpose of the national forests (actually, the forest reserves) was to provide a continuous supply of timber.

He left out the good parts. The first purpose of the Organic Act was to protect the forests, the second was to protect the sources of water, and the third was to provide a continuous supply of timber. There were other aspects

---

3. The exchange ultimately took place.—*JWT*

that he left out as well, such as cutting only dead and mature trees that are individually marked.

He knows well—he is or was a lawyer—that many laws have been passed since the Organic Act that supersede the act and impose new standards. Foremost among them are the Endangered Species Act and the National Forest Management Act.

Stevens would not allow any of the Forest Service witnesses to answer any of his questions. He came directly at Undersecretary Lyons. Then he demanded to know what the "Forest Service camera" was doing filming the hearing without his permission. He finally let me answer that question. I said I didn't think it was a Forest Service camera and that it likely was a person shooting film to be used in training sessions on how the hearing process and democracy really work. He stormed out of the hearing room still fulminating and mumbling as he departed.

As Stevens left, Senator Bumpers asked who was responsible for the second camera that was filming the proceedings. One of the aides spoke out and said, "The Republican National Committee." That was good for a laugh, even on the Republican side of the dais.

The chair, now occupied by Senator Bennett Johnston of Louisiana, recognized Senator Mark Hatfield of Oregon, who had come in during the tirade by Senator Stevens. Mr. Hatfield looked more and more pained the longer Mr. Stevens carried on. Hatfield said he would submit questions to be answered for the record. He was obviously embarrassed, or at least put off, by Senator Stevens's outburst and wanted to distance himself from the process.

Senator Johnston now remained alone in the room. He made the point that it is likely that the Forest Service's 1997 budget request will be trimmed by at least 10 to 20 percent. He encouraged us to be ready to work with the committees of the House and Senate to do that in the best possible manner. The hearing ended on that happy note.

One other point concerning the exchange between Mr. Lyons and Mr. Stevens bears mentioning. Stevens had obviously heard that I had told the regional foresters that the Forest Service would strive diligently to achieve whatever salvage targets were set forth in law, but that we would follow the standards and guidelines in the forest plans to safeguard the environmental values of soil and water. Stevens lashed out at Lyons, saying that the

Forest Service would obey the law or he would guarantee retribution, even if obedience to the law required damaging the resource.

He pounded that home several times, and finally Lyons said, "Senator, we will obey the law." What did Lyons mean by that? This new salvage law—whatever it turns out to be? This new law and the other germane statutes? What? I believe Lyons is clever enough to know what he is saying and exactly what it means to him. But the press will have a field day with that statement. Likely that is exactly what he intended and the message to the American people is that Congress is ordering the Forest Service to maximize salvage timber yields at any cost, including environmental damage. Inadvertent slip or clever ploy? Intended or not, it turns out to be clever. This will further reduce the chance that Senator Gorton will be able to get the Senate to go along with a rider to the Rescission Bill that sets a salvage target in hard numbers with the implied instruction to get out the target salvage no matter what the environmental consequences.

The exchange was a bad mistake on Senator Stevens's part and a good move on Lyons's part. I think we may well see a turn in Mark Rey's strategy as a result. If he can't get the hard targets, he will be forced to get some help from the administration. The newspapers in Washington State are pounding on Stevens about the targets, and this goof by Stevens will exacerbate his quandary.

It is well to remember that Senator Patty Murray, the freshman senator from Washington, came within two votes of beating Gorton on this issue a few weeks ago, and the administration gave her no help. This deal is falling apart on Rey—I can smell it. Thank God for Senator Stevens's tirade and his macho contest with Jim Lyons. Stevens goofed and Lyons trapped him. I believe this is a turning point in the game.

7 May 1995, Washington, D.C.

As I predicted, Mark Rey has been forced by the situation in the Senate concerning the setting of salvage targets in law to change tactics. Either Senator Slade Gorton has become convinced that he can't get the votes in the Senate to set the targets in law, or the building criticism in the newspapers in the state of Washington has made him nervous about his political base, or some of both.

The new Rey strategy is to offer the administration a deal wherein, through an exchange of letters, Senator Gorton berates the Forest Service (i.e., the administration) for not getting enough timber on the market and asks for a guarantee from either the secretary of Agriculture or the chief of the Forest Service that a certain level of salvage volume will be forthcoming. In that strategy, Rey and the timber industry he represents gain a target from the administration that they either can't get or don't want the onus of procuring through legislative action.

Rey has put this forward to me as an attempt on his part to avoid the "unfortunate precedent" of setting timber cut targets in legislation. I was impressed by his concern. Part of being effective in the continuing political chess game is interpreting what the political players really mean and what they are really after. The game is to make what you say palatable to the person that you are trying to persuade. It is important not to lie in an outright fashion or to be perceived as going back on one's word. That would be a serious breach of unwritten protocol and would lead to a serious loss of credibility.

And so Secretary Glickman received his letter from Senator Gorton, instituting the potential minuet. Associate Deputy Chief Tom Mills headed up the drafting of the Forest Service's suggested response. I personally worked on the final version that we submitted to Undersecretary Lyons for his consideration.

To top things off, Lyons has persuaded Secretary Glickman to sign the letter. Now there is progress. The easy, safe thing for the secretary to do is have me sign the letter (which I would be pleased to do) and claim credit if the political fallout is favorable and disavow it if otherwise. I am more and more favorably impressed with Secretary Glickman. I am beginning to believe that he really is going to follow up on his words in the confirmation hearings to be more engaged in Forest Service issues. And that has the potential to make the game with those in Congress who want to have timber interests dominate Forest Service management a bit less one-sided. Now, if only the more middle-of-the-road environmental groups could get up off the canvas, shake off the confusion, and mount just a little rally, this retreat into the past might yet be turned around. The trick is not to get overwhelmed by the initial attack. The longer the force of that attack is slowed or contained, the worse it will seem, upon close examination, to the American people.

This job is a real education, if nothing else, in the art and science of the practice of biopolitics at the national and international scale. Things simply don't work the way that students are taught in natural resources policy classes—not even close. There is simply no way that scholars of the subject can understand the ad hoc processes that go on within only loosely defined boundaries.

However all of this turns out, the history of natural resources management, on the national forests in particular, will bear the stamp to a greater or lesser degree of Mark Rey. If I were operating in the political arena, that would be my point of attack. The strategy would be to focus all the attention possible on Rey as mastermind, strategist, and leader of the Republican timber activities. This in turn would focus attention on the key senators as pawns—willing pawns—in a sophisticated game that they comprehend only poorly. I wonder if anyone will pick up on that.

9 May 1995, Washington, D.C.

Today there was a daylong meeting at the Fish and Wildlife Service's Patuxent, Maryland, Wildlife Research Center. The meeting was attended by about twenty-five persons, including top brass from the Fish and Wildlife Service, Bureau of Land Management, National Marine Fisheries Service, Council on Environmental Quality, Department of the Interior, and Department of Agriculture. The focus of the meeting was the potential revision of the Endangered Species Act, both the act itself and the regulations issued pursuant thereto.

There seems to be full recognition at this late date that there are severe problems with the processes, delays, appeals, and lawsuits engendered by the multiagency application of the Endangered Species Act, particularly as to how it is applied on public lands. The policy makers had already agreed to a ten-point program as to what any revision of the act should consider so far as public lands are concerned. Eight of the points were innocuous, but two were dramatically significant. The first was that the burden of any consequences of protecting a listed species should, to the extent possible, be absorbed on public lands. The second was that all actions that are feasible to head off listing of additional species should be undertaken in an interagency manner.

These two items in combination would have a profound policy effect. I made a point at the president's forest conference in Portland two years ago that, whether fully appreciated or not, the overriding objective of the management of the public lands has become protection of biodiversity. I made the plea at that time for recognition of that fact and consideration of the appropriateness of that de facto policy at the highest levels of government. That policy should be either clearly acknowledged or refuted by the executive branch or the Congress or preferably both. I made a good short speech directly to President Clinton, Vice-President Gore, and the Cabinet. My plea has gone unanswered.

However, my contention was reinforced by the marching orders given to the Forest Ecosystem Management Assessment Team, which included the order to devise a strategy that took as much of the burden of compliance with the Endangered Species Act as possible and put it on the public lands. The result was a plan that dramatically dropped the expected timber sale levels from the public lands in the Pacific Northwest from a bit over 5 billion board feet per year to about 1 billion board feet per year. One consequence was that large landholding companies have made all-time record profits and small nonindustrial private landowners (many, at any rate) also cashed in on the opportunity. Those mill operators and the associated communities that depended on timber from the public lands fell upon hard times.

Oddly, little was made in the political sense of how these decisions had cut the pie, and who had won and who had lost in the game. The press, and some hired guns for the timber industry, played up the stories of who was hurt in the process while making very little of the fact that there were big-time winners. The winners made little noise, understandably, about their good fortune, while the losers, understandably, made much noise about their pain.

11 May 1995, Washington, D.C.

There were two occurrences today that were somewhat revealing in terms of the interactions between natural resources management by the Forest Service and the constant interaction with the realities of political machinations for political gain. We seem to spend an enormous amount of time, energy, and credibility on getting the natural resources management job done and trying to be both responsive to our political bosses and keeping them from doing something stupid that hurts the agency's credibility, not to mention preventing as many self-inflicted wounds as possible.

First, Deputy Undersecretary Adela Backiel called me to say that Undersecretary Lyons, operating on orders from the White House via Katie McGinty, was relaying instructions to order the cancellation of tomorrow's auction of six timber salvage sales on the Boise National Forest. The reason was that the White House wanted time to consider whether President Clinton might announce these sales as evidence that the Forest Service could put up "very large" salvage sales under current laws through improved agency coordination. Under no circumstances was I to say what the real reason for the cancellation was.

I pushed back by saying that I would not make up a reason for that action that was not the truth. The best I could do was to simply cancel the auction with no statement of the reason why. I contended that those who ordered the cancellation should provide a quote as to why the sale was to be cancelled. Backiel suggested that I talk to Lyons about the matter, as she was only the messenger and I had her between a rock and a hard place.

I was unable to get through to Lyons, as he was in a meeting and could not be disturbed and was departing for the airport immediately upon completion of that meeting. I called in my public affairs officer, James Caplan, to find out what was on tap for the auction of the timber sales on the Boise National Forest. Within thirty minutes, he and his assistant, Alan Polk, were back.

By the time I had the information, plus ten minutes to think things over and call Adela back, the situation had turned a bit nasty. Evidently, from Adela's account, Katie McGinty had gone ballistic and had reamed Adela and called Secretary Glickman's chief of staff. Katie complained to him about lack of "strategic political thinking" in failing to use this "largest salvage sale in Forest Service history" as a vehicle for President Clinton to show success in getting the government to work and sensitivity to the needs and concerns of the people of the region. I asked Adela if it was appropriate for me to call Katie McGinty to get a firsthand account of the White House's (her) concerns. She granted permission.

Katie was indeed severely aggravated that the Forest Service was putting up for sale the largest salvage sale ever offered, and the White House was not being given a strategic opportunity to make political points for the president. When asked where she got the information that this was the biggest offering of timber salvage volume in Forest Service history, she replied that she had seen it in a Forest Service information sheet. Jim Caplan handed me the document, which read, "The six sales, when combined, make up the largest salvage volume ever offered on the forest." I pointed out that language was quite clear: it was the largest salvage volume ever offered on that particular forest but not the largest ever offered by the Forest Service. The auction she referred to was the Thunderbolt Sale Complex on the Payette National Forest, north of the Boise.

She then expressed her aggravation with Undersecretary Jim Lyons because he told her that even if the sales sold, there would be a significant delay in the actual cutting because there were problems getting clearance from the Environmental Protection Agency. So why, she asked, were we in such a hurry to put the sales up for auction as the delay would not cost anything in terms of actually being able to begin the salvage cutting?

I told her that there were no problems on these sales with any regulatory agency. It was becoming more and more clear that there was an absolute miscommunication between the White House (Katie McGinty) and USDA (Jim Lyons). That was now obvious to her as well, but she was still upset. She said to go ahead with the auction of the sales as planned. Politics and the operational aspects of natural resources management agencies simply mix very poorly.

In the second case, Secretary Glickman promised in his confirmation hearings that he would have the Forest Service take a careful look at the interaction between various environmental laws and regulations to discover conflicts and redundancies that were causing problems for the Forest Service in carrying out its mission. That report was promised for 1 June 1995, an extremely short time period.

A team from the legislative and policy branch has been hard at work on the effort for about three weeks (about 50 percent of the total time allocated for the job). They had a very carefully ordered set of instructions and were well under way.

Anne Kennedy, a very bright lady of about 30, was Secretary Glickman's assistant when he was a congressman and now serves him as an assistant in his new role as secretary. The team and I met with her about five days ago. In the discussion I pointed out that the inevitable outcome of any really rigorous analysis of this subject would reveal that the overriding purpose of the management of the national forests—and probably other public lands—was the preservation and protection of biodiversity. The application of the Endangered Species Act and the regulations on viability retention issued pursuant to the National Forest Management Act combine to produce that result. Subsequent court cases have reinforced that conclusion.

This means simply that an unplanned process has set a dramatically significant policy, and the reality of that stark policy has not yet been clearly recognized or appreciated by the administration, the Congress, or the general citizenry. Such a recognition, when it occurs, will put more focus than ever on the Endangered Species Act, which is up for renewal, and the National Forest Management Act, which Mark Rey and the senators who work for him (knowingly or otherwise) also intend to change.

That discussion evidently startled Kennedy when she realized what a hot political potato the new secretary was juggling. I don't think he had any idea what the consequences would be when he promised the Senate Committee on Natural Resources this review. Now suddenly one of his aides knows—and with crystal clarity—and that unnerves her and Secretary Glickman's chief of staff. So this afternoon, Anne met with the team and cancelled the only significant part of the charter, the part concerning the interactions of the law and regulations. There is simply no stomach to face up to the issue, as I described it, "at this time."

I was not in the meeting, but the results were immediately brought to me. There was no way that the primary portion of the analysis could be eliminated and still produce anything that would be of any use in the debate. Besides, the entire operation would not pass the laugh test: nobody could look at such an emasculated assessment and keep from laughing. Such would make a fool of the secretary and reflect poorly on the capabilities of the Forest Service.

30 May 1995, Portland, Oregon

Yesterday and this morning I met with management teams for Region 6 of the National Forest System and the Pacific Northwest Forest Experiment Station. The primary purpose of my visit was to provide the forest supervisors and regional staff an opportunity to vent frustrations. They are frustrated with the downsizing of their workforce and reductions in budget at a time when they are trying to meet the increased demands of the objectives in the President's Forest Plan for the Pacific Northwest and increased expectations for an aggressive timber salvage program.

In some cases there may well be reason to salvage dead or dying trees to improve forest health. Such actions might include significant reductions in fuel loadings so as to reduce risks from wildfires, measured as reduction in the likelihood of uncontrollable wildfire, or significant reduction in the probability of rapid fire spread, or reduction in the intensity of the burn, or any combination of the above. It may also be possible, in the course of salvage operations, to create seedbeds or otherwise enhance growing sites for selected areas with high potential for growing trees. Where salvage operations provide opportunities for enhancing forest health, those benefits should be clearly explained and be included as benefits to be gained from such action.

However, there will be situations—perhaps in the majority of cases—where timber salvage operations are carried out primarily for economic and social reasons. Those reasons include providing jobs in the woods and in the mills, giving a boost to natural resource–dependent communities, and making wood products available to the American people. Other reasons are reducing balance-of-trade problems in timber imports and substituting dead or dying trees for healthy trees in the overall harvest. These cases should be clearly identified for exactly what they are—the capture of economic opportunity. That should be justification enough provided that salvage operations are carried out in such a way that benefits—economic, social, and environmental—exceed the costs that salvage operations entail.

I want to make it clear, both inside and outside the Forest Service, that the agency will conduct aggressive and opportunistic timber salvage operations. Salvage operations will occur based on our existing plans, or in response to congressional direction or instructions from the administration.

In any case, on the explicit instructions of Secretary of Agriculture Dan Glickman, we will not, under any circumstances, violate our professional and agency ethics by failing to protect basic resource values—particularly soil and water. We can thank Undersecretary James Lyons for those instructions. Secretary Glickman made this clear in his letter to Senator Murkowski, chairman of the Natural Resources Committee, when Glickman said the

Forest Service would comply with the standards and guidelines in the forest plans.

16 June 1995, Washington, D.C.

The past week has been spent on a chief's review of the Northeastern Area of State and Private Forestry. The review team members were Deputy Chief Joan Comanor and Northeastern Director Michael Raines. The review started in downtown Chicago with a meeting in City Hall with two Chicago commissioners. Mayor Richard Daley, unfortunately, was unable to attend. We spent time in the very core of Chicago looking at several aspects of urban forestry, ranging from Chicago's GreenStreets program (a determined effort by the city to line the streets and other available areas with trees) to the reclamation of the hundreds of vacant lots that result when dilapidated properties become city property for nonpayment of taxes. These vacant lots have traditionally not been cared for and have become dumping grounds for garbage and castoffs, thereby providing breeding grounds for rats.

With a small amount of technical and organizational assistance and very modest grants from the Forest Service, Chicago's urban forestry efforts have been magnified. Some cooperators reported leveraging Forest Service grants by ten- to twentyfold in matching funds and efforts from local governments and nongovernmental organizations.

The trees seem to make a difference, both in terms of appearances and in terms of how people relate to their environment. Not so long ago, many of the neighborhood streets in Chicago were lined with American elms, the canopies of which formed arches over the streets and sidewalks. Nearly all of these trees were killed by Dutch elm disease, introduced from Europe. Yet here and there along the streets, a few of those old elm trees provide a visual respite from the sterile surroundings and afford a spot of shade in the summer heat.

We were told of old people approaching the crews cutting holes in the concrete sidewalks and planting new trees and saying, "Thank you for bringing the trees back." Block committees are being formed to care for the trees and to keep the cleaned-up and planted vacant lots free of garbage and vandalism to the trees. With that, we were told, there has in many cases been

State and Private Forestry area review, Chicago urban garden, June 16, 1995
Photo courtesy of Jack Ward Thomas.

a discernible improvement in residents' attitudes—just a ray of hope that maybe the residents can improve their lives. That seemed significant.

The next day we spent looking at the true forestland owned by Cook County, which includes the city of Chicago. Cook County is 12 percent forested and all that land is county-owned. This means that the entire population of a city the size of Chicago is within a short drive of forests. What a remarkable legacy and what a treasure for the people of Chicago.

These forests are carefully managed, including the use of prescribed burning, which to our great surprise is carried out to a significant degree by volunteers. The plan calls for burning up to 10 percent of the forest per year (usually about 6 percent is burned per year).

The next day we proceeded to central Vermont to visit a town that had undergone significant decline but was now making a modest comeback due in part to contributions by State and Private Forestry efforts. Forest Service people had contributed to these efforts with some technical advice, endorsements, and some very modest seed money that had been multiplied tenfold by local leaders. The most interesting part of the story was the local people who were taking responsibility for their own destinies, with just a little help from their friends, the U.S. taxpayers.

As we observed these good results of the very modest Heritage and Stewardship Incentives programs, it is a bit painful to realize that these programs have been zeroed out—eliminated—by the House subcommittee that deals with the Forest Service budget. Perhaps the Senate will see things in a different light, but I somehow doubt it. The subcommittee, operating under significant bottom-line budget constraints, must choose between programs. They chose programs more directly related to management of public lands, particularly those programs related to commodity production.

Too, these "eastern" programs are deemed to be the pets of once-powerful Democrats who rather ruthlessly imposed them when they were in control. That group includes Congressman Sid Yates of Illinois and Senators Robert "the King of Pork" Byrd of West Virginia and Pat "the Father of Legacy" Leahy of Vermont. Congressman Bruce Vento of Minnesota is due for his comeuppance with the elimination of two of his pet programs, International Forestry and Law Enforcement, and diminution of wilderness programs. Now, with the long-suffering Republicans in power, it is payback time. This seems a poor way of making public policy, but members of Congress are just as petty and vindictive as the rest of us, and maybe more so, given their game of choice.

Thursday was spent looking at State and Private Forestry contributions to the ongoing efforts to stop and reverse the degradation of the Chesapeake Bay. Most of those efforts are aimed at reducing the discharge of silt, nitrogen, phosphorus, and salts into the Chesapeake. It has been determined that maintaining as much of the huge watershed that drains into the bay in forest cover seems to be among the most effective and cheapest means of achieving those goals.

At noon it was time to join the celebration and cut the ribbon dedicating the Little Blackwater Demonstration Timber Bridge in Dorchester County, Maryland. This bridge was built as part of the Forest Service's Timber Bridge Initiative. With a little technical advice and some seed money, the county of Dorchester was able to get a sorely needed bridge at about half what a similar concrete and steel bridge would cost.

After a good round of congratulations and speeches, I was invited to cut the ribbon opening the bridge. There was a parade of vehicles that could not come this way before because of the inadequacies of the old bridge, including a fire truck (also made possible for the county with Forest Service help), a school bus, a truck hauling a bulldozer, a gravel truck, and an ambulance. The folks were delighted with their bridge and happy with their forest services, state and federal. I was pleased to be there.

21 June 1995, Washington, D.C.

There was yet another hearing today before Senator Larry Craig's subcommittee of the Committee on Energy and Natural Resources. This hearing concerned a bill (S 852) put together by the livestock industry and sponsored by Senator Pete Domenici of New Mexico. Interestingly, both Senator Phil Gramm of Texas and Senate Majority Leader Robert Dole of Kansas, rivals for the nomination of the Republican Party for the presidency, are cosponsors of the bill.

The title of the bill is Uniform Grazing Management on Federal Land. Given that, it is strange that Title I of the bill deals only with the administration of grazing by the Bureau of Land Management and Title II deals with the removal of the national grasslands from Forest Service management.

Careful reading makes it clear to me that the objective is to ultimately extend the bill at a convenient time to include the Forest Service. I suspect that the strategy is to begin with BLM lands, which will not draw as much attention or protest as will making a move on Forest Service lands.

The purpose of the bill is very clear, which is to make livestock grazing the dominant use on BLM lands—pure and simple. The bill codifies every aspect of grazing administration and takes every pain to ensure the "rights" of grazing permittees. This is done in large measure by establishing a number of local grazing boards that are carefully weighted to ensure a predominance of those who would protect the rights of permittees.

Senator Dorgan, Democrat from South Dakota, is the lead man on the national grasslands section of the bill. I had met with him and his staff a few months ago, and he told me what he intended to do so far as getting the national grasslands under the control of the permittees by removing management from the Forest Service. Speaking the permittees' lines, he was upset by the Forest Service's applying a rigorous multiple-use standard instead of putting grazing and permittees' welfare above all other uses.

I made some phone calls last week and turned some pressure on him from people in his own state who have broader interests than just grazing, including state wildlife departments, environmentalists, oil and gas lessees, etc. Dorgan's staffer called just before we left for the hearing. The senator's demeanor had changed. He was no longer certain of a slam dunk of the Forest Service. There was a request for me "not to take the senator on too hard," and statements that "this is work in progress" and "changes can be made" and "we want to work with you," etc. Senators ordinarily do not make such a switch unless they are nervous about the situation that is developing.

Our staff called in from South Dakota to say that the press was running against Dorgan on the issue. Evidently, planted seeds were beginning to sprout. We decided on a hearing strategy to come at Dorgan hard in the first round of questioning and then in the second round to hold out an olive branch of cooperation on alternative solutions, etc.

Again the hearing was all Republicans so far as substance. The Democrats (Dorgan, Bradley of New Jersey, Bingaman of New Mexico, and Wellstone of Minnesota), with the exception of Dorgan, made only token appearances and obviously had not done their homework and didn't show any real interest. On the other side the turnout was strong: Senators Murkowski of Alaska, Domenici of New Mexico, Craig of Idaho (who was in the chair), Campbell of Colorado, Thomas of Wyoming, Kyl of Arizona, and Burns of Montana. Burns is the faithful sidekick of Senator Craig and reminds me of Smiley Burnett, comic sidekick to Gene Autry in the old cowboy movies.

Most of the Republicans' bombast, rhetoric, and ridicule was directed at the secretary of the Interior in absentia. The surrogate for Republicans' attention was BLM Director Dombeck. Mike stood his ground and had Michael Penfold at the table with him. Penfold is a top hand and a good witness and provided toughness when it was necessary. He has announced his retirement and can afford to shield Dombeck.

Late in the hearing, Senator Kyl made what I considered a political blunder when he began to question me about extending the bill to cover Forest Service activities. That was a mistake in that it will bring a whole new set of interest groups into the game in opposition to the bill. There are a lot more people interested in the Forest Service lands than are interested in the BLM lands, and they are better organized and fight harder.

I intentionally drew out the exchange and pretended not to fully understand the question in order to get the senator to restate it in several ways in order that the audience and press would not miss the possibility of extending the bill to National Forest System lands.

Senator Craig ended the panel's questioning with a polemic statement that was even more caustic than his opening statement. At each hearing he is becoming a bit nastier and more insulting. I don't know whether he is feeling his oats as chairman or needs to respond to the hard right wing in Idaho. He had a record in the House as a bomb thrower, and he may well be reverting to that behavior.

My strategy with Craig will continue to be to act the role of the calm, cool, professional with an occasional growl just to say, "Don't push me too hard." I sense that hubris is beginning to overcome the players on this committee and may well backfire. We will see.

29 June 1995, Washington, D.C.

I was in Undersecretary Lyons's office for a late afternoon meeting. As I was leaving, I was called back. Lyons was on the phone with Leon Panetta, chief of staff to the president. The Rescission Bill was being debated in the House of Representatives. The White House was in the process of cutting a deal with the Republican majority. There was a sticking point over the section of the bill that dealt with accelerated timber salvage from national forests and Bureau of Land Management lands.

The environmentalists have been putting pressure on the White House through Vice-President Gore over the salvage sale amendment. The bill called for an accelerated salvage program over three years. The administration, fearing a Republican takeover in 1997, balked at extending the override of environmental laws past 31 December 1996. The bill further exacerbated administration fears by stipulating a salvage level that in my opinion exceeds the capability of the agencies to perform, and I could not determine where the bill's sponsors got their numbers.

The White House was negotiating with the Republican leadership and, as nearly as I could tell, bypassing the Democrats to the extent possible. Congressman Norm Dicks of Washington was the go-between, because of his position on the Budget subcommittee that deals with the Forest Service.

Panetta evidently ordered Lyons to prepare a letter for Secretary Glickman's signature, starting with the administration's position concerning the salvage section. By now it was nearly 8:30 P.M., and I was unable to locate any of the Forest Service folks from the timber shop. We were able to get our hands on the memo that the Forest Service had sent to Congressman Don Young of the Resources Committee in the House.

In the meantime, Lyons was getting pressure from Panetta and Congressman Dicks to guess at the numbers. That came to an end when I became adamant and loud about having numbers that the Forest Service could stand on. I made it clear that if a more reasoned response revealed the estimates of salvage volume to be off the mark, I would say so loudly and publicly. Assistant Undersecretary Backiel jumped in on my side. Things sobered up a bit at that point, and Lyons started hanging tough in the ongoing phone calls.

I was busy working my way through the numbers and did not get the details as to the identities of the persons Lyons was talking with. However, he was engaged in a couple of loud and frank exchanges, if not shouting matches. His theme became that he would not be party to setting the agency up for failure.

We were able to get Steve Satterfield, the Forest Service point man on budget negotiations with congressional committees, on the phone. He was able to talk us through the numbers. These numbers had to be adjusted to meet the administration's time frame of two years instead of three. At some point we were able to discern that the numbers in the bill had been altered by committee staff on the premise that the confidence interval in the Forest Service estimates was "plus or minus 25 percent." They expanded the Forest Service estimates by 25 percent!

There were also estimates of timber salvage in the bill for the Bureau of Land Management of 115 million board feet for each of three years. Evidently, no one was dealing with BLM on the issue. I was unable to get anybody connected with BLM in their Washington office. So I called Elaine Zielinsky, BLM state director in Oregon and an old friend, to tip her off. She asked me if I could prevent the letter being prepared from mentioning BLM at all. I said I would try. Lyons acceded to that request.

In the meantime, we could see the House debate on the C-SPAN network. Congressman Peter DeFazio and Congresswoman Elizabeth Furse, both of Oregon, were debating Don Young of Alaska. DeFazio and Furse were, justifiably I thought, intensely angry that they were being asked to vote on a bill that was not yet complete, and when the bill was finally configured, there would not be time to study and debate the final version of the timber salvage section. They also expressed bitterness that President Clinton was cutting a deal with the Republicans without consulting the members of his own party from the regions most affected.

By the time we finally had the numbers computed, Lyons had the letter ready with blanks awaiting those numbers. He gave Adela Backiel and me a chance to review the letter and accepted our minor suggestions. The letter was three paragraphs long, and the final paragraph was a strong and gutsy statement that the Forest Service would carry out the instructions in full, while keeping with the intent of the applicable environmental laws and the forest plans. The undersecretary had been hanging tough on those words for the past six weeks, and he had held fast on this occasion. It wasn't easy for him to do that. The Forest Service, and all concerned with good natural resources management, owe him for that.

The letter went over to Panetta at the White House; it was coupled with a similar letter from President Clinton and sent over to Speaker of the House Newt Gingrich. And so it was a done deal.

It is amazing how as a routine matter there is painstaking, detailed staff work prepared and that work is reviewed, managed, debated, and polished. Then suddenly, a moment comes when all the deliberate consideration stops and raw politics with all its attendant maneuvers, ploys, deals, and manipulations takes over and decisions are made. In this case, the debate had gone on for two days. Then suddenly, the sticking point was the language concerning a minor part of the Rescission Bill involving timber salvage.

What would have happened if I had not, almost by accident, been in the undersecretary's office when the issue came to a head? Would they have gone ahead with the guesses that they were putting together? If so, what would the consequences have been? This is no way to manage the nation's business or its natural resources.

One thing for certain is that the Forest Service has never been under such pressure to produce timber and salvage sales (albeit at reduced levels) in a manner that is inconsistent with the intent and spirit of the environmental laws. It also seems likely that this bill and its consequences, on the ground and in the law, will set off an orgy of legal challenges on an entire spate of issues. The lawyers ought to like it, at any rate.

11 July 1995, airborne between Washington, D.C., and Boise, Idaho

The House hearings on the 1995 grazing bill took place today. The two lead witnesses were Congressmen Wally Herger of California and Joe Skeen of New Mexico. Both spoke in favor of the bill, and Skeen made it very clear that there would be a serious attempt by the Republicans to transfer ownership of public lands (I think he meant Bureau of Land Management lands) to the states. However, in the meantime, this grazing bill was a good start in making the world a better place for small ranchers.

The next panel was composed of BLM Director Mike Dombeck and his deputy, Maitland Sharpe, a representative from the Statistical Services Unit of the Department of Agriculture, and me. In opening testimony both Dombeck and I expressed the opposition of the Departments of Interior and Agriculture to the bill. And then the questioning began. All the Republican committee members were present. Only one Democrat, Bruce Vento of Minnesota, showed up. Such lopsided committee hearings have become the rule. It is difficult to tell whether this is the result of the Democrats' abandoning any semblance of interest in natural resources issues and viewing such as antithetical to political survivorship, or their disgust at the stacked deck toward resource exploitation obvious in the witness lists, or sheer defeatism, or all three.

In theory, the only parts of the bill that apply to the Forest Service are the part of Title I that deals with grazing fees (which are set by a simple two-factor formula that no one can explain) and Title II, which removes the national grasslands from Forest Service management. But it is obvious that the intent is to include national forest lands in the final version of the bill. In fact, such has already occurred in the markup of the Senate version. The bill is a total mess.

In the technical sense, there is failure to recognize that the land management agencies' approaches have their roots in very different circumstances. Nearly all of the BLM lands are arid or semiarid rangelands. In comparison, the Forest Service manages primarily forestlands that support some grazing.

I think the bill's proponents may have made a strategic error in not

Central Nevada Ecosystem tour, July 1995. *From left:* Jack Blackwell, deputy regional forester; Monica Schwalbach, assistant forest supervisor, Central Nevada Ecosystem; Jim Nelson, supervisor, Humboldt-Toiyabe National Forest; Chief Thomas; Jerry L. Green, district ranger, Ely; Tony Valdes, district ranger, Tonopah; Dayle Flanigan, district ranger, Austin; Gary Sayer, deputy forest supervisor, Humboldt-Toiyabe National Forest. Photo courtesy of Jack Ward Thomas.

confining their initial thrust to BLM lands. These lands—and the agency—are much less well identified in the public's mind as valuable and important to the nation compared with the national forests and the Forest Service.

The BLM–Forest Service combination makes a much more formidable adversary, and the combination of two agency heads who are, above all else, natural resources management professionals likewise creates a different situation than would occur with political operators at the helm. Mike Dombeck and I are both professionals dedicated to good land management, and we are friends and colleagues to boot.

We both kept bearing down on the weaknesses in the bill, which are the establishment of grazing as a dominant use, the granting of a vested right in the grazing permit, the dominance of local control over federal land, the protection of graziers from nearly any control by federal land managers, and the incredible tangle of mandatory resource advisory committees and grazing advisory committees.

The Republicans kept trying to sell their wares as a protection of the "western way of life," the welfare of the small family ranches (they said nothing of the wealthy landowners and the corporations), and the control of rapacious federal bureaucrats. Any opposition was cast in the light of unreasonable resistance to the western way of life by minions of—gasp—the Clinton administration. Sometimes, in the midst of such goings-on, it is difficult not to become angry or burst out laughing at the rhetoric, but somehow everyone manages to keep a straight face.

The real stars of the show on the Republican side were Helen Chenoweth of Idaho, who was angry at ranchers' having to bear the burden of preparing full environmental impact statements (they don't), Barbara Cubin of Wyoming, who didn't have a clue as to what was going on except that she was for it, and Wes Cooley of Oregon, who wanted to defend the grazing fee formula in the bill though he obviously had no idea how it worked but had been told it was a good approach.

The only Democrat to show up was Bruce Vento of Minnesota, who was once the chair of the subcommittee before the Republican onslaught last November. He quickly demonstrated complete command of what was going on and demonstrated the foibles of the bill clearly to all who had any desire to listen and understand.

During a break that was declared to allow the congressmen to vote, an aide to Senator Dorgan (he was responsible for adding Title II, removing the national grasslands from Forest Service management) cornered me in the hall. According to her, Dorgan now realizes that he has made two serious mistakes. The first is having anything at all to do with the sponsorship of this bill; the second is the decision to carry water for the permittees on the national grasslands in North Dakota in trying to remove management from the Forest Service.

She gave me clear signals that he needs some way to save face and an excuse to withdraw his support from the bill. I didn't make it easy for her and told her that the senator had blundered, and I didn't know what we could possibly do to get him off the hook. She suggested that if we were to set up a separate management unit to handle the grasslands in North Dakota, he would be satisfied.

I acted as if that was an idea that we could pursue. I didn't tell her that we had already come up with several options for doing just that and that we considered that a good approach. I said we would think about it. The game may be to let him, and his constituents, view this as a victory, and then we do what we want to do anyway.

The conversation with Dorgan's aide was no more than over when the primary instigator who got Dorgan in this game asked if he could talk with me. The last time we visited, he was playing his trump card of removing the grasslands from Forest Service management. The reaction on my part was to declare that such would be O.K. as we had more to do than we could handle and that we had higher priorities, anyway.

Now, he wants to assure me that the Forest Service is the agency that he and his fellow association members want to manage the grasslands and that all they ever really wanted was a distinct management unit. I gave him the old "don't throw me in that briar patch" routine and let him know that though it pained me deeply and was a great concession that was being extracted from us, we would give that new idea consideration. As he walked away, I gave Bertha Gilliam, director of Range Management, the go-ahead to complete the staff work to achieve that end.

This game can become addicting. At first it seemed corrupting, and maybe it is. But more and more, it just comes across as the way that things are accomplished in this town. The main thing seems to be figuring out what the other guy needs in order to look good to those who own a piece of him or her; and that applies equally to politicians and lobbyists.

12 July 1995, airborne between Helena, Montana, and Washington, D.C.

This has been a frantic three days spent in a whirlwind effort to muster support to overturn action in the budget put forth by the House. It will affect the completion of the ecosystem planning efforts by the Forest Service and the Bureau of Land Management in the Columbia River Basin. This joint ecosystem planning effort was launched a year ago to address the emerging issues in the Columbia basin caused by the listing of all the salmon runs as threatened or endangered, the extensive mortality of trees due to insect and disease outbreaks, and the impending and increasing danger of wildfire. I was accompanied by Regional Foresters Dale Bosworth, Hal Salwasser, and John Lowe, as well as Steve Mealey, the environmental impact statement coordinator for the Columbia basin.

It was obvious from lessons learned in western Oregon and Washington and northern California that these issues can be successfully addressed only at very large scale and in planning approaches that are inclusive of ecological, economic, social, political, and cultural factors. Therefore, an interagency team of scientists (Forest Service, Bureau of Land Management, National Marine Fisheries Services, Fish and Wildlife Service, Natural Resources Conservation Service, Geological Survey, and others) was assembled in Walla Walla, Washington, to do the largest-scale (in space and in terms of variables considered) regional assessment ever conducted.

Simultaneously, two teams were assembled to prepare the environmental impact statements, with an array of alternatives to present to the public for comment, and to regional foresters of the Forest Service and state directors of BLM for decisions. Those teams are to finish their work about one year after the assessment team completes its work in October 1995.

After the Republican takeover of Congress in the November 1994 elections, those who were suspicious of the government's motives, fearful of their private property rights, and concerned about local control (i.e., county government) had the ear of several Republicans now in positions of power. Foremost among these folks is Robert Klicker, a wealthy farmer from eastern Washington, who served for a time on the board of the Blue Mountain Natural Resources Council. Klicker was a prominent figure in Congressman Nethercutt's victory over Speaker of the House Tom Foley. Klicker is a conspiracy buff of the first order and has explained his belief that ecosystem management and thus the Columbia basin assessment effort is part of a carefully orchestrated conspiracy to take over the United States. The key factor in this effort is some cartel of international bankers that operates through the "Trilateral Commission." Klicker and others, convinced that the Columbia basin planning effort represents a dire threat to private property rights, have prevailed upon Nethercutt and Senator Slade Gorton of Washington to kill it.

Nethercutt was able to accomplish this through the House Appropriations subcommittee headed by Ralph Regula. My conversations with Regula indicated that the other subcommittee members let Nethercutt have his way on that issue to make up for denying him several budget adjustments that he had requested. Regula seemed certain that this item would disappear in the full committee conference. It didn't and has made it through the final version, which has now gone to the Senate.

The budget language and allocation do the following things. The assessment will be completed with $60,000 allocated for peer review and publication. All efforts at completion of the environmental impact statements will stop, and instead seventy-four separate plans will be prepared—one for each BLM district and national forest in the Columbia basin. Also, no funds are to be expended for application of the PACFISH and INFISH[4] strategies, and the funds available to the Forest Service for planning efforts are cut in half.

The prohibition against the use of INFISH (a strategy similar to PAC-FISH, used to protect habitat for the bull trout and west slope cutthroat trout) will likely force the Fish and Wildlife Service to list those species as threatened. The bull trout had been declared appropriate for listing but was not listed on the basis of the Forest Service's adoption of the INFISH strategy.

If Nethercutt and Gorton are depending on the prohibition against listing the bull trout that Congress has instituted for the rest of the year to remove problems, they are wrong. The Forest Service would be forced to deal with the habitat of that species as a requirement of the regulations issued under the National Forest Management Act. That would lead to the development of some alternative strategy to INFISH, which in reality would be INFISH in drag; what other legal options are there? And so we come back to the same point, having paid the price of an additional listing that could have been avoided.

Now we come to the planning requirements to do seventy-two separate plans. No matter what, in the end FS and BLM will have to consult with NMFS as to the adequacy of the plans to protect the listed species and, if

4. U.S. Department of Agriculture. 1995. Decision notice/decision record, FONSI. *Ecological analysis and appendices for the inland fish strategy (INFISH), interim strategies for managing fish-producing watersheds in eastern Oregon and Washington, Idaho, western Montana and portions of Nevada.* Forest Service Intermountain, Northern, and Pacific Northwest Regions.

the bull trout is listed, also consult with the Fish and Wildlife Service. It can be reasonably assumed that the regulatory agencies collectively would insist that the plans protect the habitat of the listed species in a coordinated fashion. When that is done, it seems likely that the agencies will be back where they would have been, having gone through the process anyway.

To complicate matters even further, if NMFS is not able to tie the required recovery plan for the salmon (and maybe the recovery plan for the bull trout prepared by the Fish and Wildlife Service) to the FS and BLM plans, NMFS would be forced to impose appropriate management strategies in a recovery plan as required by the Endangered Species Act. That is an even worse and less responsible course of action.

As if there weren't enough problems, consider the potential consequences of having produced a hugely significant new body of information and having published and made such information available to all who asked for it. The National Forest Management Act and the Endangered Species Act require that new environmental impact statements be prepared when new knowledge becomes available. FS and BLM then would be required by law to immediately reexamine all affected forest and district land-use plans, and if the plans are challenged in federal court, it seems likely, given the dire straits of the salmon and the stressed conditions of the forests, that the courts would issue injunctions against any management actions that might cause damage to the habitats of listed species. That was the scenario that I presented to each group or person we visited over the two-day period.

14 July 1995, Washington, D.C.

I got into Washington at midnight last night. When I arrived at the office, it was obvious that, at last, concern had been piqued at political levels in the Interior and Agriculture departments over the grazing bill making its way through Congress. Senator Max Baucus of Montana, a Democrat, has signed on to the bill as a cosponsor. That bipartisan support is likely to ensure that the bill will escape the committee and be voted on in the Senate. There is one thing one can say about Max Baucus: he has little stomach for anything beyond his own reelection.

The "grazing reform" effort was badly botched from the beginning and the Forest Service, at my instruction, made every effort to avoid becoming enmeshed in the process. That effort was successful to a large degree. For example, the Forest Service did not get stuck to the tar baby of resource advisory committees for each management unit as prescribed in the regulations promulgated by the Bureau of Land Management.

My first action of the day was to discuss the situation with Interior Secretary Babbitt. I suggested that the better part of pragmatic valor might be for him to withdraw or significantly delay the institution of the grazing regulations, especially since the national forests and grasslands have been drawn into the issue, making the Department of Agriculture and the Forest Service undeserving and unwilling combatants in an issue we had resisted from the beginning.

He told me that he saw no political advantage in handing the Republicans a victory in forcing a withdrawal of the grazing regulations. I expressed my concern with being brought into the fray and with the consequences of the Republican-sponsored grazing bill (which seems to be gaining momentum), which would be a disaster for the overall Forest Service approach to planning and management.

Secretary Babbitt then told me that he had already discussed the matter with Leon Panetta, chief of staff to the president, and had been told not to withdraw the proposed regulations. Further, he said that he was confident that the president would veto the grazing bill if it actually passed. He then qualified that rather certain statement by saying that when standing fast on veto threats, "the White House gets a little squirrelly sometimes."

Just following that discussion, the staffs from Legislative Affairs and Range showed up with a draft letter to go, along with a letter from Bureau of Land Management Director Mike Dombeck, to editorial boards across the country. The letter was strong and pointed, but I beefed it up a bit. I'd done such before and the letters were always toned down or killed in the clearance process in the Department of Agriculture.

Assistant Undersecretary Adela Backiel met me upon arrival at the meeting and wanted to discuss my letter to the editorial boards. To my amazement, she encouraged me to make an even stronger statement. As a result, I modified the letter as the meeting went on. At the close of the meeting, we went over the letter again and agreed I would have it put in final form by staff.

I discussed the situation developing over the grazing bill with Undersecretary Lyons and Deputy Secretary Rich Rominger. I related my conversation with Secretary Babbitt to them, and I suggested a meeting on Monday with Babbitt and Glickman, appropriate undersecretaries and assistant secretaries, the Bureau of Land Management director, and me.

My bottom line was that I was willing to load my last clip, fix bayonet, and charge over this issue, but only if the two secretaries would recommend a veto if, in spite of our efforts, the grazing bill passes. I said I was not ready to put the reputation of the Forest Service and the credibility of the chief on the line without full support.

Undersecretary Lyons at that point instructed me to delay my travel to the Toiyabe National Forest, scheduled for Sunday night, by one day to allow for such a meeting. I reluctantly agreed.

15 July 1995, Washington, D.C.

Tonight it was my pleasure to take Carol Applegate to supper while she was in Washington on a teacher's short-course. She and her husband, Dr. James Applegate, a professor at Rutgers, are old friends and spent a sabbatical with me at the Forest and Range Sciences Laboratory in La Grande. Jim hunted with me in the high Wallowas in northeastern Oregon last elk season. Carol delivered several 8"×10" photos to me of our hunting trip.

As I examined the photographs, I was dressed in a pinstriped suit, a crisp, starched white linen shirt, and a regimental striped tie. I recognized the Jack Thomas in the photograph dressed in wool pants and suspenders, winter pacs, a black wool shirt, and an old pile army cap. I pondered the photograph, and I suddenly didn't recognize the person holding the picture. It was a most strange feeling and one of great melancholy.

I don't really belong in Washington. I don't belong in these clothes. I don't belong in the Cosmos Club sipping expensive wine that I can't tell from cheap wine and listening to a harpist play. I think about the ride home on the subway and the high-rise condominium in which I live.

I am tired of trying to be someone I am not, tired of pretensions of power, and tired of games. I am tired of simultaneously fighting off assaults by the political right on the parts of the Forest Service budget that make for a balanced approach, and fighting off assaults on the management and ownership of the people's lands and on the laws that protect the environment.

I am also tired of struggling to get the administration to keep the word

Elk hunt, Minam River Drainage, Eagle Cap Wilderness, 1994. *From left:* Robert Nelson, director of Fish and Wildlife, Forest Service; Chief Thomas; and James Applegate, Rutgers University. Photo by Will H. Brown, courtesy of Jack Ward Thomas.

they gave to me and to Forest Service employees that my appointment as chief would be converted from political to nonpolitical. I gave my solemn pledge to my fellow employees in the Forest Service that I would accept the chief's job only on the assurance that such a change would be expeditiously accomplished. It has been eighteen months since those pledges were made. I feel that all concerned have made a good-faith effort to make good on the promises. It now seems obvious that the Office of Personnel Management will not under any circumstances allow this conversion to occur.

So the time has come for me to ask to be replaced by a nonpolitical appointee. I must move now in order to give the secretary time to get a replacement in position before the silly season of the upcoming presidential election year begins in January 1996. I will propose two courses of action to the secretary, in order of preference to me: Return to my previous assignment as a senior research wildlife biologist (ST-17) at the Range and Wildlife Habitat Laboratory in La Grande; or retire at a time of the secretary's choosing between the first of September and the end of 1995.

Filling the chief's job with a nonpolitical appointment is too important to the Forest Service as an organization, to the vision that Forest Service employees have for themselves, and to the future of the management of the national forests and grasslands to be jeopardized by a broken promise. And it is critical to Forest Service people that chiefs stand on their word. I see no alternative and can only trust that the circumstances are for the best.

17 July 1995, Washington, D.C.

I had been scheduled to be in Salt Lake City last night but was ordered to delay my travel for one day to be available for a meeting with the secretaries of Interior and Agriculture regarding the grazing reform bill. This meeting was to be arranged by Undersecretary Lyons. When I had heard nothing by 11:00, I called and made an appointment to see Jim at lunch. He had made no contact with Mr. Babbitt, and it was obvious that no meeting with the secretaries would occur today.

Nothing transpired for the rest of the day so far as any action on the grazing bill. Undersecretary Lyons knew that I was departing for Nevada late in the afternoon and voiced approval. Just before I departed for the airport, I received a call from John Lowe, regional forester in Region 6, who was passing on information that as a result of our "lobbying trip" to Idaho, eastern Washington, and Montana last week, the county commissioners had, almost across the board, agreed to back the Forest Service position that we be allowed to complete our operations in the Columbia basin.

On the way to the airport, I called Assistant Undersecretary Adela Backiel to pass on the word about the county commissioners' support for the Columbia basin project. She was quite upset and agitated that I was leaving town. She thought I should remain in Washington to deal with the upcoming markup of the grazing bill in the Senate Natural Resources Committee. I told her that it was also important for me to keep my appointments of long standing in Nevada, and that I was holding to my schedule. She wanted to know if I had made phone calls to the senators and representatives who had been assigned to me. The answer was no, as my instructions were to wait for instructions to proceed. She wasn't happy when the call ended. My wonder at the lack of coordinated action in government above the agency level seems to be ever expanding.

This situation has been on a clear trajectory for at least ten days. Now, less than thirty-six hours before the Senate committee markup on the grazing bill, the administration chooses to get excited. It is simply too late to get a full-court press on the swing votes in the committee. The bill will likely, according to our count, be passed out of committee by one vote. The key is Mark Hatfield, who will see the problems with the bill and likely vote against it in the full Senate. However, as Bob Dole, Senate majority leader, is a cosponsor of the legislation and Hatfield is a part of the leadership, Hatfield will likely defer to Dole. This probability is enhanced by the ration of crap that Hatfield took in being the critical vote that prevented the vote needed on the balanced budget issue earlier in the session. It is simply too late. However, it is my bet that there will be a strategy meeting called at the White House tomorrow by Katie McGinty, with an equal chance that it will never occur. We will see.

My original list of senators and representatives to call about the grazing bill included Senators Dorgan, Conrad, Kyl, and Craig and Representatives Vento, Pomeroy, and Roberts. When I got to the hotel, instructions awaited me to also call Secretary Babbitt; Senators Domenici, Bingaman, Campbell, Hatfield, Wellstone, and Bumpers; and Representatives Richards and Williams. That is sixteen calls in a six-hour period to very-difficult-to-reach senators and representatives. Not a chance, but I will do my best.

21 July 1995, airborne between Salt Lake City and Houston

This morning, we (Regional Forester Dale Bosworth, Humboldt-Toiyabe National Forest Supervisor Jim Nelson, and I) met with Nevada's state forester, the Nevada state director of Natural Resources, and the Nevada deputy director of Fish and Wildlife. Beyond pleasantries, our primary subject was the grazing bill working its way through Congress. To my relief, we were in total agreement that this is an atrocious piece of legislation with significant ramifications beyond dealing with grazing on federal lands. We concurred that the bill was very cleverly constructed to achieve the following ends:

(1) Make planning by BLM and the Forest Service a mere support to county planning. This de facto transfers these lands to the counties with financing by the federal government.

(2) Establish resource advisory boards for each national forest to ensure local resource users' dominance of management, which was established by the planning. These resource advisory boards are essentially appointed by the governors and have direct access to the secretary.

(3) Give graziers with permits to graze on federal lands a property right in those permits as opposed to keeping the permits a privilege. This is a huge giveaway of a valuable commodity owned by the people at large to a select few—mostly large corporations and millionaire landowners. If these valuable grazing rights are made property rights and transferred to the private sector, they should, at the very least, be auctioned off to the highest bidder. The bill would take grazing rights that have sold at far below market value for nearly a hundred years and, as a final gift to the recipients of that largesse, make them into a property right for the winners of an ownership lottery—i.e., he who is fortunate enough to possess the permit at the moment wins the property right. What a deal!

(4) If the permittee has a property right to the grazing, does that convey a water right for the water necessary to support the livestock? Quite likely. Therefore, the granting of an ownership right in the grazing permit would likely carry a water right. Why? Because unless the water was available, the grazing right would be impaired. At the least, other allocations of the water—say to fish or wildlife or recreation— would likely be a "taking" and subject to compensation.

(5) Then, in order to prevent end runs by federal land managers, a grazing advisory board is established—essentially by the governor of the state where the grazing land is situated—and the board is so structured to assure domination by livestock interests and absolute domination by local interests. Though the wording is carefully contrived and "multiple use" is inserted in the appropriate places, the overall effect is to make grazing a dominant use. Why are there no hunting advisory boards, fishing advisory boards, timber advisory boards, etc?

(6) Management action deemed not likely to benefit the permittee in the short run is further blocked by provisions that so tangle permit compliance as to make effective management impossible. There are provisions that prohibit inspections of allotment conditions without notifying the permittee. No permit action to reduce livestock numbers can be taken absent a determination of "irreversible resource damage." No action can be taken to reduce livestock numbers if there is a loan outstanding on the property right in the permit. Land grant colleges will be relied on for advice concerning permit actions—i.e., is irreversible damage occurring. Some faculty members in some land grant universities (New Mexico and Nevada come to mind) are notorious pimps for the livestock industry.

All in all, the effect of the grazing bill is to turn over management of the federal lands to the tender mercies of the counties and the states. That, in itself, is a very clever ploy. Polls indicate that there is currently a very strong feeling that the closer to the people the government is, the better it is. That is a complete reversal of the sentiment of several decades ago.

This bill, after all, will bring the management of the federal lands close to the people. You bet it will! The local people won't have a prayer to counter the power of large corporations and wealthy landowners, not in grazing or any other issue, such as timber. People dependent on those industries for a living will understandably go with their pocketbooks; and when the rip-off is complete, they will be on their own. We have seen it before.

This bill would be the ultimate rip-off. The exploiters get the benefits and the taxpayers at large carry the costs. What could be a sweeter deal than that?

Later in the morning, we met with the U.S. Attorney for Nevada (Department of Justice). We have not been happy with the Department of Justice's handling of cases that we have turned over to it for prosecution. In our opinion, Justice has been too hesitant to prosecute cases that seemed solid to us, in order "to avoid confrontation." In other cases, it has decided to pursue violations as civil rather than criminal cases in order "not to make martyrs of the defendants."

Our officers have been advised not to issue citations without consulting with the Justice Department, if the cases "potentially might involve a county supremacy issue or a state's rights issue." These instructions from the Nevada U.S. attorney's office have had a chilling effect on our law enforcement efforts. We believe that this caution breeds contempt and leads to more flouting of the law and ultimately to decreasing respect for the law and the agency and increasingly dangerous situations for our employees.

It has also become clear to me that the Department of Justice is the most significant agency in government in terms of setting natural resources management policy. I don't think the employees of the Department of Justice pack the gear to make such policy and often don't even realize that they set policy through their actions.

It also seems that our law enforcement people are more subject to direction by U.S. attorneys than by agency officials. The law enforcement people in the Forest Service work for the chief of the Forest Service and not for the attorney general and certainly not for the U.S. attorneys.

I went over all of the above points in very plain English with the U.S. attorney. She handled herself well and changed her demeanor during the conference toward a more conciliatory position. She began to say that these were misunderstandings and that, certainly, she didn't mean we should not issue citations for violations. She stated that we had a communication problem and surely we had misunderstood our "consultations" with her lieutenant.

It seemed the best way to come to an amicable solution was for me to accept that our people had misperceived the instructions. That seemed to release the pressure. Then I thought it would be a good idea for the Forest Service people to meet with her and her lieutenant to get a clear understanding of their advice. She agreed.

I gave her several heads-up warnings to consider in the meantime. These were (1) I will request meetings with the head of the FBI, Lois Schiffer of the Attorney General's Office, and the president or vice-president to discuss the situation; and (2) it is likely that we will have to deal with a flagrant livestock trespass case some time in the foreseeable future, and the chief, the deputy chief for National Forest System, the regional forester, the forest supervisor, and the district ranger will participate. She got all of the points I wanted to make. She seemed to want to get to a better relationship. I think we made some progress.

The best thing about the meeting was that it gave me an opportunity to exhibit to folks on the line that they had my attention, my full support, and my willingness to go on the line with them. The word will get around. It always does.

30 July 1995, Washington, D.C.

The debate on the grazing bill continues to heat up, and I believe the tide is turning against the bill's proponents. The press is running about 9 to 1 against the bill, and even the press in New Mexico is giving Senator Pete Domenici a very hard time—putting him into a fury and even exacerbating his vendetta against Interior Secretary Babbitt.

The markup of the bill in the Senate demonstrates a considerable retreat from the original version, though the language is now even more confusing. After it became obvious that the Forest Service would adamantly oppose the bill, a considerable set of adjustments were made, including (we think) removing the Forest Service from the requirements for resource advisory councils and grazing advisory boards for each national forest, removing the reference to grazing permits as a right rather than a privilege, etc. In addition, regarding Title II of the bill concerning the national grasslands, it has been made clear that the Forest Service is to remain the managing agency.

My conversations with Senator Byron Dorgan of North Dakota, the daddy of Title II, indicate that he has withdrawn as a sponsor of Title I and that he really wants to pull out on Title II. Dorgan asked me to ask the regional forester for Region 1 to visit Dorgan's constituents in North Dakota concerned with the grasslands. He realizes that he has spent considerable political capital on a nonissue that has been made into a real issue. He obviously realizes that we have not turned loose the real opposition yet.

The western Republican sponsors of the bill are shaken and are threatening to hold hearings on the Bureau of Land Management and, to some extent, the Forest Service for doing inappropriate lobbying to defeat the bill. They wouldn't even consider that unless they detected that they were in serious trouble with the bill. Again, their hubris and arrogance are pushing them to make another strategic error. Such a hearing would only bring additional attention to this real stinker of a bill from some special interests, and the more attention the bill receives, the greater the stink that will permeate the political atmosphere.

2 August 1995, Washington, D.C.

I met in the morning with top staff and our attorneys from the Office of General Counsel to discuss policy suggestions from the Forest Service on the forestry-related matters in the Rescission Bill. The first issue was the matter of the 318 sales. Some of these sales have been cancelled or postponed or modified as various plans for the Pacific Northwest have come and gone and legal processes have played out. The Rescission Bill gives some rather clear direction to either release all the 318 sales or supply the buyers with substituted volume of like kind and amount. As the Forest Service has neither the volume nor the personnel to put the sales on the market within the prescribed time frames, the obvious—and probably only feasible—decision is to quit the argument and follow the law. At the same time, we should point out the consequences of cutting those sales on the efficacy of the President's Forest Plan.

The second decision was that, regardless of the industry interpretation, the law's instructions on previously sold sales were confined to the original 318 sales only. That will draw a lawsuit from the industry side, and our lawyers give the odds as even.

The third decision was that the land management agencies should have final decision authority as to whether to proceed with a sale after receiving advice from the regulatory agencies. Such is the clear intent of the law. We can't be expected to be responsible for meeting the salvage volumes described in the law if we do not have the authority commensurate with our responsibility.

BLM Director Mike Dombeck and I met with the undersecretary late in the afternoon prior to Lyons's attendance at yet another séance of undersecretaries and assistant undersecretaries at the White House to determine procedures for carrying out the Rescission Bill.

The discussion centered on one issue, the issue of authority. Dombeck and I both insisted that we should not be held accountable without the authority to make final decisions when inevitable differences of opinion arise between management and regulatory agencies. Lyons told us that, realistically, such would not be granted and asked for Plan B.

I suggested as an alternative that all the involved agency heads be given the same charge to meet the salvage targets in the legislation in compliance with the intent of the environmental laws. If that result is not attained, the heads of the agencies should be cited for unsatisfactory performance and similar ratings should be passed down the line to all so deserving. It would be interesting to see if holding regulatory personnel responsible for performance along with the land managers would make a difference. Lyons liked the idea.

3 August 1995, Washington, D.C.

This was the third day of negotiations by political appointees trying to come up with a memorandum of agreement—a MOA—as to how to carry out the salvage provisions of the Rescission Bill. The last version that existed last night represents almost no change from the existing process. It is as if the Rescission Bill had never been passed. In fact, the National Marine Fisheries Service has, as it does at every opportunity, slipped PACFISH-PLUS back into the requirements for avoiding a "jeopardy call" on any proposed action. The Forest Service and the Bureau of Land Management have consistently rejected the National Marine Fisheries Service–developed additions to the PACFISH strategy as a mandate, insisting that they be considered guidance in decision making. NMFS representatives have, one more time, slipped the PACFISH-PLUS standards into the proposed memorandum of agreement. As it is obvious that the Forest Service is being rolled by the other players at every turn, I deemed it necessary to write a memo to the secretary of Agriculture saying that, operating under the proposed MOA, the Forest Service will be unable to produce the anticipated salvage volume of 4.5 billion board feet by 31 January 1996. Our best estimate is for a volume of 1.5 billion or less, about the standard program level. I said that the program had been changed dramatically from the volumes estimated previously. Those estimates were made using the assumptions that the Forest Service would have decision authority in any dispute, that environmental assessments rather than environmental impact statements would be used, that there would be accelerated processes and no appeals. Under the current circumstances reflected in the MOA, we cannot produce as promised and should clearly say so in order to give industry, communities, and the politicians an idea of what to expect.

I showed Undersecretary Lyons a draft of the memo, and he was highly upset, to say the least. We argued for nearly an hour and rather intensely. He so firmly wants to believe that the warm fuzzy agreements by Washington-level brass will lead to marvelous enhanced performance in the field. Experience has told us to be very doubtful of that assumption.

I did not convince Lyons of my position, and he demanded that the Forest Service provide him a briefing tomorrow on how we could have possibly come to that conclusion. Fortunately, there is a detailed paper trail that marks the path to the conclusions that I espoused in my memo. I promised to arrange the briefing for tomorrow morning.

We then met with staff who are preparing Lyons's testimony for next week's hearings on Senator Stevens's bill to put timber first and foremost in Alaska. Regional Forester Phil Janik and staff were in attendance via conference call. I thought the draft testimony was very pointed and strong in blanket opposition to the proposed legislation. Yet Lyons asked that the testimony be made even more pointed. We discussed with Janik what Janik should be prepared to cover in his testimony. Key to that preparation was a list of the ten worst things about the legislation, with details, so that among the three of us on the witness stand (Lyons, Janik, and Thomas), we are certain to make the points as the opportunities present themselves.

From there, Lyons and I answered a summons to the White House to visit with Katie McGinty on the salvage situation. To my surprise, Katie was much more interested in my view of the overall situation concerning performing on the President's Forest Plan than the new edicts on salvage. She made it very clear that she did not want to assume any role at all in the salvage situation.

I think she may have been spooked a bit by the request from Senator Murkowski's Natural Resources Committee to appear before an oversight hearing on the salvage situation. If she is going to be the decision maker in this sticky situation, she needs to stand up and take responsibility. She really doesn't want nor need that and, appropriately enough, is getting some distance between herself and the situation.

She asked me about the morale of Forest Service troops, and I told her that morale was suffering from a combination of government bashing (including that by the president), constant upset due to the dragging out of reinvention, continued downsizing, and the chaos in the budget situation. I made it clear that I thought downsizing had already gone too far and it was time to examine that presidential mandate. She seemed sympathetic, for whatever that is worth.

At her invitation, I expressed my concerns that the Forest Service would not be able to meet the projected salvage levels of 4.5 billion board feet by 31 January 1996, due to the difficulties involved when four or five agencies have essentially veto power in the process. I put forward my conviction that it was essential to put someone in charge who has final decision authority. That archaic view of management was dismissed summarily with a smile. I then put forward my Plan B, in which all concerned agency heads are held jointly responsible for meeting the projected salvage levels in compliance with the spirit of the environmental laws and in compliance with the standards and guidelines in the forest plans. The agency heads should be jointly evaluated on their performance in this regard by a panel of undersecretaries and each will receive the same, identical rating. The agency heads will, in turn, jointly pass similar judgments on agency employees at the next level, and so on. It is time that the agencies jointly share responsibility for getting the entire job done. She agreed—in principle.

I hate these group gropes and being second-guessed by political appointees with no experience in and no knowledge of natural resources management or making a large bureaucracy function. It doesn't have to be this way. I am becoming more frustrated by the day.

4 August 1995, Washington, D.C.

Additional tensions are building over the fact that the undersecretaries and assistant secretaries in the concerned departments are the arbiters of the negotiations but expect the agency heads to be the signatories to the agreement. The intent is to deflect the political heat from where it belongs. The old law of gravity seems to rule here: crap runs downhill. I don't mind taking both responsibility and heat for matters under my control but I am getting weary of taking the heat for the fumbling and stumbling of the politicos.

This administration's lack of will or sheer intestinal fortitude to make tough decisions and then hang tough afterward is endemic. That, coupled with the unwillingness to put anybody in clear charge of anything, is producing one convoluted mess after another.

Lyons summoned Gray Reynolds, deputy chief; Robert Nelson, director of Fish and Wildlife; Dave Hessel, director of Timber; the associate deputy chief of State and Private Forestry (who is handling the staff work on the memorandum of agreement negotiations); and me to his office to explain my memo to the secretary. That memo made it clear that since the agreement being negotiated did not meet the assumptions made when salvage volume estimates were made, the estimates could not be met. The projected levels of 4.5 billion board feet are now more likely to be about 3 bbf.

Lyons did not like that and wanted to argue at every turn that everything would be OK, that things were working and it was reasonable to assume that the original estimates would hold. We held fast to our position, and as time passed, he became more and more agitated. Finally he said, "I consider what you have done unprofessional, and this memo to the secretary is a cover-your-ass document."

At that, to my later chagrin, I exploded. That was too damn much to take. It was an insult to my people, who were struggling to do a top job, and it was an insult to me in front of my subordinates. My rather vigorously delivered retort was, "I deeply resent that and I won't silently accept it. That memo was not meant to cover our ass. It was meant to cover your ass," I said, pointing at Lyons, "and Secretary Glickman's ass. The president and the secretary are now set up for huge embarrassment. Are you going to let them walk into that political buzz saw without warning?"

That calmed things down and essentially brought the meeting to a close. The final shot was that we should not have assumed in the analysis we provided to Congressman Charles Taylor that the land management agencies would have the "discretion" to make final decisions. We should have conferred with Lyons prior to having made that assumption. It was futile to explain that we had made the assumption we were asked to make.

5 August 1995, Washington, D.C.

I received a call from Associate Chief Dave Unger in the wee hours of this morning to tell me that a bomb had been detonated in front of the home of Guy Pence, district ranger at Carson City, Nevada. The bomb was either placed or thrown under the Pences' van parked in the driveway. The blast did extensive damage to the van and blew out the windows in the front of the house. At the time, Guy was on a horse trip in the wilderness with the Toiyabe forest supervisor, Jim Nelson, and the U.S. attorney for Nevada.

Guy's wife, Linda, and their two daughters were at home but were not injured in the explosion. They were moved by law enforcement officers and spent the rest of the night at another Forest Service employee's home. Forest Service law enforcement officers and numerous agents of the Federal Bureau of Investigation are on hand to gather evidence and provide security for the Pence and Nelson families.

A helicopter was dispatched at daylight to find the party in the wilderness area and pick up Pence and perhaps Nelson to return them to Carson City. Everything seems to be proceeding in good order.

Manny Martinez, director of Law Enforcement, called at about 7:00 A.M. to update me on the situation. I asked him to discern from the regional special agent in charge if it would be a problem if I showed up early in the week. Manny came by the house in late morning to give me a further update. He told me that the special agent in charge did not need my presence to complicate the situation—i.e., he would have to worry about my security, and he didn't want to divert officers to that purpose. I concurred, though my instinct is to make an appearance just to make a statement of support and concern.

I spoke to Linda Pence on the phone. She seemed in perfect control. I told her that we were concerned about her and her daughters' safety and encouraged her to let me know if there was anything the Forest Service could do to make her life a bit easier.

In pondering the situation after I visited with Mrs. Pence, I seemed to recall from visiting with Jim Nelson on my recent trip to Nevada that Pence had put in for a job somewhere and was awaiting a decision. That led me to call Deputy Chief Gray Reynolds to suggest that, as these bombings seemed to be aimed at Guy Pence personally, it might be well to promote him into another job well away from Carson City. I made it clear that this suggestion was meant to make life a bit safer and less stressful for the Pence family and should be done to reward good service (i.e., a promotion should be involved). Gray concurred and said that he would go to work to address the situation.

Somehow, dealing with attacks on Forest Service personnel and property was not something that I had conceived of as being part of the job of chief of the Forest Service.

8 August 1995, Washington, D.C.

A memorandum of agreement is being formulated under the auspices of the Council on Environmental Quality office in the White House to guide how the timber-related parts of the recently signed Rescission Bill will be carried out. These negotiations are being headed by undersecretaries and assistant secretaries from three departments, Interior, Agriculture, and Commerce, and five agencies—Fish and Wildlife Service, Bureau of Land Management, National Marine Fisheries Service, Environmental Protection Agency, and Forest Service. Agency heads have been precluded from the negotiations, and staffers are doing the grunt work and providing "input" from agency heads.

The environmental community has really worked over the White House through CEQ Director Katie McGinty, and it is becoming evident that the Forest Service and the Bureau of Land Management will be allowed none of the leeway that Congress clearly intended to bestow in the Taylor Amendment. In fact, the agreement thoroughly involves the regulatory agencies to a greater degree than ever before.

The Forest Service people working on the agreement are Bill McCleese, associate deputy chief, and Robert Nelson, director of Fish and Wildlife. They tell me that Undersecretary Lyons is doing his best to maintain some flexibility for the Forest Service. However, he is outnumbered by those with an agenda of maintaining all possible provisions of the environmental laws, and the prerogatives of the Fish and Wildlife Service and the National Marine Fisheries Service to make "jeopardy calls" if they, alone if necessary, deem such to be appropriate. In short, that means that these agencies can stop any proposed management action by the Forest Service when a threatened or endangered species is involved.

The drafts of the memorandum that we have reviewed are an absolute mess—the language is obfuscatory, confusing, and cannot be used to guide field operations. It will be necessary to turn what we think this means into clear, crisp instructions to cover field operations. Of course, that will require further negotiations among all the involved agencies. In all probability, this will evolve into jointly issued instructions by five agencies to field troops. How much time and negotiation will that take? Who knows? But one thing is certain, the clock is ticking.

My plea has been to put somebody, anybody, in charge with decision-making authority. This "management by committee" with appeals to yet other committees has all the seeds of disaster planted and watered. Worse yet, it looks bad and will draw scathing reaction from Congress.

I have asked that a clause be inserted in the document that makes all the agency heads jointly responsible for achievement of the goal—production of 4.5 billion board feet of salvage within the environmental laws in the two calendar years of 1995 and 1996. Further, I have made it clear that, as they have had no part in the negotiations, agency heads should not be required to sign the document. If agency heads are required to sign, a statement should be added that "the undersigned agency heads understand and will carry out the instructions in this memorandum." I sent word I would not sign the document under any other circumstances, as I believe that the memorandum is not in keeping with the law that the president signed and certainly not in keeping with the intent of Congress.

I *don't like* this new law and recommended that the bill be vetoed so long as the salvage rider was included. Yet when the bill overwhelmingly passed Congress and was signed by the president, it did become the law. The president's disclaimer that he didn't like the salvage rider doesn't alter the fact that he signed the bill. As a result, I want to make it clear that the Forest Service, and I, did not make the decisions that are contained in the memorandum, but we will carry out the orders issued to us by the administration for whom we work.

The word has gotten to me that the White House is displeased with Jim Lyons and me, as we are discerned as dragging our feet in carrying out the wishes of the White House. I'd like to meet "the White House" some day. Who is the White House, anyway? Somewhere there must be a person who is responsible, but maybe not. It is both frustrating and maddening to have responsibility without adequate authority.

I had breakfast this morning in the office of Interior Secretary Babbitt with the secretary and BLM Director Dombeck. Our first course of business was to discuss the status of the grazing bill that is working its way through Congress. It is Babbitt's opinion that the bill will not pass the Senate, as the proponents were unable to rush the legislation through before opposition built up. In failing to get the legislation passed by both houses prior to the congressional August recess, the chances of failure have increased greatly.

Babbitt said that he had met with the bill's primary sponsor, Senator Peter Domenici of New Mexico, and that Domenici recognized that the hubris that built up in the wake of the sweeping changes in the last election led some Republicans to believe they had support for anything they wanted to do in the name of taking down government. In the case of environmental issues, they clearly misjudged the situation. This is coupled with the misjudgment of letting the livestock industry write the legislation and not seeking any review. Domenici has been taking real heat in New Mexico, some sponsors have withdrawn, and one voted against the bill in committee.

Though the bill may go down, there will be payback in the form of oversight hearings on "lobbying activities" by government employees against the legislation. The hearings will likely be aimed at Mike Dombeck and the Bureau of Land Management, as BLM is in the Department of the Interior, and the Republicans can get at Babbitt in that fashion. I have certainly been as active as Dombeck in rallying support against the grazing bill but suspect I have been a bit more circumspect and I have left no paper trail. It is quite legal to tell people and groups about pending legislation and to discuss the political ramifications of such legislation. The people or group addressed can decide what they want to do or not do about such matters. In such circumstances it is not necessary to lobby—just telling the truth is adequate.

I discussed with the secretary the necessity to hold all the associated agency heads jointly responsible for meeting the aim of the Rescission Bill's salvage provisions to offer and award 4.5 billion board feet of salvage volume during the calendar years 1995 and 1996, in compliance with the environmental laws. He agreed and asked what he could do to help. I suggested, as soon as possible, a meeting between the secretaries of Interior, Agriculture, and Commerce; the agency heads of the Forest Service, Bureau of Land Management, Fish and Wildlife Service, National Marine Fisheries Service, and Environmental Protection Agency; and the director of the Council on Environmental Quality. The purpose of this meeting should be very narrow: lay it on the line that responsibility for achievement is joint and rests squarely on the agency heads, and that they will succeed or fail together. He said he would make that happen. If it does happen, it might be helpful.

Secretary Babbitt was very complimentary of Dombeck and me for our "steadfastness" in standing shoulder to shoulder in all of the issues that have faced the land management agencies. That was nice to hear.

9 August 1995, Washington, D.C.

In the morning there was a hearing scheduled before the Senate Energy and Natural Resources Committee chaired by Senator Frank Murkowski of Alaska. The purpose of the hearing was to ram through a bill sponsored by Murkowski and Senator Ted Stevens of Alaska to mandate a timber cut level on the Tongass National Forest, for the purpose of "maintaining 2,400 direct timber jobs from the Tongass." Without going into detail, this action would essentially make the primary purpose of the Tongass providing wood to the timber industry, all other laws notwithstanding.

Most significant are the provisions that will ensure the welfare of the Ketchikan Pulp Company, including the company's pulp mill at Ketchikan. It has been noticed, and will likely increasingly be, that Senator Murkowski holds (or has held) an interest in the Ketchikan Pulp Company and a stake of over $1 million in the bank in Ketchikan.

The administration's testimony, written by the Forest Service for Jim Lyons, was strongly opposed to the bill and couched in much more direct language than I have seen. The secretary's office actually asked several times that the testimony be strengthened. The Office of Management and Budget cleared the testimony with little comment. Finally, we have been able to stand up and resist the Alaska juggernaut.

Regional Forester Phil Janik and staff came into Washington two days ago to make sure that the Lyons-Thomas-Janik testimony team was together and well prepared. By hearing time, we were better prepared than I have ever seen. Oddly, I had the feeling that we were eager for the inquisition and prepared to fight back, for a change.

Then we got the word from Mark Rey, chief of staff to Murkowski's committee, that the formal hearing had been canceled. Instead, there would be a "workshop" of all the participants who had been scheduled to testify. Our first question was, "What's happened here?" This action is unprecedented, and Rey (Murkowski isn't shrewd enough to figure that out) had to be maneuvering to avoid something. Then aides showed up with the testimony of all the witnesses. The statements, including that of the governor of Alaska, were overwhelmingly opposed to the proposed legislation, and most rather vociferously so.

This workshop is obviously a ploy to keep those statements from appearing in the *Congressional Record*, and to save Stevens and Murkowski from having to sit at the dais and be blasted by their own folks. We wondered what the reaction of all those folks coming all the way from Alaska, flying all night on community funds, would be upon being told they were going to get to make a three- to five-minute statement off the record and then be dismissed with Senator Murkowski's appreciation.

When we showed up at the workshop, the hall was filled with some very angry Alaskans. The hearing room was set up so that the workshop participants sat elbow to elbow at the dais (where the senators usually sit), and Senator Murkowski and his chief of staff sat at the witness table.

Murkowski opened the meeting with a statement to the effect that strange, strange things were going on. Indeed. Murkowski then said that obviously there was great confusion as to what the proposed legislation would do. That was not too astute as he told all these irritated people that they were too dense to read the proposed legislation and understand it. Senator Stevens arrived and was rather obviously not a happy camper. He made his usual surly performance and then announced that he had other more important duties to perform and departed. Another public relations triumph had taken place.

Murkowski introduced the first two speakers and then left. As there were no other senators present, Mark Rey (former timber industry lobbyist and now the committee's chief of staff) took over the workshop: yet another public relations coup.

I made a very brief statement in strong opposition to the proposed legislation, making two emphatic points. First, the Appropriations Bill being debated on the Senate floor at that moment dictated a forest plan for the Tongass, i.e., much of what is being opposed in the legislation under discussion is taking place in another form and with no public input and no debate. Second, the statements by Murkowski and Rey that they were just giving us the flexibility we said we needed was just the opposite: a dramatic reduction in flexibility was proposed. I turned my place over to Regional Forester Phil Janik and left to get ready for the oversight hearings tomorrow on the timber salvage provisions in the Rescission Bill.

In my opinion, this fiasco has been Mark Rey's first major screwup, and a first-class, humiliating, and dramatic advance to the rear. The most devastating thing in this town is to make oneself look ridiculous, and

Murkowski looks ridiculous in this case. That means, I think, that he will
be even more outrageously aggressive in tomorrow's hearing.

The afternoon was spent in a final review of the memorandum of agreement for the salvage operations. The meeting took place in the Department
of Agriculture with Jim Lyons in nominal charge. There was a cast of dozens:
undersecretaries, assistant undersecretaries, lawyers from the Office of General Counsel, lawyers from the Department of Justice, staff from various
agencies, and at last, two agency heads—Rollie Schmitten of the National
Marine Fisheries Service and me.

Three significant things emerged from the meeting. The regulatory agencies will not, under any circumstances, give up their prerogative to call "jeopardy" to a threatened or endangered species from a proposed management
action; i.e., no matter how much we talk about teams and team approaches,
the regulatory agencies retain the right to have the last word. In my opinion that is exactly contrary to what Congress intended.

There is no decision as to what to do about the 318 sales. I tried to lead
them through a decision process to make the necessary policy decisions.
They were not buying any of it. In my opinion most simply don't want to
obey the instructions in the Rescission Bill. That is the next long drawn-
out decision, and we have thirty-one days left to execute the contracts. We
will, obviously, be sued by both sides in the next few days. Mark Rutzick
will file on behalf of the timber industry to force the release of all sales bound
up in geographic area of the 318 sales, and the Sierra Legal Defense Fund
will sue to prohibit sales in conflict with the President's Forest Plan for the
Pacific Northwest or the Endangered Species Act.

I refused to sign the memorandum of agreement, making the point that
I did not agree. I said I would sign only if it was clear that all that was agreed
to was that the instructions were understood and would be carried out.
Lyons left the room to get that cleared by somebody (I assume the "White
House") and came back in saying that the requested clause would be reluc-
tantly added. As an aside, he said, with some weariness, "You owe me one."
My reply was that it did not compute that way to me: the "powers" could
accept or not, and there were no favors involved.

This ain't no way to run a railroad. Tomorrow's hearing will be a turkey
shoot for Senator Larry Craig and whatever sidekicks show up.

10 August 1995, Washington, D.C.

The oversight hearing conducted by Senator Larry Craig's subcommittee of the Committee on Energy and Natural Resources went as expected. As usual, no Democrats bothered to show up, leaving the stage to the usual cast of Republicans. Alaska Senator Murkowski, appearing as the chair of the full committee, led off with a typical tirade replete with finger pointing and fist shaking and loud protestation. The first time or two I witnessed it, the performance was impressive in its own way. Now it is hard to sit through with a straight face, although it is unclear what one's expression should be. The choices that come and go are boredom, disdain, disgust, and amusement. Then the cold realization comes over me that this clown is a very powerful man, and that is a sobering thought, indeed.

I remember little of what he said, spending most of my energy in analyzing the performance in terms of body language and gestures. I think testosterone poisoning will do as a diagnosis until something better comes along.

He would be relatively easy to take on in a fair debate where he would be required to think and deal with facts. I think he knows that, and therefore almost never stays around after his opening tirades. And sure enough, as soon as he completed his statement, he departed with a flourish.

That left Senator Larry Craig alone at the dais to read the script put before him by Mark Rey. Ordinarily, Craig does much better with his script than the others on the dais. As he launches into his tirade, I find myself wondering where his trusty sidekick Conrad Burns of Montana is and hoping that he will show up. I enjoy Burns's performances, particularly the downhome humor that he interjects in his chosen role as the cowboy senator from Montana.

The senators aim their slings and arrows almost entirely at the chief witness, Undersecretary Lyons. The senators have a good case. They drive home several main points over and over, and these are good points. First, the administration has refused to give the management the authority that Congress intended to carry out the intent of the legislation. Second, the process set out to achieve the objectives is complex and there is no clear authority in charge. Third, the course of action prescribed is not in keeping with the intent of Congress and almost certain to fail to produce the promised 4.5 billion board feet of salvage by the end of 1996.

Lyons, playing with a very weak hand, did a brilliant job of dealing with the unrelenting attacks. His key point was that the legislation in question gave the secretaries of Agriculture and Interior "discretion" to give decision authority to land management agency heads and how they will comply with the environmental laws. The course of action prescribed in the memorandum of agreement is a proper use of that discretion.

The questioning made it clear that the policy decisions did not involve the agency heads, that the decided-upon process to achieve the objective involved the regulatory agencies deeply in federal land management, and that the "White House" was the wizard behind the curtain pulling the strings. That last observation will have a spillover effect on the fate of the Columbia basin assessment efforts. Those efforts are already high-centered on the absolute conviction of some that the situation will be manipulated by the White House no matter what the professionals find and the land management agency heads put forward. They saw that, or thought they did, with the President's Forest Plan for the Pacific Northwest and they see it again here. Is it unreasonable to be suspicious that the same will occur with the Columbia basin assessment?

The hearing may have had at least one good effect. The point was made that the Forest Service had the confidence of the chairman of the subcommittee (a clever ploy) and that the other regulatory agencies would override the Forest Service at their peril. This was made crystal clear in one sequence of questioning. Craig asked me if the Forest Service had competent biologists. Answer: Yes, probably the very best. Craig asked if I trusted them. Answer: Yes. In turn, Craig asked each of the witnesses if they trusted Forest Service biologists. Each witness answered yes. The final question in that sequence was to Undersecretary Lyons as to the competence of the Forest Service biologists. Lyons replied, "They are the best." Craig asked if he trusted Forest Service biologists. Answer: Of course.

In the afternoon I met with Marvin Rolienson of the Sierra Club, Marty Hayden of the Sierra Legal Defense Fund, and Mike Francis of the Wilderness Society. The subject of the conversation was the Forest Service's execution of the timber salvage provisions of the Rescission Bill. I gave them copies of the just-signed memorandum of agreement to give them a vision of the government's *modus operandi* in carrying out the law. I made it very clear that the Forest Service would do its best to meet the mandate for timber salvage within the environmental laws. They wanted me to understand that they would monitor our actions from beginning to end. I assured them that they were not only expected but were encouraged to do so.

Hayden informed me that the Sierra Legal Defense Fund would file suit against the Forest Service tomorrow to enjoin any move to cut the 318 sales. My reply was, "Good, the more the merrier." I went on to inform them that the timber industry's primo legal eagle, Mark Rutzick, had filed suit today to force the Forest Service to release the 318 sales, and all other delayed sales within the geographic area of the 318 sales, posthaste. So here we are, immediately after the latest attempt to legislate a fix for the festering situation, right back in court to get our management orders from a federal judge.

In parting, I pointed out that (outside the disposition of the 318 sales) the salvage mandate is a minor-league issue. However, planned or not, it is serving beautifully as a ploy by the supporters of the timber industry to distract the attention of the environmental community away from the grazing bill, the Tongass bill, and the land disposal bills. They listened, but I don't expect much change. They are simply too disjointed to do much more than focus on local situations that merely meet the needs of their local constituencies. They simply have not come to grips with the present; they are stuck two years in the past.

15 August 1995, Washington, D.C.

There was yet another meeting held today in the Old Executive Office Building concerning the administration's position on the execution of the timber portions of the Rescission Bill. Again, the meeting, though ostensibly chaired by T. J. Glauthier of the Office of Management and Budget, was notable for the absence of any discernible process and any clearly identified decision maker(s). I hope I am never queried in a congressional hearing or in a court proceeding as to the process of how "policy" in this matter has been determined.

I came to the meeting with the hope of instilling some rational, ordered process to determine the course of action. This hope was represented by a flow chart of decisions to be made considering the germane factors. Our work was not given any weight whatsoever.

The meeting was absolutely one-sided and was, in my opinion, an indulgence in the fantasy that the Rescission Bill's language is vague and can be interpreted in various ways. So in rapid succession, with the Forest Service in a singular minority, one "consensus" after another was reached. I was reminded of the queen's assertion in *Alice in Wonderland*, that she was quite capable of believing a number of impossible things—before breakfast.

The Fish and Wildlife Service contingent, headed by Curt Smith and Dan Barry, frankly warned the group that their field biologists would vocally and vigorously oppose any adherence to the instructions in the Rescission Bill. The statements went something like this: If 318 sales are released, biologists would be forced to render "jeopardy opinions" on any future Forest Service and Bureau of Land Management sales that involve mature or old-growth timber. The Fish and Wildlife Service would not be able to proceed with any habitat conservation plan for private lands. In addition, it is likely that restrictions would have to be placed on net fishing in coastal waters, etc.

This led to the conclusion that it will be necessary for the land management agencies to provide "substitute volume of like amount and kind." I explained over and over that the Forest Service has no such substitute volume to offer, as all such volume was part of that scheduled to supply the timber promised as part of the president's plan. Furthermore, even if there were such volume available, the Forest Service does not have the necessary personnel to prepare such sales and stay on line with the president's plan. The attendees also insisted on ignoring the bill's instruction that any substitute volume be in addition to—not in lieu of—the scheduled "green sale" volume expected under extant plans.

I was stunned, and said so, that the Forest Service was never asked about the consequences of having to adhere to the provisions of the Rescission Bill. The language in the bill is clearly a political trap engineered to bite the administration no matter what decisions are undertaken. Every paragraph is a lose-lose decision framework, and the land management agencies take the brunt of the anger and frustration produced by one side or the other and likely both.

16 August 1995, Washington, D.C.

I was briefed this morning by the minerals staff on the progress of the environmental impact statement being prepared by the Forest Service and the state of Montana on the proposed New World Mine near Yellowstone

Park. The proposed mine has become a cause célèbre for the environmental movement, which bills the mine as a threat to Yellowstone Park. The park superintendent and Interior Secretary Babbitt have spoken out loudly against the proposal. The *New York Times* editorialized against the project.

This is my second review of the ongoing assessment, and the truth seems dramatically different from the press reports. The proposed mine location is in a valley that is 90 percent privately owned and 10 percent national forest. The area is hardly pristine, having been heavily mined in the past, and it has severe mine drainage problems from tailing piles and abandoned shafts.

The valley is separated from Yellowstone Park by several miles and several ridges. The valley drains away from Yellowstone into the Clarks Fork of the Salmon River. It seems reasonable to await the completion of the environmental impact statement before condemning the proposal.

Former Senator Birch Bayh, now a lobbyist, called and asked if he could visit to discuss the New World proposal. He left a number of letters exchanged by George Frampton, undersecretary of the Interior, and the United Nations. In those letters, Frampton invited the Heritage Program of the United Nations to visit the area, at the expense of the United States, to evaluate whether or not the environmental assessments were being appropriately done and covered the pertinent questions. Such is of concern to the Forest Service, as we are the agency doing the environmental impact statement. Mr. Frampton has not bothered either to discuss the matter with me or to send me a copy of the correspondence.

That makes me wonder about several things related to this matter. Is this a deliberate slight by Interior? Or is it another example of the right hand not knowing what the left hand is doing? The other wonderment is at the lack of political astuteness in inviting a United Nations review team into the hassle. This is not, repeat not, the time to involve the United Nations in an internal United States matter. I can hardly wait to see what Senators Murkowski, Craig, Burns, Thomas, et al. will do with that—not to mention giving Congresswoman Helen Chenoweth and Congressman Don Young and others raw meat to feed upon. If the lobbyists for the mining industry have the correspondence, it can be reasonably assumed that the above-named politicos have it as well.

Mr. Bayh was very complimentary toward the Forest Service and the people working with the mining industry. Forest Service employees were praised as thoroughly professional, hardworking, competent, and not involved in politics or prejudgment as to the appropriate outcome of their work. I assured Bayh that this would continue to be the case. Again, I am distressed by the penchant of some in the government to disobey or circumvent laws with which they disagree. The 1872 Mining Law is, in my opinion, an anachronism that should have been repealed decades ago. However, so long as it is the law, federal agencies should obey that law faithfully.

6 September 1995, Washington, D.C.

Upon my return to the capital city, the number-one crisis that seems to require high-priority attention is, "What happened to the C-130s?" There was a fire on Long Island during my absence. The local authorities had difficulty fighting the fire. Evidently, there are over twenty fire jurisdictions on Long Island, all with autonomous authority and no experience in fighting wildland fire.

It turns out that Senator Alphonse D'Amato of New York has a home on Long Island and inserted himself into the issue, even to the point of being visible on the firelines wearing a fire captain's coat. Either out of a desire to be helpful or the desire for some publicity, D'Amato decided that the answer to the fire problem was a couple of C-130 aircraft equipped for retardant drops. So the good senator calls President Bill Clinton on vacation in Wyoming and demands that the president order C-130s to the fire. The president, for whatever reason, agrees and gives the order. That order is relayed to Undersecretary Lyons, who—unable to get either the associate chief or me on the telephone—calls someone he knows on the Forest Service fire staff in California to get advice on what to do. He is informed about how to properly activate the Incident Command System: the state of New York makes the request. That done, Lyons proceeds to New York at the instruction of the "White House."

The Incident Command System kicks in and works expeditiously to get the *right* equipment—including aircraft—and the right overhead team and the right crews to the fire and puts out the fire. But the C-130s don't show up when expected; and when they do show up, the pilots have exceeded their maximum flying hours and must take the mandatory rest period

Salmon fishing out of
Ketchikan, September
1995. Photo courtesy of
Jack Ward Thomas.

before they can fly missions on the fire. D'Amato goes berserk and starts
making all kinds of noise about incompetence and demands an investigation, etc. That, of course, embarrasses the president. Nobody seems to
notice that the objectives were to put out the fire, protect property, and
save lives, and those objectives were accomplished in a professional, expeditious manner.

The only thing that seems to matter is that the C-130s did not make the
evening news shows. Never mind that C-130s were the *wrong* aircraft for
the situation.

So an investigation team has been formed to find out "what went
wrong." Jim Gilliland, head of the Office of General Counsel for the
Department of Agriculture, has been named by Secretary Glickman to be
the watchdog over the review team's work, in order to ensure an outside
view. I think that translates as Gilliland is to make certain the Forest Ser-

vice doesn't whitewash the situation. Gilliland asked me to accompany him to the meeting of the review team to exemplify my (our) concern over the situation.

Talk about concern! The review team is outnumbered at this meeting by a cast of characters, besides Gilliland and me, that includes Secretary Glickman's chief of staff, Undersecretary Lyons's chief of staff, two Office of General Counsel attorneys (James Perry and Tim Obst), and a couple of others I don't recognize. With that much pressure in the room, I felt it necessary to ask for clarification as to the team's mission in order to take the heat off the team to get after "the truth." My question went something like this: "I assume we want the team to go after the whole truth, even if that means that they might say that political folks involved themselves who shouldn't have done so; and that poorly thought out orders were given and in a way that confused the issue." The overseers nodded and said that was O.K. What else could they say under the circumstances?

As I was leaving the meeting, several team members came up to tell me that they appreciated my speaking up and giving them the cover they needed to do the job that needed to be done. Before I left, I spoke to team leader Mike Rogers and told him, "Tell it exactly like it is. If there is a problem, I will take the heat." He is a solid and gutsy man and would do no other. But it is well that he knows he will get backing if it comes to that.

Picking at this is a mistake for the politicians. It will come out that they should not have inserted themselves into the picture, beyond ordering federal assistance to the state of New York. Ordering C-130s before there was any evaluation of what was needed, what was available, and the logistics involved was clearly a mistake. Ordering the C-130s made the issue one of saving face for politicians—deliver the "ordered" C-130s for a dramatic appearance on TV—rather than the proper issue of expeditiously and efficiently protecting life and property by putting out the fire.

However, there is a good reason for this review, and that has to do with the problems in getting a fast call for firefighting assistance to the right people and in the correct fashion. That needs to be fixed, and it will be.

7 September 1995, Washington, D.C.

Today is my 61st birthday. It has not been a happy birthday.

I met with Undersecretary Jim Lyons preliminary to meeting with Secretary Glickman in the afternoon. The efforts to convert my appointment from a Schedule C (political) appointment to the Civil Service's Senior Executive Service have essentially come to a dead end. The Office of Personnel Management has a rule against converting such an appointment—that's it.

When Lyons told me the news, he was obviously heartsick. I have no doubt that he made the commitment to the Forest Service and to me in absolute good faith that such a conversion would occur, and I have no doubt that all possible avenues have been explored to make good on that promise.

However, it is now time to make good on my pledge to the Forest Service workforce that I would not be part of a situation that leads to future chiefs' being political appointees. Therefore, I told Lyons that I would leave the chief's job at a time of his choosing, but not later than December 1995. I made that agreement in response to his request for a transition period to get my replacement in place.

Lyons said that I could return to my research position in La Grande if that was my choice. I expressed appreciation for that opportunity, and that may be my choice. However, next week I will check out some tentative job offers in a university or private foundation setting.

He asked me for a list of five or six potential replacements from within the Forest Service and an evaluation of each person's attributes for the job. Of course, I said that I would comply within a few weeks. And so the decision seems to be made.

Strangely, I feel nothing—not anger, not remorse, not loss, not relief, not regret. I have kept my word and have not flinched from duty and have done my absolute best. Now it is simply time to find something productive to do with the rest of my life. It will be for the best. I won't think about it anymore today.

There was a meeting in the afternoon with Secretary Glickman, Deputy Secretary Rominger, Undersecretary Lyons, Assistant Undersecretary Gaede, et al. to discuss Forest Service issues. The two primary issues were the C-130s and funding for the Quincy Library Group's plan for its area in northern California. There was new nothing in the conversation about the C-130 issue.

The Quincy Library Group—folks with various views on forest management—began meeting at the library in Quincy to see if they could come to some consensus on how the national forests in that area should be managed. Glickman and Rominger (California is his home territory) have made the political decision to give them what they want. It simply didn't buy much when I tried to introduce consideration of the consequences of shifting such large amounts of money from elsewhere to that area of northern California. I nearly choked when Lyons said we would just shift "discretionary" funds, no problem. They simply are not in contact with the realities of the budget: there is no way to pay Peter without robbing Paul. My arguments fell on deaf ears. That's what the secretary wants, and Lyons said it's a good idea and we will do it out of discretionary funds. So much for careful planning.

14 September 1995, Washington, D.C.

I met in the afternoon with Secretary Dan Glickman, Deputy Secretary Rich Rominger, Chief of Staff Greg Frazier, and Undersecretary James Lyons concerning the status of my appointment as chief of the Forest Service. I quickly ran through the story to put everyone at the same level of understanding. Essentially, I accepted the chief's job as a Schedule C (political) appointee on condition that the appointment be quickly converted to regular Civil Service within a reasonable time. That was promised in writing by Undersecretary Lyons and conveyed to all Forest Service employees. Now, twenty-one months later, that promise has not been fulfilled. Investigation has revealed that the Office of Personnel Management director has simply said that such conversion is not possible. My purpose for the meeting is to make arrangements for an orderly transition to a new chief.

That seems not to be an attractive prospect, and I was asked to respond to a proposal to vacate the chief's job for ninety days while I take on another assignment so that the position can be advertised. The intent would be for me to apply and, if all goes well, be returned to the chief's position as a Civil Service senior executive. I declined that offer on the basis that it wouldn't stand the "smell test." The Forest Service should not be subjected to such undignified and manipulative devices. It is better, in my opinion, to simply move on to a new chief. The agency does not need any such shenanigans to add to its already bruised dignity. I declined the offer.

We discussed a list of six candidates whom I proposed as my successor

Chief Thomas with Vice-President Gore, September 14, 1995. White House photo, courtesy of Jack Ward Thomas.

and my assessment of the merits and weaknesses of each of those candidates. I reiterated that it was my wish to make the transition as smooth as possible, but it should be complete by the end of December.

The secretary said that left two choices: accept the situation or elevate it to the office of the president. To elevate the situation would, without doubt, produce strain between USDA and the Office of Personnel Management. Top USDA staff will discuss that and make a decision.

In the evening, I attended a reception at the home of the Vice-President and Mrs. Al Gore. When I visited with the vice-president, he already knew of the situation and said straight out, "We do *not* want you to leave. We will keep our promise. No ifs, ands, or buts"—just those words. I told his aide that I needed an answer by Wednesday of next week, to which she agreed. So things continue to drag out.

15 September 1995, Washington, D.C.

Greg Frazier, chief of staff to Secretary Glickman, called and asked if I would accompany him to a conference of staff aides trying to reach compromise language for Alaska Senator Ted Stevens's amendment to the Appropriations Act concerning a dictated solution to forest planning in Alaska. There were staffers from the Appropriations committees of both houses, aides to the Alaska delegation, T. J. Glauthier of the Office of Management and Budget, Greg Frazier, and me.

The Alaska delegation's staffers laid out their case, replete with distortions and outright errors. They have spouted this rhetoric so long that I think they believe it. I carefully put out a view of truth as we saw it and explained technical matters. Finally, an aide of Congressman Don Young's who was seated to my immediate right launched into a tirade against the Forest Service, the administration, and me, replete with table pounding and a raised voice. I answered back with some vigor and my own table pounding, saying that there were alternative explanations of what had happened. He said, somewhat under his breath, that one-half of the timber jobs in Alaska had been lost under my tenure. Then he said, "You son-of-a-bitch, you will get a chance to make your explanations to our committee."

That did it. I was infuriated and got in his face with an offer to kick his ass. We were both on our feet and nose to nose before good sense began to overrule emotion. I felt more than a bit embarrassed, but I have tolerated more than enough bullying from the Alaska delegation and I was certain that taking such from a congressional aide was beyond tolerance—my tolerance, anyway.

The outburst had, if nothing else, a remarkably sobering effect on the tenor of the meeting. There was a sudden switch from rhetoric to getting something done. Senator Stevens's aide was asked to handle the negotiations. To my great surprise, there was no argument as to letting the Tongass Land Use Management Plan process continue to completion if Alternative P was instituted for two years.

Mark Rey was concerned that none of the alternatives being developed would include the timber yields anticipated under Alternative P (about 425 million board feet). I told him that I was sure that the alternatives would bracket that annual sale quantity level. He seemed satisfied with that and seemed to intimate that if such were the case, Congress would simply legislate whatever alternative the Alaska delegation deemed appropriate.

My final suggestion, whispered to Greg Frazier, was to go for language stating that Alternative P of the 1991 plan would be in force until a new alternative is put in place by the TLUMP process; i.e., don't give away two years under Alternative P, which the Alaska delegation will claim means that we should have put 425 mmbf of timber up for each of those two years. We anticipate bringing TLUMP to a close in midsummer 1996, cutting the operations under Alternative P to under a year.

I made it very clear (and must remember to follow up in writing) that saying we will operate under Alternative P does not mean that the Forest Service will put anything near 425 mmbf on the market for at least three years. Rey said that they need a stable, predictable timber supply so that a buyer can be found for the Wrangell sawmill. This action won't do that, unless Rey anticipates legislation that will ensure a similar timber sale level once the alternatives arise under TLUMP. That is why he wanted to ensure that the alternatives being considered contained levels on each side of the current Alternative P.

As the meeting ended, I apologized to Young's aide and to the assembly for my outburst of temper. My antagonist did likewise. But maybe saying, through action, that enough is enough was useful.

Upon arriving at the office, I learned from a phone conversation with Jim Lyons that Senator Stevens was pushing an amendment to the Appropriations Act that would cut off funding for the USDA undersecretary's Office on Natural Resources and the Environment. This is the same course of action that Senator Stevens had threatened me with earlier in the year if I did not see things his way. I reacted with some anger and nothing happened. Now he seems determined to take on Jim Lyons.

26 September 1995, airborne,
en route from Lafayette, Louisiana, to Washington, D.C.

I have spent the past two days at the annual meeting of the National Association of State Foresters. These folks are the Forest Service's primary partners in the State and Private Forestry area. Most of them are facing the same problems within their states that the Forest Service is facing at the national level: downsizing, budget reductions, and trying to satisfy conflicting political demands.

Given these problems, it is not too surprising that a certain amount of petulance directed at the Forest Service arose from time to time during the meeting. For example, Conrad "Connie" Motyka, state forester from Vermont, gave the opening address, a well-thought-out wakeup call as to the deteriorating situation of government forestry programs at all levels across the United States. In the process, he took a few gratuitous shots at the Forest Service. The more significant shot was that State and Private Forestry had taken a 20 percent reduction in budget compared with an overall cut of 10 percent for the agency, and somehow the Forest Service was responsible for that situation.

The State and Private Forestry cuts are particularly painful for the state foresters, because the cuts were concentrated in the pass-through areas of the Stewardship Incentive Program (money used to provide incentives to private nonindustrial forest landowners to execute beneficial forestry activities) and the Legacy Program (used for the purchase of development rights to maintain property in forest uses). This federal largesse is the meat and potatoes of some state forestry organizations.

The frustration is understandable. Laying the blame at the door of the Forest Service is less so. The Forest Service recommended modest increases in the budget for State and Private Forestry, including the Stewardship Incentive Program and the Legacy Program. Those recommendations were carried forward in the president's budget submitted to Congress.

The appropriations committees in both House and Senate did not see things the same way. These committees were limited by the rules of operation to a prescribed level of spending, and that was less than the president's suggested budget. The overall cut for the Forest Service was determined to be 10 percent. Then the committees began to set their priorities within that ceiling—i.e., they began to shift money around between line items to meet their individual and collective objectives.

Areas of emphasis were commodity production items, particularly the cutting of salvage and green timber sales. The offset was largely concentrated in "subsidy" programs to private landowners—the Stewardship Incentive and Legacy programs. This attack on subsidies was in keeping with the philosophical positions of the newly dominant Republicans on the subcommittees with purview over the Forest Service budget. Also caught in the pinch were the Forest Service efforts (which were originally mandated by Congress) in the International Forestry arena. The president's budget recommended a $10 million program, which was an increase of $6 million over the 1995 budget level, and Congress reduced funding to zero.

The state foresters were warned about what was happening in Congress. The political clout to take on that issue, almost by definition, was their arena. Their efforts were only marginally successful. Now it is intimated that this failure should be laid at the feet of the Forest Service.

In response to their criticisms, I told them what had happened and said that the Forest Service would not, even if we could control the budget process of the Congress, recommend across-the-board budget reduction approaches. That approach is a no-brainer, and if repeated enough times, merely results in an across-the-board ineffective program. In my opinion, I said, it is better to set priorities and end up with stronger, if fewer, programs. I was frank in saying that we are undertaking strategic planning sessions looking at further budget decreases of 30 percent or more. And in those exercises, the conclusion was that the Forest Service would drop back to the stewardship of those lands entrusted to us by the American people— the national forests and grasslands.

That did not play well with the audience and I was closely questioned on the relative value of state and private lands and the national forests and grasslands. I did not retreat from the obvious reality that the Forest Service's core mission, when all else is stripped away, is the stewardship of the lands in our care. Asked to justify that conclusion, I responded that the owners of private or state land (or any land) are primarily responsible for that land, including financial responsibility.

The point is that the state foresters and those concerned with the national forests and with Forest Service research should concentrate on ensuring that the overall Forest Service budget does not deteriorate further. Talking about how to divide up significant reductions is a losing game over the long run. It is just one more wedge being driven between factions in the conservation monolith. From the response of the state foresters, this wedge has the potential to create further tension (as opposed to cooperation and mutual support) between state governments and traditional federal partners.

Some of the state foresters have decided to at least nibble at two lures that are being cast by the masters of the wedge technique. The first of these is the glittering idea of the block grant. In this approach, the portion of the Forest Service budget that goes to State and Private Forestry should be parceled out to the states in block grants with few or no strings attached.

The problem with that approach is that the block grants will be gradually—or suddenly—terminated as the old lesson is learned again: that no politician will transfer money, and the power and political credit that go with largesse of the public purse, to another politician for long.

The second idea or threat part of the equation is that the Natural Resources Conservation Service might better serve as the conduit of money and expertise from the federal level to the state foresters. The assumption in that proposal is that NRCS would be more attentive to the state foresters than the Forest Service. Besides, there is an undertone from some state foresters that the Forest Service is becoming less concentrated on "multiple-use management" (interpreted as emphasis on timber production) than they deem appropriate. That idea may have merit as NRCS has the unadulterated mission of influencing conservation of all kinds on private lands of the United States.

But the state foresters would likely be minor players when mixing it up in the political games with NRCS's traditional constituencies of agriculturists and graziers. Besides, as one state forester said, it may be best to "dance with the one who brung you to the dance."

30 September 1995, Washington, D.C.

This has been one of the more hectic weeks of my tenure as chief. The early part of the week was spent visiting with key congressional people in an effort to dissuade them from carrying out the Senate Budget Committee's resolution to "defund" Undersecretary Lyons's oversight of Forest Service activities. Those efforts, by a number of folks, proved successful in the conference between the House and Senate budget committees, and the provision was deleted. I am left wondering, however, how much political capital was spent in the effort and what sorts of promises have been made to protimber senators and congressmen to fight off the attack. I smell that Mr. Lyons was saved by some commitments by the secretary to take on more personal oversight of the Forest Service. If that hunch is correct, we will begin to see the secretary's immediate staff (Greg Frazier and Anne Kennedy) have more and more direct contact with Forest Service activities.

We were also successful in holding on to $4 million to complete the Columbia basin assessment. I interpret the language in the Budget Bill as saying that local planning (i.e., one record of decision for each national forest, as opposed to the much more efficient course of one overarching document) is the appropriate course of action, regardless of vastly increased costs in money and time. It is clear that another intent is to delay and thwart any significant land-use management decisions until after the 1996 presidential elections. This is based on the fear of Republican senators and congressmen that the Forest Service and Bureau of Land Management officials will not make the decisions. They suspect that, as in the case of the President's Forest Plan and the memorandum of understanding for carrying out the forest salvage and 318 sales release provided for in the Rescission Bill, the decisions will be made by White House political appointees—the undersecretaries and assistant undersecretaries in the departments of Interior, Agriculture, and Commerce. I believe those fears to be unfounded. However, the distrust and fear are palpable, and maybe those fears are justified.

Then, on Wednesday, a regional director of the General Services Administration instructed the Forest Service to take possession (i.e., ownership) of all air tankers acquired under the provisions of the Antique Aircraft Act by the Forest Service and turned over to contractors that provide firefighting services.

As there are no pilots or service personnel in the Forest Service to fly and maintain these aircraft, this action would remove about half of the air tanker fleet just as the firefighting season is coming on in California. The regional director of the General Services Administration evidently acted on instructions from a regional official of the Justice Department without consulting with its national offices or the Forest Service. I told our people not to comply until they received orders from me. The director of the General Services Administration was taken aback at the information and got more so as the pragmatic and public effects of the decision were explained to him.

By the end of the week, the action had been put on a more deliberate course, and our air tanker capability had been retained for the moment. More information surfaced as the week went on. Evidently, a federal grand jury had returned a true bill on charges against some air tanker contractors and perhaps one or more Forest Service employees for improper actions in the procurement and contracting procedures.

I thought this problem had been completely dealt with just as my watch as chief began in late 1993. My staff, at the moment, has no idea as to what the problem(s) may be. Once these screwups occur, they assume a life of their own and distract from the ability to move past old mistakes and address the problems of the moment.

5 October 1995, Washington, D.C.

There was a meeting today at the Council on Environmental Quality to consider whether to appeal a federal judge's recent decision on the meaning of the timber salvage provisions of the Rescission Bill. The government argued that the language was ambiguous and the bill did not intend to conclude *all* sales that had been made "within the geographic range of the 318 sales" and then withdrawn or otherwise held up for environmental reasons. The judge came to the inevitable, in my opinion, conclusion that the language was clear, pointed, and difficult to misinterpret. That was clear to me, and I so argued, before the case went to trial. It is even more clear to me now.

The Department of Justice lawyers are very explicit that there is very little chance of the government's prevailing on appeal. A week ago, the departments of both Agriculture and the Interior had recommended that the decision not be appealed. Obviously, something has happened between then and now—that did not involve either the Department of Agriculture or the Forest Service—to change that picture. The Department of the Interior, backed by Commerce (which includes the National Marine Fisheries Service) has reversed the decision not to appeal.

Undersecretary of the Interior George Frampton led the discussion. The arguments are strictly political: it is essential to keep faith with "our constituents," which, it was clear to me, was the environmental community. I said that it was difficult for me to understand how Congress can overwhelmingly pass legislation that is then signed by the president—with stated reservations—and the executive branch can then argue that it should not be carried out. Crossing one's fingers and yelling loudly that you don't agree with what you are doing as you sign a contract doesn't make the contract any less binding or legal. The arguments being made here are to the contrary. When I pointed out the very clear contradiction, the cold stares were dominant. Mine was not the politically correct posture.

Before the debate could even develop, George Frampton sought to cut off any in-depth discussion by declaring that he had visited with Secretary

Babbitt only a few hours earlier and had been informed decisively that the secretary had decided to appeal the judge's decision.

I could feel my color start to rise at that edict. Here we were again in a position where the Department of the Interior was making decisions for the Forest Service (and the Department of Agriculture) with absolutely no consultation.

Before I could say anything, James Gilliland, the Department of Agriculture's general counsel, spoke up in a low southern voice and firmly made it clear that Secretary Glickman would have equal voice in that decision, and Glickman had not yet decided. Now there was a change in the political calculus—finally. In two years of struggling with these issues, this was the very first time that I have seen the secretary of Agriculture not simply defer to the "White House" or the secretary of the Interior. Gilliland went further and said that appealing a court case in the absence of a good legal position was, in his opinion, both bad legal and political practice. Such would produce a bad result and was likely to be considered irresponsible by the court.

Now the debate began in earnest. The overriding objective was to protect the efficacy of the President's Forest Plan. My jaw dropped at that. Before the president signed the Rescission Bill, I had made the point, as loudly as I could, that the 318 release language would imperil the efficacy of the president's plan. *Now* the other players get excited. *Now* the National Marine Fisheries Service has decided that release of these timber sales will cause "jeopardy" to threatened fish species. *Now* they come to this conclusion. That had to be obvious before the president signed the bill.

This poses a most interesting legal question. Can a regulatory agency charged with enforcement of the Endangered Species Act call jeopardy on the consequences of a law passed later? Did the president understand from the Fish and Wildlife Service and the National Marine Fisheries Service the likelihood that in signing the Rescission Bill he was putting the president's plan at risk? Either he did know and signed it anyway, or he didn't know and should have. Neither looks good.

Frampton then fell back to the political argument that it was essential to appeal to keep faith with "our friends" or "our constituents." Such an effort was deemed essential to demonstrate, even in an almost certain losing situation, resistance to the timber industry and its friends (Republicans) in Congress.

That was a bit too much for me, and as in previous meetings, I pointed out that the Forest Service had been selling timber to industry for nearly a hundred years and that people in the timber industry are our customers and part of our constituency. I went on to emphasize that the Forest Service was by custom and law dedicated to the multiple-use doctrine, and such included the cultivation and cutting of timber. That seemed to be something of a faux pas, committed in mixed company.

The argument then evolved into one of the degree of "irrevocable damage" that the release of these timber sales would cause to one threatened species or another. Will Stelle, northwest regional director for the National Marine Fisheries Service, had submitted a formal written response to some question posed by the Justice Department, which had requested an oral reply. The question concerned the effect of the 318 sales on threatened species of anadromous fish. Stelle's carefully written response was that most of the sales would have significant negative consequences on the survival of listed species and the efficacy of the president's plan. That letter, once leaked to the legal eagles for the environmentalists, will help ensure that any legal action predicated on the efficacy of the president's plan after the 318 sales are cut will succeed. Nice move, or stupid blunder?

Stelle is a lawyer and it is difficult to believe that he was not fully aware of the potential consequences of his written reply. Peter Coppelman of the Department of Justice was a tad upset, but that won't change anything.

The meeting ended, as most such meetings do, with a definite decision to have another session on the subject at hand. As is too often the case, it is not clear who is in charge, who will make the decision, and who will stand responsible for the consequences of the decision.

6 October 1995, Washington, D.C.

Today was my day off and I just came by the office to sign some documents before leaving for New York for a long weekend. Anne Kennedy, assistant to Secretary Glickman, caught me in the office and insisted that I be in on a 2:30 P.M. meeting in Secretary Glickman's office.

I arrived at the appointed time for a meeting whose purpose was unclear, at least to me. Key players present were Secretary Glickman; Assistant to the Vice-President Katie McGinty; Undersecretary Lyons; Undersecretary of Commerce Doug Hall; Undersecretary of the Interior George Frampton; General Counsel for Agriculture James Gilliland; Lois Schiffer, Department of Justice; and ten or so others.

To my surprise, Mr. Glickman turned to me and asked me to bring the group up to speed on the status of the timber sale programs (both green sales and salvage) in the Pacific Northwest and across the country. By sheer serendipity, I had in my possession the required data, which I was delivering to Anne Kennedy in response to a week-old request.

The gist of the briefing was that the Forest Service was on track to meet the green and salvage levels projected for 1995 and 1996. However, I warned that the Forest Service now believed that the sale level for the President's Forest Plan for the Pacific Northwest would likely be 10 percent below anticipated for fiscal year 1997. This shortfall would result from the alterations made in the record of decision to Option 9 put forth by the Forest Ecosystem Management Assessment Team. These changes included increases in buffer widths along ephemeral streams from 50 to 100 feet and including the buffers as protection for territorial vertebrates associated with late-successional forests. This meant, to me at least, that it was unlikely that these buffers would ever be modified, or if modified, only slightly and after much trouble. This meant that the yield estimates in the original Option 9 were now at least 10 percent too high.

My repeated warnings to that effect at that time were ignored, as some wanted desperately to believe that a free lunch was forthcoming. Besides, back then the day of reckoning was at least four years away. Now that time is two years away and the cold truth is beginning to dawn.

However, Undersecretary Lyons has the Forest Service at work analyzing the situation and making recommendations as to what might be done to ensure that the timber sale program would be on target for 1997. I said that such an analysis was almost complete.

At that point, George Frampton abruptly shifted the conversation to the decision to appeal the court's decision on the release of the 318 sales. Nearly all the arguments that had been put forward in the 5 October meeting were rehashed. Frampton made it clear to the secretary of Agriculture that Secretary Babbitt had made the firm decision to appeal.

Mr. Glickman interrupted Frampton and said that he had not made up his mind on that matter, and it would be necessary for him to talk to Mr. Babbitt before a joint decision was made. Oh, happy day!

Now the conversation shifted back to the 318 sales situation. I pointed out that based on conversations with Boise-Cascade Corporation officials, it was likely that modifications could be worked out in most of the sales of real concern that might make them much less damaging from the environmental standpoint. I emphasized that these negotiations would have to be conducted in a collegial fashion. Industry is convinced that they have a winning hand in court and are not inclined to negotiate under the gun. I suggested that, as the advice from our attorneys was that the government had almost no chance of winning an appeal, it might be well to work with the purchasers of the sales to get a better job on the ground.

Then, trying to appeal to the political instincts of the assembled under-secretaries, I pointed out any political benefits to be realized from an increased flow of timber to Pacific Northwest mills would have to occur between now and November 1996. If there is any hope of a positive political effect, it is well to release the 318 sales, end that controversy, and get more timber into the mills. Oddly enough, that timber will flow into the economies of Washington, Oregon, and California, three must-carry states for President Clinton. When all other arguments fail, point out the political ramifications of a pending decision—whether good or bad—which will have a decided propensity to get attention on the issue. Once that attention is focused, it is much easier to get attention on the technical programmatic issues and to see that those are considered.

This meeting, which was a good one in my opinion, if for no other reason than having Secretary Glickman enter the fray, ended with the decision to meet yet again on the issue of the appeal some four days hence. I thought—or rather, hoped—that I saw a significant change take place in the dynamic of policy decisions involving National Forest System management. Finally, the secretary of Agriculture has claimed his prerogative to be a player on matters concerning the agency. If that perception is correct, then there may be some shift toward a more pragmatic and commonsense approach to federal land management than has recently prevailed.

10 October 1995, Washington, D.C.

As an adviser to the secretary of Agriculture, I seem to be on a real losing streak. This morning, special assistant Anne Kennedy called to tell me that the decision has been made to go with the negotiated settlement in the

Mexican spotted owl case (Silver v. Thomas). The Department of Justice, working with Forest Service people in Albuquerque, negotiated the settlement with the victorious plaintiffs.

In my opinion, and those of my legal advisers, the proposed agreement contained three real clunkers to which we strenuously objected. The Department of Justice is, in my opinion, almost always too eager to settle legal actions, particularly when plaintiffs are of the environmental persuasion. It was a shock to my system to find that the Department of Justice does not consider the Forest Service a client. They have little concern as to the desires of the Forest Service or any other agency. They set their own course and in doing so are de facto setters of policy. Somehow, that seems to be a serious flaw in the system. But for now, at least, it is the system.

The first error is in agreeing to language that the Forest Service "admits" that the case is lost and that consultation with the Fish and Wildlife Service concerning threatened species must take place at the forest plan level. There is no reason to admit to anything. I would "recognize" but wouldn't demean the agency by admitting to anything.

The second error is in committing to no management activities on some designated lands for a period of fifteen years. That agreement is, in my opinion and that of my counsel, a violation of the National Forest Management Act, which requires land allocation to be made through a carefully prescribed land-use planning process. The attorneys for the Department of Justice disagree, arguing that such deals have been made before in the Southwest.

The third error is to obligate the Forest Service—through the Rocky Mountain Experiment Station—to conduct a "workshop" of appropriate scientists and experts on management of old-growth forests of the Southwest. This seems quite reasonable on the surface. But what lies beneath the surface? In reality, this would cause an enormous effort to be put in motion to address all currently listed species (over thirty) plus all other species thought to be associated with old-growth forests. This, given the Pacific Northwest experience, would quickly evolve into a large and complex undertaking that will certainly produce "new information" that would then have to be melded into revamped forest plans. That is complex and difficult and, given present circumstances of personnel and funding, could not be quickly accomplished. That likely failure would, in turn, provide the rationale for yet another sequence of legal actions that would likely bring land management activities to a halt.

Do we have to repeat the entire sequence of events that is still being played out in the Northwest? Insanity has been defined as doing the same thing over and over and expecting a different result. But in this case, we are headed right down that path. I think the problem is one of no institutional memory. There are very few people around who went through the saga of the Pacific Northwest. Even most of the government people who went through it never grasped the big picture. But I guarantee that the environmental lawyers who successfully fought those legal battles are still in business at the same places, and I further guarantee that they understand perfectly the big picture and the game to be played.

Anne Kennedy listened to my reservations and informed me, once again, that the decision has been made to accept the negotiated settlement that promises peace in our time. She was a bit agitated that I continued to press for a better job of negotiation.

Understandably, she was concerned for the welfare of the people whose jobs are on the line if the injunction continues. I believe that she and whoever else was involved in the decision are quite sincere in that concern. And of course, these are the political ramifications of the continued injunction.

The issue of jobs has evolved into the perfect political weapon in all conflicts over allocation of natural resources. The potential consequences of any action are measured in the currency of gains or losses in jobs.

This, however, is the first time I have seen the ploy used to justify a decision in favor of the environmental side of an argument. In this case, we settle the case on less than favorable terms in order to protect some jobs in the short term.

This decision, which buys very little in the short term—a few jobs for a few months—will come back to haunt the Forest Service and the Southwest over the next decade as the path laid out in the Northwest extends into other regions. But the short-term political considerations override those concerns. And maybe they should.

14 October 1995, Washington, D.C.

Yesterday I met with Greg Frazier, chief of staff to Secretary Glickman, concerning the hot issue of the Forest Service's moving into roadless areas in Montana for purposes of salvaging timber and letting leases for oil and gas exploration. None of the areas under consideration for entry were identified in the roadless area review exercises—RARE I and RARE II—as suitable for wilderness classification or wilderness study areas. Those reviews were conducted during the Carter administration, and areas not identified as wilderness study areas were allocated to the multiple-use management category, for eventual roading and timber harvesting. Many such tracts have since been roaded and logged—they are not roadless.

All of the Montana "roadless" areas were included in the multiple-use category through the processes described in the National Forest Management Act. Then, each roadless area to be entered was the subject of a full-blown environmental impact statement with full opportunity for public involvement and political participation. Most of the forest plans have been in place eight to ten years.

The politicians have, under both Republican and Democratic administrations and Republican- and Democrat-dominated Congresses, had every opportunity to address the question of the classification of these roadless areas. The politicians have either shied away from the issue or been unable to effect a lasting political solution.

Nearly eighteen months ago, I sent out instructions to regional foresters either to remove contentious roadless areas (essentially all are contentious) from the timber base via forest plan amendment or to proceed forthwith to enter those areas for purposes of timber management—i.e., initial roading and logging. I felt it necessary to issue such instructions because most Forest Service managers had, reasonably enough, shied away from the controversy inherent in entry into roadless areas. Instead, the managers met their timber targets by cutting at a rate projected to be sustainable but doing most of that cutting in already roaded areas. In other words, it was assumed that eventually some manager would enter these roadless areas to meet planned timber outputs. Those decisions were put off time after time for the "night crew" that would have the misfortune of being on watch when the day of reckoning finally arrived. Now that time has come.

When the Forest Service made moves to enter roadless areas in Montana that were clearly and legally identified as part of the timber base, Congressman Pat Williams objected strongly in a letter to Secretary Glickman.

The secretary reacted positively to Mr. Williams without, I think, understanding the situation fully and without appreciating the consequences of such a decision. Senator Max Baucus, who is facing a tough reelection fight, backed Pat Williams's request for a delay of entry into roadless areas. But that was done in the form of a personal note on plain bond paper (not letterhead) and with no public announcement.

I went through all of this in detail with Frazier, who quickly grasped the larger problem and the legacy left to the night crew. To his credit, Frazier agreed that any decision to halt ongoing moves into roadless areas for management as requested by Congressman Williams was clearly a political decision and should be openly identified as such. He asked me to draft two letters for the secretary's consideration.

The first of the letters would comply with Pat Williams's request. The letter would point out the consequences of such a decision on the amounts of green and salvage timber that would be sold over the next few years, the investments in sale preparation that would be forgone, effects on timber supply as projected in the forest plans, and effects on standing decisions to proceed with oil and gas leasing. This moratorium on entry will be for a certain period (which was not specified at this point). Then, if congressional action had not taken place by that certain time, entry into the roadless areas would proceed. Further, the secretary would not go forward with this decision unless at least one of the Montana senators (Max Baucus or Conrad Burns) and Congressman Pat Williams came out *in full public support* of the secretary's decision. I doubt he will get that support. Senator Burns is philosophically committed to expeditious exploitation of natural resources and has no truck with environmental types: he has nothing to gain. Senator Baucus, given his tough reelection fight, seems unlikely to publicly support any action that can be painted as costing jobs in Montana. Besides, given the likelihood of his opponent's being a hard-right conservative Republican, he has no need to court environmentalists: they simply have nowhere else to go.

Pat Williams is a tough man and seems quite willing to take a stand on principle. Baucus seems more of a political opportunist with very few core values beyond reelection to the Senate. It seems likely that he will straddle the issue by privately encouraging us to defer and saying nothing publicly.

The second letter to be prepared will announce the decisions will be left to the chief of the Forest Service (actually the regional forester) to be dealt with as prescribed by law, regulation, and applicable land-use plans.

My advice, and that of the regional forester, is to proceed with the second letter. I elaborated by telling Frazier that the secretary was getting pulled more and more into making decisions that should be left to the chief of the Forest Service. Making such land-use decisions at Cabinet level is nearly always a political loser over the long term and results in bad natural resources management as well.

Elevating such decisions above the chief's level sets a precedent that leads to a rush to elevate more and more such decisions to the political level, thereby bypassing the decisions being made on technical and legal bases. If this continues, the secretary will become mired in such decision making with the attendant problem that other politicians would much rather have such decisions made by a fellow politician than by a natural resources professional.

A bow to Mr. Williams's request would exemplify why the timber industry (and other industries) become so peeved at the extant system. Nearly all such decisions outside the prescribed system favor the environmental advocates. The best that commercial interests can expect the Forest Service to do is to follow the approved forest plans.

Environmental interests, on the other hand, can and do exert pressures (political and legal) that result in temporarily or permanently not proceeding with the parts of plans having to do with outputs of products. That reality is producing a more and more volatile situation.

17 October 1995, Missoula, Montana

The University of Montana's dean of forestry, Perry Brown, and Acting Boone and Crockett Club Professor Dan Pletscher were my luncheon companions. The subject of conversation was my interest in the Boone and Crockett Club Professorship and, if I am interested, under what circumstances and in what time frame.

This professorship is a real plum and, in all modesty, well suited for my talents, tastes, and experience at this time of my professional life. The University of Montana is the right size for my tastes, and Montana as a place to live and work suits me well.

I pointed out the circumstances concerning my professional status and the conversion of my appointment from political appointee to the career Civil Service. I told them that I expected a final decision from the secretary of Agriculture some time this week. If the answer was that no conversion was possible, it was my intent to resign (retire) effective before the first of the year. They indicated to me that it might be possible to accelerate the selection process to accommodate that schedule if I was the selection of the search committee.

In the midst of the conversation, a waitress came to the table to tell me that the secretary of Agriculture was on the phone. Mr. Glickman told me that there was simply no way that my appointment could be changed that would not bring the possibility of ridicule on the Forest Service, that the process could become the focus of disagreement by the Office of Personnel Management or the Association of Senior Executives or both. In short, there appears to be no way that the promises made to me and to the Forest Service employees at large by the administration (i.e., Undersecretary Lyons) can or will be honored.

Secretary Glickman then told me that he had discussed the situation with President Clinton and Vice-President Gore. They both felt that my departure, particularly at this time, would be a political blow to the president. Further, the president said that such would be most unfortunate for the Forest Service in both the short and the long term.

Then Mr. Glickman said, "The president has asked that you remain in place as chief until the election. At that time, whether the president wins or loses, a career senior executive from within the Forest Service will be named the fourteenth chief." The secretary said that he was sympathetic to my desire to keep my word to Forest Service employees that I would not be a party to permanently converting the chief's position to the status of a political appointee. However, he considered it my duty to look at the situation in a different light. The present mess did not result from bad faith but from the hubris of those newly in power and the absence of good staff work. There was absolutely no fault that could be attributed to me.

He now considered it critical for me to put ego aside and do what is best for the Forest Service and the country. He asked, "Where is the greater good?"

Damn! I was reminded of a problem from some basic philosophy text.

When it is not possible to do right, how does one do the least wrong? I really resented being placed in such a position. Yet I know that this present situation did not result from any chicanery. So now what is the right thing to do?

30 October 1995, Washington, D.C.

There was a meeting today at the Council on Environmental Quality concerning the environmental impact statement on the New World Mine. The EIS is being jointly prepared by the Gallatin National Forest (which is adjacent to Yellowstone National Park) and the state of Montana.

The entire issue is a political hot potato and an increasingly visible test of the balance between development and environmental protection. Interior Secretary Babbitt and associates in his department, particularly the National Park Service, have been vocal in opposing the mine.

The Forest Service has been consistent in insisting on strict adherence to the system prescribed in law and regulation. This has positioned the Forest Service, as usual, as the odd man out in the top echelon of government that places political agendas over adherence to prescribed processes.

This meeting represented no change. It was quickly obvious, to me at least, that the stage was being set to reject the New World Mine. As discussion began to develop, there was a drift toward what decision should come out of the ongoing EIS process. At that point, I balked.

I noted that the forest supervisor on the Gallatin National Forest was the designated decision maker and that he should and will be given that authority. As long as I am chief, there will be *no* political messing around with the prescribed system. If the forest supervisor is not allowed to make that decision, it should be openly acknowledged and the decision maker clearly identified.

The room became very quiet. Though no one chose to debate the issue, my impression was that no one agreed with me. With that, the discussion ended. I don't think that this is the end of it. This issue is a political litmus test that will come fully to fruition during the upcoming presidential election year.

3 December 1995, Washington, D.C.

I have been unable to write in this journal for about the past month because of arthritis in the thumb of my writing hand. So the entries for last month are only approximate in terms of the dates and not necessarily in the correct sequence.

[nd] November 1995, Washington, D.C.

I met with Undersecretary Lyons today on several matters, including my performance evaluation. Although no ratings of individual performance elements were below "fully satisfactory," the evaluation was the worst I have received in a twenty-nine-year Forest Service career. I was sorely disappointed, did not agree with the rating, and bluntly said so. My relationship with the undersecretary continues to deteriorate. We simply do not see eye to eye on how the Forest Service should be managed and on the evaluation of individual performances of key senior staff. He announced that he was removing two of the senior executives from the list suggested by the Forest Service of those to receive bonuses. Further, he was substituting two others of his choosing in their places, who are certainly well qualified for bonuses.

One of the people removed from the bonus list, Deputy Chief for Programs and Legislation Mark Reimers, has been a target of Mr. Lyons's for all the time I have been chief. The animosity seems to stem from efforts Reimers ostensibly made to protect Chief Dale Robertson and Associate Chief George Leonard from being removed prior to my appointment. Reimers was probably guilty of that, as were hundreds of other Forest Service people (including me). Lyons continually comments that Reimers and his staff "do not serve [Lyons] well." Nothing the chief and staff say in this regard shakes Lyons's determination to go after Reimers and his top associates, particularly Steve Satterfield, who handles budget affairs. He decided not to override the recommendation of the Forest Service for a pay level adjustment for Satterfield, probably to throw a bone to me, as I was becoming livid at that point in the conversation.

He then made the point that "our travel" (his and mine) should be coordinated so as to achieve maximum effect in getting "our message" across. I did not argue the point, though I wondered about how to coordinate travel between us and maximize "our message" when I have no idea what his travel schedule is or what may pop out of his mouth in front of various audiences.

I know what the Forest Service message is, and there is no advantage to the Forest Service or to its message by being associated with Undersecretary Lyons. He simply has lost his support on both sides of the aisle in Congress and within the Forest Service.

Now that the administration and Mr. Lyons have produced a thoroughly convoluted situation for management of the national forests with the dramatic mistake of agreeing to the Rescission Bill and its timber salvage provisions, it is apparent that the Forest Service will be set up to take the brunt of criticism from both extremes in the issue. I resent that, and I think the rank and file resents it as well.

Mr. Lyons seems back in the saddle after maintaining a low profile for the past several months. He informed me that he wants a full-time public affairs officer (and he has a candidate from the Department of Defense) and a full-time speechwriter and a liaison staffer to replace Gary Larson, who is going to the field. The problem is that no one in the Forest Service will volunteer to work in that office. Another problem is that the rank and file wonders why, in view of Washington office downsizing, the undersecretary's office is building up its staff and hiring from outside the agency in the process. Those are good questions.

Jim said it was his intent to "repair things" on the Hill and with the Forest Service. I wish him luck but I think that it is far too late.

[nd] November 1995, Washington, D.C.

There were back-to-back hearings on two days. The first was a joint oversight hearing before the Senate and House committees on natural resources. As has become usual in these hearings, the lineup of inquisitors was all conservative Republicans; there were no Democrats present. The regular protagonists carried water for the Republicans: Senator Larry Craig and Congressman Wes Cooley from Oregon and Congresswoman Helen Chenoweth from Idaho. As near as I could discern, the purpose of the hearing was to continue the "big lie" propaganda that the administration was failing, and deliberately so, to meet the intent and desire of Congress in regard to timber salvage.

Cooley and Chenoweth reported on the field hearings held by the timber salvage task force (which were boycotted by the Democrats). They purported to have heard from the Forest Service field staff that the agency wanted to meet the intent of Congress but was being thwarted by the requirements of the memorandum of agreement between departments and agencies. Actually, the testimony did not support those conclusions.

They did all in their power to ridicule the idea that the Forest Service was on track to meet the 4.5 billion board feet (plus or minus 25 percent) by the end of 1997. In fact, the Forest Service is ahead of that schedule.

Wes Cooley claimed emphatically that the target was 6.5 bbf. I replied that my marching orders from the secretary of Agriculture were to produce 4.5 bbf (plus or minus 25 percent) by the end of 1997. We both repeated ourselves several times—with increasing emphasis—just to be certain that the press got the point.

We have explained the difference in the numbers to Wes Cooley several times. The original estimates prepared by the Forest Service, in response to appropriate inquiry by Congress as to the amount of salvage that might be available under the Taylor Amendment to the Rescission Bill, was about 6.5 bbf. However, it was assumed that the bill would pass and go into effect by April-May 1995 (it did not go into effect until August) and would extend through December of 1997 (in the final version the authority extended only through December 21, 1996). The original estimate covered a period of thirty-three months and the operative version covers only twenty-one months. Most significant, the final version lopped off one full field season (of three) from the original assumption. And there is the difference between 4.5 bbf and 6.5 bbf.

Cooley insisted that the bill called for 6.5 bbf. I told him what the bill said—that the administration would endeavor to increase salvage above programmed levels so as to reduce the backlog of potential salvage. He insisted that he was correct and would not back off even when the operative sections of the bill were read to him.

Cooley angrily told me that the Congress would hold the Forest Service accountable for 6.5 bbf. I read him the letter from the secretary of Agriculture to the speaker of the House, Newt Gingrich, specifying 4.5 bbf (plus or minus 25 percent).

Additional intimidating shots across the bow were fired from time to time about the total incompetence of the Forest Service, and that if we didn't straighten up and obey Congress it would be necessary to dismantle the agency. This seemed a strange counterpoint to similar shots across the bow by some in the environmental community that the Forest Service acceded too much to commodity interests and should be disbanded on *that* account.

These games are beginning to bore everyone, including the participants. Nothing, absolutely nothing, is being said that has not been said several times before. The only reason that I can discern for the hearings is the media out West—and that (from the press clippings) seems to be dwindling rapidly—or the satisfaction of the political action committees that feed money to the Republicans on these committees in return for promising some break or another for commodity interests. Oddly, these same political shenanigans seem to have convinced the administration that standing tough on environmental and natural resources is a political plus. President Clinton, a rather indifferent environmentalist, has now begun to stiffen on such issues.

How can the two sides read the political tea leaves so differently? The Republicans from Idaho, Montana, Washington, Oregon, northern California, Nevada, Arizona, New Mexico, and Utah stand to gain even while alienating voters from the rest of the country. The Democrats see their gains in the urban portions of those states and a gain in all other states. Nobody seems to be backing off. That will make the presidential election year of 1996 an interesting experience.

The second hearing was under the House Committee on Resources. As nearly as we could tell in preparing for it, the title could have been "Potpourri" or "A Little Something for Everybody." Again the lineup was all Republicans save one; Congressman Bruce Vento of Minnesota (the ranking minority member) did show up and did battle on an even keel.

Congressman James Hansen of Utah, who has not the killer instinct of the rest of the committee, was in the chair. The primary Republican tormentors were (who else) Cooley and Chenoweth, joined by Doolittle of northern California, and Pombo, who carries the anti-Endangered Species Act mail. Other bit players were on the scene but their performances lacked luster.

Cooley and Chenoweth repeated their performances from the previous day. They are both so rude, arrogant, overbearing, obnoxious, ignorant, and filled with hubris that they can be played easily. If they, particularly Helen Chenoweth, did not exist, it would be to our advantage to invent them. I had trouble keeping a straight face during exchanges with Cooley

and Chenoweth. Someone could sell a book composed entirely of her stupid statements. Among those that came out in this hearing (and the one
the day before) were the following.

"What's the problem? A species goes extinct every six seconds. That
means a new one appears every six seconds. Right?"

Waving a thick volume of laws that apply to the Forest Service, she said,
"Chief, these are your regulations." Actually, they are laws passed by Congress. After reading selected passages from several laws, she asked, "Why
don't you obey those laws?" After I explained that we had to obey all the
laws and as interpreted by case law, she said, "Judges, liberal judges,
shouldn't be making law and I am going to introduce a bill to set terms on
federal judges." She obviously does not understand that such would require
amendment to the Constitution.

She then suggested that the Forest Service should follow the selected passages of laws that she read aloud and ignore the others and any federal court
orders to the contrary. I replied that such would constitute contempt of
court, and I did not choose to go to federal prison for contempt. She said,
"That will never happen." I replied that I did not choose to find out.

Cooley and Chenoweth continued to fire away at Forest Service "incompetence" and threw out other insults that began to include the field troops.
Finally, I had had enough and came back hard with something like the
following.

"Mr. Chairman, I have remained silent long enough. I *resent,* I deeply
resent, this trashing of my employees and the Forest Service. This agency
is, in my opinion, the best conservation organization in the world with the
most skilled and dedicated workforce. They are out there day after day trying to make things work under very trying and very confusing circumstances. You want to chastise the administration? Fine—that's fair. You want
to chastise the chief and my top staff? Fine—that's fair. But it is not fair to
castigate the field. I am asking you to stop these attacks and slurs." That
seemed to change the tone of the hearing as even Chenoweth and Cooley
changed the tenor of their attacks for the remainder of the hearing.

Chairman Hansen professed confusion over ecosystem management—
again. I carefully went back over ground plowed a dozen times before in
front of these committees. Chenoweth, on schedule, jumped into the fray,
asking, "Where in the law does it say you do ecosystem management?"

My answer was that evolutions in science and case law and capability have brought us to this point. Ecosystem management is simply a concept whose time has come and it is an approach that will be increasingly applied, no matter what is or is not ordained in law. One might as well stand on the shore and order the tide not to sweep ashore.

She did not like that response and demanded again to know where we believed there was any permission to practice ecosystem management. She maintained that such an approach to natural resource management was illegal.

My reply was that, to my knowledge, there had only been one legal test of application of ecosystem management to Forest Service and Bureau of Land Management planning. That was the case before Judge William Dwyer in Seattle concerning the legality of the President's Forest Plan for the Pacific Northwest. One of the questions before the court was the question of the legality of using an ecosystem management approach.

Judge Dwyer ruled that not only was ecosystem management legal, such is mandatory if all laws applicable to land-use planning and management are to be simultaneously obeyed.

The hearing, at least our portion of it, ended with an exchange between Cooley and Vento where the two talked over one another for some ten or fifteen minutes. I'd love to see a transcript of that exchange. Bruce Vento can hold his own against the entire bunch of Republicans. It makes one wonder if the debates might be a bit more productive if some Democrats would show up and take part. I have yet to understand why they do not attend any of the committee meetings or, so far as I can tell, play any role in committee business.

[nd] November 1995, Washington, D.C.

I was summoned to a meeting with Senators Kerrey of Nebraska and Bond of Missouri. Ron Stewart from the Research staff and Steve McDonald from the State and Private Forestry staff accompanied me.

The subject was to push back on the Forest Service's decision to reduce the Research Work Unit in Lincoln, Nebraska, which is focused on agroforestry, from six scientists to four. Bond was on the attack. Kerrey didn't seem to be too concerned.

The details of the meeting are not important but the lesson was worth learning once again. In the words of the former speaker of the House, Tip O'Neill, "All things are political and all politics is local."

Here we sat with two U.S. senators who voted to slash our research appropriations by an additional 10 percent this year, making it necessary to go to a reduction-in-force action. And these two senators want to protect two research positions at a particular location.

After Senator Bond saw that I was not going to accede to his wishes, he dropped the usual threat of transferring the agroforestry program to another agency. He said, "I'll bet the Natural Resources Conservation Service would be glad to have the program." I responded, "Perhaps."

Bond countered with the promise to earmark money to Lincoln. I said that if that were done, we would, of course, comply with the direction. Bond knows that the Budget subcommittee chair, Senator Slade Gorton, won't allow that to happen lest it set off a stampede of senators protecting this unit or that. They know we have to make cuts and close locations and they have no problem with that—until it occurs in their bailiwick and one of their constituents pitches a bitch.

It seems that the politicians have two choices in dealing with the politics of downsizing—let us make the decisions and take the blame, or involve themselves through micromanagement of the budget. Such involvement is a zero-sum game: one politician can score only at a direct cost to another politician. Or the game may be merely to be able to say to constituents that they brought in the agency head and told him what-for and in no uncertain terms—but of course he wouldn't budge; you know how those arrogant bureaucrats are.

All politics is local and there is more than one way to mollify a constituent.

[nd] November 1995, Washington, D.C.

There was a meeting of the Board of State Foresters in the Forest Service conference room. This was my first meeting with this group since the fiasco in Louisiana several months ago in which I managed to aggravate a number of the state foresters by saying that if the Forest Service went down by another 30 to 40 percent, we would be pulling back to the "green line" on the map—the national forests and grasslands. Some of the state foresters took dramatic exception to that statement, taking the comment as the Forest Service's abandoning leadership for forestry across the United States.

After that meeting, I sent a letter to all state foresters explaining my statement (which was quite correct) and emphasizing that the issue becomes moot if we can stop the downslide in the Forest Service's budget. I used the letter to point out that I had obtained the biggest increase in history for State and Private Forestry in the president's budget. And I had made attempts to heighten the profile of State and Private Forestry within Forest Service reinvention efforts. The result was that we—the Forest Service and the state foresters—were not able to hold those increases and avoid significant cuts.

The representatives seemed somewhat mollified by the letter, and the purpose of this meeting was to build on that beginning. The tempest may have served a useful purpose in making it clear that the Forest Service and the state foresters need to work *together* to hold the overall Forest Service budget.

By the time the meeting was over, we seemed well on the way to healing the break that I inadvertently started in Louisiana.

3 December 1995, Tucson

Today marked the beginning of the long-planned Ecological Stewardship Workshop. This effort is designed to get the technical foundation under the ecosystem management concept and derive some agreed-upon approaches to bringing the concept into context and hence into land-use planning. We worked hard, with the significant help of Amos Eno and the Fish and Wildlife Foundation. For symbolic and practical purposes, over half the costs of the effort have been contributed from nongovernmental sources. The workshop has finally come to fruition despite considerable internal resistance. This resistance has been largely centered on having this "expensive workshop" at a time of agency downsizing and budget reduction and fear of the luxurious five-star hotel at which the workshop is being held.

I remain wed to the idea that the Forest Service can and should be the conservation leaders for the nation and the world as the 21st century looms. That dream cannot be realized with a bunker mentality and a clinging to the status quo. Therefore, I have held tight to the concept of ecosystem management, despite criticism and attack from hard-right Republicans and the conservatives in natural resource management. No matter what happens in the short run, I am determined to move us so far along the path of more comprehensive management (in terms of space, time, and variables considered) that there can be no going back.

It was my honor to speak at the opening plenary session of the workshop. My comments, titled "Guiding Principles and Workshop Overview," were essentially as follows:

Ecosystem management is an idea whose time has come. While some view this approach to land management as revolutionary, others—correctly, in my view—recognize this approach as part of an evolutionary and ongoing process. We have come to a point where there is a juxtaposition of streams in philosophy, science, awareness, and technological capability that makes ecosystem management approaches inevitable.

I view ecosystem management as a concept of land management that must always be placed in context. That context differs from past approaches in that spatial aspects are larger, temporal scales are longer, and the variables considered are much expanded—certainly including humans as part of the system.

Having said those things in front of a Senate committee, I was requested to explain the concept in more simple terms. I asked the senator if his mother had ever told him the fairy tale about the goose that laid golden eggs and asked him the moral of the story. He answered, "If you wanted golden eggs, it was necessary to ensure the health of the goose." Exactly.

On the battlefields of the Potomac there is an ongoing battle raging between those who wish to turn back the clock for natural resources management to what existed at some idealized time in the past, when timber flowed at desired rates unimpeded by other considerations. One thing is very clear to me. The present situation in which public land management exists is not tenable over the longer term. I vote for moving forward to a future that incorporates a combination of new knowledge and understanding with old wisdom.

I believe that the *context* for this potentially watershed workshop includes the following items:

· Ecosystem management is a journey and not a destination.
· Ecosystem management is evolutionary and not revolutionary in approach.
· The objective is to promote the long-term sustainability of ecosystems by

ensuring their health, diversity, and productivity. This will involve development of integrated strategies to deal with concerns for ecosystem sustainability, biodiversity, forest and rangeland health, maintenance of aquatic and riparian systems, and effects of introduced plants and pathogens within a more overarching concept.

· This larger strategy is the development of a comprehensive framework and use of the framework as a base to address implementation and strategies between agencies and other entities.

· This effort is the next logical step to follow previous efforts undertaken by agencies to use ecological approaches to natural resources management.

In summary, I challenge you to think beyond the barriers of agency custom and culture, beyond the limits of how things are, beyond boundaries, and beyond the present. To paraphrase, I ask you not to dwell on the present confusion and conflict. I encourage you not to be paralyzed by the situation. Instead, look for what can be and ask, "Why not?"

Land management agencies must move forward or risk a move backward to management paradigms of the past. For we now stand on shaky ground, where we cannot remain. I want to move forward and I consider ecosystem management a way to accomplish that. I need and ask for your help.

12 December 1995, Washington, D.C.

The Forest Service national leadership team met to discuss courses of action if the Congress fails to pass a continuing resolution to allow the Department of the Interior and the Forest Service to continue to operate in spite of the fact that there is no budget. Oddly enough, the Forest Service budget comes under the Interior and Related Agencies budget.

The Congress has passed a budget that is unacceptable to the president because of inadequate funding for programs for Native Americans and "antienvironmental" riders. Among these riders are a mandated timber cut level in Alaska and limitations on completion of the Columbia basin assessment. So the president vetoed the bill.

Staff informed the national leadership team, based on latest estimates, that there was enough carryover from the FY 1995 budget plus the money in special accounts (such as the salvage fund and the Knutsen-Vandenberg Fund) to continue operations until Thursday, 21 December. As usual, the team was split down the middle as to whether to shut down immediately or to continue to operate through Thursday. I was persuaded by the arguments of Kathleen Connelly, deputy chief for Administration, to go to an orderly shutdown by close of business on Tuesday, 19 December. Her point was that the accounting system was not designed to give information on budget status on a day-to-day basis *and* the year-end numbers are estimates and always wrong. Therefore, it was advisable to be cautious in order to avoid problems with the Anti-Deficiency Act.

And so the orders went out to close down Forest Service operations. There were exceptions. "Emergency personnel" stayed on the job. The term emergency personnel now replaces the term essential personnel—and the flip side of "nonessential" personnel. Those folks identified as nonessential were, understandably enough, offended by that tag, and it led to considerable confusion in the public's mind and ridicule in the press: Why would the government employ people who were not essential?

Law enforcement personnel will still be on the job as required to protect health and safety. Folks involved with supervising and preparing salvage sales will continue to work, with funding from the salvage fund.

All other persons will be furloughed. There is great irony in this fiasco. Given the political ramifications of furloughing hundreds of thousands of federal employees at Christmastime without pay, the politicians on both sides of the aisle are proclaiming that it is the fault of the other side and assuring furloughed employees that they will be paid anyway. That means that all these people will be paid for not working while emergency personnel will be working—whether they choose to or not—and being paid exactly the same as their nonemergency personnel brethren who are not working. There is something quite ridiculous in all of this. Merry Christmas! Alice is alive and well in Wonderland!

I am beginning to understand with increasing clarity the observation by (I believe) Will Rogers, "The Republic is in danger—Congress is in session." It now seems likely to me that Congress will go into recess this week without either a budget in place for Interior or a continuing resolution. That means that the shutdown will continue until several days after January 10.

That means a "vacation" of at least twenty-two days, of which fourteen days are workdays, for nonemergency personnel.

In the afternoon, I met with the team that is revamping the planning regulations issued pursuant to the National Forest Management Act. I ordered several minor changes and told them to prepare the final draft for the final approval process in the Department of Agriculture and the Office of Management and Budget.

This has been a long road. Over two years were spent on revisions during the administration of President George H.W. Bush. The rule was ordered held during the 1992 election year, as it might have injected some controversy into the election debate. After President Clinton was elected, the effort began anew with my instructions to formulate the operation around the concept of ecosystem management.

The release of the new draft rule set off a storm. Industry (or at least parts of it) came unglued over the inclusion of "ecosystem management" and the maintenance and restoration of ecosystem function as guiding principles. Industry believed that the overriding goal should be multiple-use, with emphasis on "use": i.e., the management of the national forests should be focused on timber production and, of course, the jobs that the timber industry provides.

The timber industry, through its former mouthpiece, Mark Rey (now chief of staff to Alaska Senator Frank Murkowski, chairman of the Energy and Natural Resources Committee), has insisted that action on new regulations be delayed, as the Republicans have every intention of revamping the National Forest Management Act, the Endangered Species Act, and other assorted acts. The industry wants to delay until the laws are changed and the undersecretary of Agriculture overseeing the Forest Service is appointed by a Republican president, which is assumed will happen in 1997.

The hardcore environmentalists are equally upset because the appeals process is slightly more constrained and accelerated. Also, their favorite nebulous requirements for "diversity," "viable populations well-distributed" and "sensitive species" management have been more specifically stated and require more rigor for application. They see that as the loss of a primary tool to thwart Forest Service management actions with which they disagree.

So as usual, the Forest Service is squarely in the middle. Perhaps, at least in this case, such is the appropriate place to be.

I am now convinced that the new planning regulations have been completed in full compliance with the law, there has been a more than adequate effort to involve the public, extensive changes have been made in response to public comment and internal Forest Service employee input, and the new planning rules are a vast improvement. Further, it is essential to press these new rules into place as soon as possible to cut off the increasing political temptation of the politicians to avoid controversy in an election year by again delaying revision.

Therefore, I instructed staff to prepare an action plan to get the rules adopted as soon as possible—within the month, if possible—and said I would be fully involved in accelerating the process. That personal involvement will accomplish two things. The process of approval will be accelerated and those who are vehemently opposed will center their attacks on me rather than on the Forest Service. As a result, this will be one additional "sin" that I can take with me on my departure from the chief's job.[5]

20 December 1995, Washington, D.C.

The work session to prepare the briefing on air tankers for Secretary Glickman was an interesting experience in and of itself. The problem was to bring some order to the presentation and tell the secretary what he needed to know without overwhelming him with detail and trivia.

By the time my briefing outline was complete, I understood the situation for the first time. The short version includes the following facts.

The Forest Service acquired the surplus military aircraft through the Antique Aircraft Act. The idea was to put aircraft in the hands of contractors who would fit them with the equipment necessary to make them air tankers. In addition, enough aircraft were made available for spare parts to keep the tankers flying. These aircraft were to be used for no other purpose than firefighting.

An argument quickly developed over who owned the aircraft. Did they belong to the Forest Service or to the contractors? That issue was confused by a court case in which a contractor borrowed money from a bank

---

5. The proposed planning regulations were held in limbo until after the 1996 presidential elections. Following Clinton's reelection, this version of sixteen regulations was withdrawn and the process begun anew with the addition of a committee of scientists. As of June 2002, no new regulations had been adopted.—*JWT*

using the aircraft as collateral and then defaulted on the loan. The bank went after the collateral and the government claimed ownership. The judge ruled that the bank had lent the money in good faith and was therefore entitled to the aircraft. As a result, it is now unclear who owns the remaining aircraft.

Several years ago, an investigation by the Office of the Inspector General found that the process of obtaining the aircraft was improper. The investigation also questioned the propriety of the Forest Service person who negotiated the deal and then racked up four-engine flying time on the traded aircraft to upgrade his commercial flying license. This led the Forest Service to fire the negotiator. The Labor Relations Board deemed that action too severe and ordered the person reinstated. The Department of Justice declined prosecution at that time.

At that point, the Office of the Inspector General and the Office of General Counsel insisted that they review and approve any future contracts for the air tankers. That review and approval process has been followed through three subsequent contracts, including the contract now in question.

During Operation Desert Storm, several C-130 aircraft obtained under the Antique Aircraft Act turned up hauling cargo to Saudi Arabia. This led to a legitimate complaint from other contract aircraft providers. Congressman Charlie Rose went public with accusations that the entire aircraft exchange involved the Central Intelligence Agency. I do not know if there was any such connection.

This set of circumstances produced a flap in the media that was spun into a "scandal" and a "coverup." The flap has died down for the past year and a half or so.

Then, the Department of Justice branch in Tucson informed the Forest Service that it was building a case against the Forest Service employee mentioned earlier, a middleman negotiator, and two employees of the Department of Defense. The Department of Justice requested that the Forest Service do nothing to resolve the issue of ownership of the aircraft.

A short time later, the Department of Justice requested the director of the Southwest Region of the General Services Administration to order the Forest Service not only to declare ownership of the aircraft but to take actual possession. That order was given.

As the fire season was still on, I refused to obey the order and asked General Counsel Gilliland for advice and help. Gilliland advised me not to comply until he instructed me to do so. As a result, the Forest Service has yet to take possession of the aircraft.

The potential source of embarrassment to the secretary lies in the perception that these "valuable" aircraft (which were worth only scrap in a mothballed state) were placed in private hands and are now being used to make huge profits for the contractors currently in possession. In addition, these aircraft were being used for other purposes than firefighting, including undercover operations for the Central Intelligence Agency.

However, the worse embarrassment would be to go into fire season without a federal air tanker capability. Therefore, given that fact and the overriding obligation to protect health and safety, there is no alternative but to sign the contracts.

That was obvious both to the secretary and to the general counsel. Still looking for cover for the secretary and sympathetic to the position of the Department of Justice, Gilliland proposed sending a letter to the contractors. The letter would state that ownership is still in dispute and that by signing the contract, the Forest Service is not making a decision in that regard. The significance of such a letter was lost on me, but it was enough to get the Department of Justice not to object to signing the contract.

I also pointed out that this controversy, with the attendant potential embarrassment, would continue until some clarifying action took place. The Forest Service has proposed, for each of the past three years, that the Department of Agriculture seek legislation that would accomplish two necessary things. The first is to clear up the ownership dispute, preferably by transferring title to the contractors. The second is to prescribe a method by which the Forest Service can acquire surplus military aircraft for conversion to air tankers. The first may be difficult because of the negative publicity that surrounds the current situation. Gilliland suggested that this approach may not be politically feasible and suggested a clarification through court action. If the matter is forced into court, the Department of Justice can negotiate a settlement that would clarify ownership.

Achieving legislation to clear the way for additional aircraft was termed "acceptable" by Gilliland and "doable" by the secretary. I made a mental note of those statements and determined to pursue drafting of such legislation and a plan for achievement as soon as personnel return to work.

22 December 1995, Washington, D.C.

Forest Service operations have ground down to a virtual stop. The phones have ceased to ring and electronic mail messages are infrequent. At noon, I issued orders to keep one actual or acting deputy chief per deputy area on duty, with one clerk to answer the telephone, and sent all other personnel home for the holidays—with or without pay is the question on everyone's mind. There are fewer than 20 people in the Auditor's Building on duty out of a normal workforce of over 700.

I have read all the mail, all the accumulated documents, and cleared my electronic mail. I have cleaned off and out my desk. Now, for the first time ever, I am writing in my journal during business hours. What a fiasco. What a truly stupid fiasco!

28 December 1995, Washington, D.C.

While I was sitting in my office, being one of the ten Forest Service employees on duty answering the phone and standing by for emergencies, an "emergency" arose. Steve Satterfield of the forest policy staff came in to tell me that the Department of the Interior had complained to the Office of Budget and Management that all the ski areas on National Forest System lands were being allowed to continue to operate in spite of the government shutdown caused by the failure of Congress and the president to come to terms.

Interior had made the decision, without consulting the Department of Agriculture (i.e., the Forest Service), to close down all national park operations, including several ski areas that are within park boundaries. It is inevitable that this is bringing the Park Service criticism for causing inconvenience to the public and significant lost revenues to concessionaires that operate in the parks. It is likely that one of two things has occurred, or both. The Department of the Interior (i.e., the Park Service) is concerned that the government is not taking a consistent position in response to the shutdown. Or the decision to take a hard line on park closures was meant to have a maximum effect on local economies and the recreational pastimes of affluent Americans and foreign tourists. Or, of course, both.

The Forest Service had *planned* to keep ski areas and campground concessions open, keep timber sales, miners, and grazing permittees operating, and continue with preparation of salvage sales. This intent and the process for accomplishing it were clearly spelled out in the required shutdown plan that was submitted to and approved by the secretary of Agriculture and the Office of Management and Budget.

Now the Department of the Interior complains about our course of action, not to the chief of the Forest Service or to the secretary of Agriculture, but to the Office of Management and Budget. I find it hard to imagine that the Forest Service or the Department of Agriculture would ever comment on an action taken by the Department of the Interior, but the reverse is more and more common.

The Forest Service decided to keep the above-mentioned operations open based on careful consideration of several factors—philosophical, legal, pragmatic, and political. The most significant factor in my recommendation to keep operating was the consideration of the Forest Service mission to "care for the land and serve people." It made no sense to close down ongoing contractual operations that require oversight by very few Forest Service personnel; those operations should be continued as long as possible. Putting people out of work and hampering marginal economic enterprises did not strike me as consistent with the mission of serving people.

I was also concerned about the legal ramifications and the liability for losses caused to contractors if their operations were closed down suddenly. Pragmatically, assuming the shutdown does not go on for a prolonged period, it is *far* cheaper and requires *far* fewer "emergency" personnel on duty to allow the ski areas to keep operating than it would require to shut them down and, I suspect, to deal with the fallout of the shutdown.

Given the fact that we have, with some difficulty, beaten off attempts by conservative Republicans in Congress to privatize ski areas (i.e., turn over federally owned ski slopes to ski operators), it seems quite ill-advised to essentially shut down the ski industry. That action would almost certainly guarantee another effort to place the ski slopes in private hands.

In addition, I don't think the people in Interior have really thought their way through the entire matter, and their position is not consistent across the board. Have they shut down livestock grazing, mining, hunting seasons, etc.? No? Why not? Politics?

1996

6 February 1996, Washington, D.C.

There was an oversight hearing today before the Energy and Natural Resources Committee in the Senate. Senator Craig Thomas of Wyoming was in the chair. The announced subject of the hearing was a report by the General Accounting Office on the trends in public land ownership, methods of acquiring land, and plans for additional acquisitions.

The first witness was Governor Jim Geringer of Wyoming. His point was that the federal government is the dominant landowner in Wyoming, with well over one-half the land. He said that the secretary of the Interior had more to say about things in Wyoming than did the governor. He encouraged changes in the federal estate, such as consolidating ownership and giving more attention to the management of the federal lands to benefit the citizens of Wyoming. I thought it was interesting that he did not call for any reduction of federal land ownership in Wyoming.

Senator Thomas, in an opening statement, reacted pointedly to the submitted written testimony of Interior Secretary Babbitt, in which the secretary did not address the subject of the hearing but launched an attack on the potential sale or giveaway of the federal lands. I was a little surprised when I read the testimony, and puzzled over the strategy. The secretary essentially chose to create a new hearing, one focused on the devolution of public lands rather than on the current status of the federal estate. Senator Thomas and Senator Larry Craig took the bait and devoted most of their time to attacking Babbitt and his "mischaracterization" of the committee's intent. They both stated that they had no intention of removing any land from public ownership. Thomas stopped with that. Craig continued to raise the specter of having the states take over management. The other Republicans present—Burns of Montana (who dozed through the hearing and said nothing significant) and Kyl of Arizona—gave no indication of backing Craig's statements and made it clear that they do not intend to reduce federal land ownership.

For a change, two Democrats actually showed up for the hearing—Bumpers of Arkansas and Bingaman of New Mexico. Bumpers made some serious statements in favor of retaining federal lands. Bingaman asked some questions of the General Accounting Office witnesses that came out as generally favorable to federal ownership.

Secretary Babbitt was clearly the focus of the hearing and by his testimony kept the focus on himself. My testimony was strictly to the point of the hearing. The questioning by the Republicans (the Democrats left the dais after the GAO testimony, and before Babbitt and I came to the witness table) was strictly a politically acrimonious exchange between Craig and Thomas on one side and Babbitt on the other. Almost no questions were directed to me and those few were technical in nature.

Toward the end of the hearing, Craig held up two videotapes and asked if I was aware of the "public service announcement" videos starring General H. Norman Schwarzkopf, in which the pitch was "keeping public lands public." These videos were released to television stations in the Boise area under the Forest Service logo. When the regional forester, Dale Bosworth, brought these videos to my attention two weeks ago, he had already removed the tapes from circulation. I affirmed his decision and instructed that the videos not be shown again. While I agreed with the message, using such videos identified as Forest Service products was clearly inappropriate lobbying on a political issue.

So when Craig questioned Forest Service use of those videos, I said simply that the action was wrong and inappropriate, and further, that the videos had been withdrawn from circulation. Craig seemed a bit taken aback by my answer and replied, "Thank you, chief. Thank you very much." That ended that, for the moment, at least.

The moral of the story is that when someone in the agency messes up, the best thing to do is take personal responsibility, admit the error, and make certain it doesn't happen again. It is amazing how few people can bring themselves to do those simple things. A statement of "I'm sorry" doesn't hurt either.[1]

As the drama played out, I thought I understood Mr. Babbitt's strategy. His intent was to divert the hearing from a carefully crafted Republican strategy. That strategy, through a long series of hearings, has been to reach a point where it is possible to turn over, or devolve, some aspects of public land ownership or management to the states.

Babbitt, buoyed by public opinion polls and Congressman Ron Wyden's victory over State Senator Gordon Smith in Oregon's special election to replace Senator Robert Packwood, decided to jump into the fray. Babbitt forced the committee to deal with the issue on which he had seized the high ground, when the committee was unprepared to discuss it. The Republicans reacted by protesting that they never, ever, had any intention of taking land out of federal ownership.

I couldn't believe that they took Babbitt's bait, but they did. Now, suddenly, the Republicans on the committee are on the defensive, after pursuing a carefully crafted offensive game plan through twelve previous hearings. Damned clever if intentional, and serendipitous if not. I am even more convinced now of the conclusions I reached after the last hearing. The Republicans have lost momentum and are now in retreat, even to the point of proclaiming themselves proponents of public land ownership.

At the close of the hearing, I leaned over and whispered to Secretary Babbitt, "I don't know if your opening statement was an intentional ploy or not, but one way or the other, you flushed the hardcores out. They were falling all over themselves to deny that they personally ever had any ideas about transfer of federal lands into other ownership." He just smiled.

As I left the hearing room, I noticed that Babbitt was in the corridor surrounded by cameras and reporters. The senators left by the back door. Mr. Babbitt had taken a real tongue-lashing but he won the battle for media attention with his spin on the issue.

---

1. General Norman Schwarzkopf was displeased with this decision, feeling that his help on the issue was not respected. I wrote the general a letter of appreciation and apology, saying that Forest Service personnel had misused the video for a political purpose.—*JWT*

7 February 1996, Washington, D.C.

Today the Forest Service received the 1997 budget "passback" from the Office of Management and Budget. This is the result of OMB's making adjustments in the budget submitted by the Forest Service. Under this administration, it is actually the response to the Forest Service's suggested budget as modified by the undersecretary. Mr. Lyons changed the budget mechanism to keep the Congress from asking questions in the budget hearings as to what the Forest Service budget recommendations were. In other words, political twists are already built into the Forest Service's budget requests.

OMB provides additional political guidance on the budget. I have no idea how OMB operatives come to the conclusions that they do, and frankly the adjustments make very little sense except politically.

The Forest Service, with the approval of the political handlers in the Department of Agriculture, has one opportunity to appeal OMB's decisions. But OMB bureaucrats, with very little knowledge of what the Forest Service is required to achieve, make the final determinations as to what is submitted to Congress as the administration's recommended budget for the Forest Service. This amounts to political modification of budget recommendations by professionals in the natural resources field. Of course, it is probably fair enough for the administration to get a political twist to the budget before the House and Senate budget committees put their political twist on the (usually) final budget.

In this case, OMB made several obvious changes with which I disagree. The first was to increase the Stewardship Incentive Program from its likely present level of about $4 million, to $14 million, in the face of pointed cuts by the budget committees last year. This is money that essentially goes to provide incentives to private landowners to practice good forestry. The political attractiveness is that it is money largely at the disposal of the state foresters. The danger in this action is that the congressional budget committees will be agitated that the administration did not get the message last year. A good lobbying effort by the state foresters and the Forest Service might overcome that agitation.

The second move was to beef up the timber program by about $4 million more than is required to hit the anticipated timber objectives. Downsizing has produced a circumstance whereby the limitation on timber sale preparation is the number of skilled people available, and not money.

The third major disagreement is the decision that the Forest Service absorb the new $12 million cost of administering the subsistence hunting and fishing program for natives in Alaska. The state of Alaska, in startling contrast to the constant tirade against federal interference in Alaskan affairs, has chosen to force the Forest Service to handle subsistence hunting and fishing. The decision to absorb the cost, which is a new imposed cost, in a budget that is already too low, seems ill advised. This should be new money and an add-on to last year's budget level.

The fourth and last area of disagreement is the continuing instruction to absorb the cost of the AmeriCorps program in the Forest Service budget—i.e., the program will continue without making the expenditure obvious to Congress. The national leadership team, with the concurrence of the regional foresters, puts a very low priority on the AmeriCorps program, preferring to spend increasingly scarce resources on the regular workforce to accomplish the agency's mission. Clearly, the AmeriCorps effort is a high priority for President Clinton, and there is little doubt that the Forest Service will be told to absorb the cost. The argument is whether the budget documents submitted to Congress will make that expenditure obvious.

So the decision was made by the national leadership team to resist OMB changes on the above four items. In the case of the Stewardship Incentive Program, the proposal will be to increase the program to $6 million rather than to $14 million. I suspect the department will not agree.

We propose to scale back the increase in the timber program significantly. The department will likely agree.

We will suggest that the $12 million required to administer the subsistence program be requested as an add-on to the 1996 budget. The department will probably agree. OMB's reaction will be a toss-up.

In the best-case scenario, the 1997 budget will not show any increase over that for 1996. That means a budget cut by the rate of inflation and whatever the increase in salary costs turns out to be—a likely 2.5 to 3.5 percent decrease in an already too-tight budget.

9 February 1996, Washington, D.C.

There was a breakfast this morning at the Department of Agriculture involving Secretary Glickman, Undersecretary of the Interior George

Frampton, Assistant Undersecretary of Agriculture Brian Burke, Assistant to the Secretary of Agriculture Anne Kennedy, BLM Director Mike Dombeck, and me. The session was, in general, a discussion of the status of a number of issues that Interior and Agriculture jointly face.

There was recognition, at long last, of the hodgepodge of policy that is being derived through uncoordinated negotiated settlements of dozens of lawsuits across the country. This situation results because the Department of Justice exercises its power to handle these cases. DOJ does not consider an agency in a lawsuit its client. Rather, DOJ maintains an independent posture and considers itself above the fray and hence acts out its own view of appropriate settlement terms. This process is often executed by relatively young and inexperienced attorneys who know little about natural resources management issues, whether technical or in matters of policy. They are particularly insensitive to the long-term impacts on policy of their immediate case-by-case actions.

Some of the settlements are, in my opinion, almost bizarre and go far beyond the circumstances of the case. For example, in several cases, agreements have been made that make land-use decisions (which are required to be made in forest plans or amendments to such plans, and which require environmental assessment) outside the required processes; these decisions will be in place for years. Worse yet, each of these settlements is independently derived, without any overall policy guidance.

I have been upset by this process and, in my opinion, the inappropriate delegation, whether purposeful or inadvertent, of such policy-making authority to DOJ rather than to the secretaries of Agriculture and Interior or their delegated agency heads.

Now, at last, pounding on the issue has focused the secretaries' attention on the problem. I will take advantage of their momentary attention to ask Jim Gilliland, general counsel for the Department of Agriculture, to prepare a white paper on the subject. That paper can be used to open a dialogue with DOJ. That may have some effect. One thing is certain, the heads of agencies simply do not have enough clout to make any impression on DOJ. I hope to get a joint clearinghouse set up to bring some consistency to settlements in policy terms to be established. I hope that such national policy for settlements will include at least the following: (1) no long-term land-use obligations; (2) no concessions that are not directly connected to the issue in question; (3) DOJ will give agency concerns more weight, i.e., there is more of a client-attorney relationship; and (4) a review panel to examine all proposed settlements prior to entering them.

21 February 1996, Washington, D.C.

I met today with Secretary Glickman's chief of staff, Greg Frazier, and assistant to the secretary, Anne Kennedy, accompanied by Assistant Undersecretary Brian Burke. The subject of the briefing was timber theft and was scheduled by Anne Kennedy. She was charged by the secretary (after a meeting with Dan Devine of the Government Accounting Office, who was described as an acquaintance of the secretary) to find out about the situation surrounding the timber theft issue and the continuing whistleblower complaints by disaffected members of the disbanded Timber Theft Investigation Branch.

Present from the Forest Service were Deputy Chief for National Forest System Gray Reynolds, Director of Law Enforcement Manny Martinez, Assistant Director of Law Enforcement Hank Kashdan, and me.

Martinez conducted the briefing and, one by one, refuted the charges made by the few disaffected former members of TTIB. In my opinion, the only weak spot was that because of budget considerations plus downsizing, the training and consulting cadre that had been initially proposed had not been activated, although every law enforcement officer was receiving a mandatory forty hours of training per year. That training includes a module on timber theft detection and prevention.

Kennedy took the position that somehow the Forest Service was wrong and she was the special prosecutor. She most obviously has no understanding of the workings of the Forest Service, and when facts were presented to her bearing on the point in question, she retorted that the Forest Service was, once again, being resistant to what the politicals (i.e., the administration) wants to occur. During one of these exchanges, I let my frustrations out and told her that we were merely trying to give her the information needed to appreciate the situation.

Martinez explained that we were woefully understaffed in law enforcement and that further significant downsizing was under way. Further, we were shifting personnel slots from special agents (investigators) to uniformed patrol officers. This is being done at a time when illegal immigration across Forest Service lands, marijuana gardening, timber theft, threats and violence against employees, political antagonism, militia activities, numbers of visitors to the national forests, and threats to visitor safety are increasing.

Kennedy asked him what law enforcement needed to cope adequately with these problems. Manny replied that about a $17 million per year increase in funding would be a good start. She looked at me and asked me why I didn't just make that a priority. I explained once more that there was no free lunch and that I had no place left to shift resources except from other areas that were likewise short of resources. From the look on her face, it was clear that once again, she felt the Forest Service was being intransigent.

Damn, this is getting harder and harder to tolerate. And it is obvious that the politicals are equally frustrated. It seems clear where the disconnect and the inability to communicate come from. The Forest Service professionals view themselves—correctly, in my mind—as people who have been getting the job done for ninety years and will continue for the next ninety. The Forest Service professionals, in other words, have a track record and a long-term vision and close connections with the land and various constituencies.

On the other hand, the politicals see things as separate and isolated events and emphasize quick fixes to political concerns. Even more common is their eagerness to satisfy the desires of politicals above them in the hierarchy. Such are the criteria of their success and the means to maintain or increase their power or influence. Further, most of the politicals are not encumbered by much knowledge of the issues at hand or by any long-term memory or long-range view. For them, everything lies in the present and the very immediate future, and the goal is to triumph in the next election.

Given these circumstances, interaction between the professionals and the politicals is bound to produce friction, and it obviously does.

5 March 1996, Washington, D.C.

There was a meeting today with Senator Hank Brown (Colorado) concerning the Forest Service's position on his amendment to the Farm Bill. The impetus for his amendment was the Forest Service's position that all permit renewals on impoundments on national forests include provisions for bypass flows to maintain aquatic communities and recreational opportunities.

Senator Brown and Colorado Congressman Wayne Allard seized upon this action as a political winner and sought to pass legislation to prohibit such action. Brown is retiring after this session and Allard will seek his Senate seat.

The situation has been exacerbated by a particular set of political maneuvers among the political appointees in the Department of Agriculture. The last secretary in the Bush administration, Edward Madigan, set out the "Madigan policy." Simply stated, Madigan directed that the Forest Service would seek no water rights or make any decision that impinged on the water rights of state or local entities. After the election of President Clinton, the Forest Service (with Regional Forester Elizabeth Estill in the lead) courageously set out to guarantee bypass stream flow authority when water facility permits were renewed. This required overturning the Madigan policy, which the Forest Service had requested on a periodic basis.

Secretary Mike Espy did decide to overturn the Madigan policy. However, for some reason unknown to me, the Forest Service was not made privy to that decision. In fact, every time we (I) inquired as to the status of that decision (which we strongly supported), the answer was that "no decision had been made." I would pass that word to Estill, who in turn continued to tell Brown, Allard, and others that the Madigan policy remained in force, which was not true.

So when it became clear that the Madigan policy had been revoked by Espy early on, Brown and Allard assumed that Estill had lied to them, and they were furious at her. She and I were blameless, as we had told them what we thought was the truth.

Brown and Allard then demanded all the Forest Service and Department of Agriculture documents that in any way pertained to the issue, including personal notes. Those records plus testimony before Congress clearly revealed that the Madigan policy had been overturned, and the Forest Service and the Congress had been misled. Either Brown and Allard (or their staffs) never reviewed this material or they did not catch that fact, or for reasons of their own they decided not to make an issue of the circumstances. At best, the situation clearly cost the administration, and the Forest Service, in terms of credibility.

Brown and Allard chose to address this issue through an amendment to the Farm Bill, which would preclude any hearings on the issue, as full hearings would probably lead to defeat of the legislation. For a while it appeared that the amendment would have clear sailing, but the consistent Forest Service position and the subsequent attacks and high-profile tirades by Brown and Allard brought the situation to public scrutiny. And the moribund environmental groups, which were reinvigorated by the salvage rider and this amendment, became more active. Trout Unlimited, for example, began printing full-page advertisements showing dried-up trout streams.

Then, the administration began to see the environment as a potentially significant issue that could tilt the balance in a few critical states in the upcoming presidential election. As a result, the administration began to speak out on this and other environmental issues and against "stealth legislation" passed in the form of amendments to must-have legislation. This built to a point where I was told clearly, "The administration is unalterably opposed to any legislative language in the Farm Bill related to water rights on public lands."

I had visited with Senator Brown, at his request, via telephone on several occasions last week. In those conversations he expressed concerns that the "effects statement" that the Forest Service had submitted on his amendment to the Farm Bill was filled with factual errors. I told him that if he could demonstrate that any statement was incorrect, we would quickly attest to our error. I asked Deputy Chief Gray Reynolds (who had worked on Brown's staff during a training assignment) to visit with the senator and his staff to go over the effects statement line by line.

That was done and the Forest Service analysis held up to examination. About that point, it became clear to Brown that the amendment was in trouble. That point was emphasized when Allard withdrew his companion effort in the House. Allard then made things even worse by aggravating the key members of the Agriculture Committee, sneaking the amendment back in as a "technical adjustment."

When I met alone with Brown today, he was very friendly and seeking some way out of an increasingly embarrassing situation. First, he stated that he was willing to delete significant portions of the amendment to gain administration acceptance. I told him that my instructions were to tell him that the administration was opposed to any inclusion of such an amendment to the Farm Bill. And I was not in any position to negotiate with him in any case.

We discussed technical aspects of water management for some time, and I pointed out to him that many aspects of his amendment begged critical questions to which answers were not readily available. It was my intent to give him a way out of the situation by asking for a study and evaluation of the situation, which is needed to head off the next assault. By the time the conversation was over, he had that idea firmly in mind. Most encouraging of all, he was characterizing his problems with Estill as "miscommunication" rather than being lied to or misled.

The next meeting (with Undersecretary Jim Lyons, Assistant Undersecretary Brian Burke, Deputy Assistant Undersecretary Mark Gaede, and staffer Stephanie Hague) was even more frustrating. The first subject was the budget to be presented to Congress. Lyons and the Forest Service leadership team are in absolute disagreement on several points, and we were supposed to reconcile those differences in this meeting. That is unlikely. We simply see the world through different lenses. My lens is that of a natural resources professional with a science background who uses and trusts staff in a collaborative decision process. Lyons's lens is that of a person with degrees in natural resources, no field experience, and a professional career as a political operative, who distrusts all staff and is arbitrary and dictatorial by nature.

Therefore, we discussed the issues and I lost every argument. The big area of disagreement was the amount of funding needed for roads to meet the timber targets. Lyons's considered opinion (based on his instincts) is that the Forest Service is requesting far too much. He pretends that if I have staff do further analysis, he will change his mind. Not likely. My position is that we should take the advice of staff people who have spent careers building roads and putting up timber sales.

The second issue is that the Forest Service is in a terrible position vis-à-vis the range program. Lyons agrees that the program is in sad shape and getting worse. But he is simply unwilling to transfer any money from elsewhere in the budget. The Forest Service had suggested $10 million from the Stewardship Incentive Program in State and Private Forestry. That amount was cut by Congress last year and restored by Lyons in the budget to cover his political base with the state foresters. I fear that Congress will see the restoration of these dollars to last year's budget proposal, after Congress cut the program, as an in-your-face maneuver and will just cut out $10 million from the Forest Service budget. Lyons argues that the nation will get more timber this way. Maybe, in fifty or so years.

The Office of Management and Budget overruled us on the transfer of $10 million from land acquisition to the range program. Land acquisition is popular with the environmental community. Lyons is not willing to argue with T. J. Glauthier of OMB over the issue. These dollars are also vulnerable, as Congress made clear last time around.

Lyons believes the Forest Service is concentrating suggested cuts in the programs he is interested in. These discussions always come around, sooner or later, to this point. I tell Lyons he is paranoid and he tells me I am hopelessly naïve.

Brian Burke watched this exchange as if it were a tennis match: back and forth, back and forth. I think he was a bit taken aback by our frankness and contrary positions. He may realize more with every additional day on the job that the disconnect he was sent to straighten out might be a bit more difficult than he originally perceived.

6 March 1996, Washington, D.C.

A call came in from the deputy secretary of Agriculture, Rich Rominger, at 11:30 A.M. for me to be in his office at 1:30 P.M. The meeting was to concern two subjects: the Quincy Library Group and my pending travel to Russia.

The first item discussed was the Quincy Library Group. These folks have reached some consensus on how the national forests in their region of California should be managed. While such an achievement is remarkable enough, it should be noted that the group is self-appointed, and to say that they represent the community or even the interest groups within that community is a real stretch.

However, this group was able to convince Secretary Glickman that they had such a good idea and such a good approach to management that, without consulting the Forest Service, he agreed to recognize the group (though not as an advisory group under the Federal Advisory Committee Act). Furthermore, Glickman obligated $4.5 million in additional funds to help carry out their views of forest management. He also promised that no other national forests would have their budgets reduced to provide the funds. In other words, it was to be a free lunch. Hardly—there is no free lunch. In order to comply with Glickman's order, I transferred $2.5 million in forest health funds from other forests in California and another $2 million from forests outside California, Oregon, and Washington. That $2 million is a transfer from the least well fiscally endowed regions to the second best-funded region, and comes on top of our already having made multimillion-dollar shifts from those less well-off regions to support the President's Forest Plan for the Pacific Northwest.

In my opinion, this spur-of-the-moment decision made in the absence of full understanding of costs and benefits and political volatility was a technical and political blunder. The idea from the politicals in the Department of Agriculture is to take it off the top and then send that money out for special projects and programs—i.e., politically expedient expenditures. I objected, saying that one did not take it off the top. Instead, money is taken off the bottom. Rominger politely and quietly kept explaining the worthiness of the project and the free lunch that comes with direct transfer of funds from Washington.

I politely and quietly disagreed, in detail, several times. I finally said, "May I cut to the chase?" With Rominger's assent, I summarized the issue as follows:

(1) The secretary has committed to the Quincy Library Group. The Forest Service is to make it work. There is to be no further discussion.

(2) The chief will, in the next fiscal year, hold enough money off the top for use in projects like the Quincy Library Group. This will be a free lunch and there will be no indication that any unit budget went down because of that action.

I said I understood. It's legal, but is it ethical? The next item did not take long. My long-planned trip to Russia is cancelled because of election-year sensitivities.

7 March 1996, Washington, D.C.

I met in the morning with Forest Service Deputy Chief for Policy and Legislative Affairs Mark Reimers and staff concerning congressional activities to carry out the president's avowed intent to revoke the timber salvage rider. There are two moves going on in the Senate by Patty Murray (Washington), Mark Hatfield (Oregon), and Slade Gorton (Washington) in the form of amendments to the Appropriations Act. This of course means that Congress will address a critical natural resources issue without committee hearings and without more than superficial understanding by Congress and the public of the matter at hand.

The suggested amendments have only one thing in common. Both contain provisions to allow either buyouts or use of substitute volume or both to close out the outstanding timber sales under Section 318 of the 1990 Appropriations Act that present environmental problems or conflicts with the President's Forest Plan.

Murray's amendment says that any buyouts or additional above-budget timber sale preparation costs may be paid from the settlement fund (i.e., money set aside to pay costs of government settlements in legal cases). The Hatfield-Gorton amendment says that all costs must be handled within agency budgets, and that none of the costs can come from commodity programs (i.e., timber, mining, grazing, and some recreational programs). This would also result in massive internal transfers in the Forest Service budget from other areas in the country to the Pacific Northwest, which is already hugely disproportionately funded compared with the rest of the country.

Murray's amendment gets worse in the terms of pragmatic execution. She provides for an accelerated process to put up salvage sales that is no improvement over standard procedures. Appeals of salvage sales are reauthorized, and that includes all salvage sales, even those that have been awarded and are currently being cut. The legal liabilities could be enormous. The new definition of salvage would be almost impossible to make work.

Reimers and staff have a preliminary effects statement completed, which will be forwarded to the Office of Management and Budget once it has cleared the Department of Agriculture. That could take some time if the normal amount of delays and revisions take place. Whatever happens, it will be too late for this afternoon's meeting in the office of the director of the Council on Environmental Quality, Katie McGinty, to discuss the Murray Amendment.

When I arrived at McGinty's office for the meeting, which was for principals only (code for having no technical civil service present, only political appointees), the discussion of the amendment was well under way. There were a number of Justice Department lawyers in the room, headed by Peter Coppelman. The others included McGinty, Don Barry (Interior), Jim Lyons (Agriculture), and T. J. Glauthier (OMB), and they were the only ones who spoke.

Shortly after entering the room, I handed Lyons the as-yet-uncleared effects statement prepared by the Forest Service. In my mind, the meeting was decidedly premature as no specific analysis of the consequences had been, or would be, considered by the group. I was the only one in the room with any experience in on-the-ground forest management. Yet there sat the decision makers.

It was immediately clear that the group—or at least McGinty—liked the Murray Amendment as the way for the administration to back away from the salvage rider. As I started to explain the consequences of passing the Murray Amendment, which had obviously been written by hardcore environmentalists, it was clear that they really did not care about technical practicalities. The only interest I could detect was the political expediency of satisfying environmentalist-type constituents as the swing vote in the critical urban areas of Oregon, Washington, and California. The lessons learned from Ron Wyden's election to the Senate in the special election in Oregon have not been lost on these folks: the election was carried by voters in the urban Portland-Eugene corridor, which favored Wyden, while all the rural (timber) areas went to the conservative, Gordon Smith. The difference was perceived to be the proenvironment vote. As Oregon, Washington, and California are the key to a Clinton-Gore victory in November, and it is clear that the timber interests will not vote for them regardless, the strategy has now shifted to a proenvironment (antilogging) stance.

The obvious trick for the politically minded is to avoid being critical of either side in the conflict. Any upset from either side will be handled by attacking the Forest Service as too zealous or too bureaucratic or whatever. This technique was recognized as a successful gimmick some years ago by then-Governor Neil Goldschmidt of Oregon. The ploy has been widely applied by other politicians since, knowingly or unknowingly.

The capstone of the meeting was a retort to my statement that the Forest Service had not "discovered" the timber to be used as substitute volume to replace 318 sales. Rather, I said, this was the volume that had accumulated during the first four years of the president's plan, as each year's timber sale program increased toward the targeted level. The timber that was not cut was targeted to be cut, above the promised level, during the last three years of the decade. In other words, there was no free lunch here.

The reply was, "Before anybody figures that out, President Gore will be in the White House." I was a bit taken aback at such an attitude. This was not about good natural resources management. This was not about whether the salvage rider was being properly applied. This was about the presidential election in November, that and nothing else. These are not, I truly believe, unethical people. We simply operate from a totally different set of ethical values.

I came to Washington believing that some things, such as good natural resources management and sustainable use of those resources, should be above manipulation for political advantage. I now know better. Nothing is above manipulation for political purposes, absolutely nothing. Such is the essence of the acquisition and exercise of political power. My response was to chuckle when I heard former Speaker of the House Tip O'Neill of Massachusetts quoted as saying, "All things are political and all politics is local." He was dead on, and anyone in this town who doesn't understand that can never be effective.

And so the corruption of principle and idealism continues, and it comes not in great confrontations but little by little by little.

13 March 1996, Washington, D.C.

Secretary of Agriculture Dan Glickman called this morning to express his exasperation over an editorial in some California newspaper. The editorial concerned the transfer of funds from other national forests to satisfy the secretary's $4.5 million commitment to the Quincy Library Group for "collaborative" management of the Plumas National Forest. This was portrayed in the editorial as a broken promise by the administration that the money would not be taken from any other national forest.

Of course, there was no other place to get the money, since all but $15 million had been distributed to the field before the order was received from the secretary. The secretary intended that the money be taken off the top and then distributed, thereby camouflaging the fact that other national forests' budgets were being reduced to satisfy the demands of the Quincy Library Group. There is only one way to accumulate dollars at the Washington Office level: take it off the bottom—any resources held at the Washington level are resources that cannot be distributed to the field.

The secretary was quite exercised and said, "Why is it so hard to find $4.5 million out of a $4.5 billion budget to carry out my priority programs? I've heard from several people that you are not supporting my decision."

I became angry and replied, "Dan, that is bullshit! I told you my reservations and you decided to go ahead. We are executing your orders to the letter, using the mechanism available to us."

The secretary calmed down at that point and said he was just venting. I allowed that I could understand that, but I was getting more than a little irritated with being called on the carpet about the Quincy Library Group.

There was a meeting today of Forest Service representatives (Gray Reynolds, deputy chief for National Forest System; Mark Reimers, deputy chief for Programs and Legislation; and Steve Satterfield, program leader for Budget) with Brian Burke, assistant undersecretary for Natural Resources and the Environment, and Anne Kennedy, assistant to the secretary. The subjects were, once again, funding for the Quincy Library Group and AmeriCorps.

The atmosphere, once again, was hostile to the Forest Service, and we were accused of not being responsive to the priority programs of the president and the secretary. Reimers and Satterfield did a masterly job of explaining the intricacies of the budget, particularly the necessity to maintain the integrity of spending money within the limits of seventy-two separate line items. The problems were compounded by not yet having a final budget and the difficulty, due to government shutdowns, of attaining any reliable figures on the carryover of Washington Office funds from last fiscal year.

Quincy Library Group and AmeriCorps are now funded off the top. Until this week, it was not feasible to carry out that instruction of the secretary. Burke and Kennedy seemed relieved that we could now announce that none of the money going to the Quincy Library Group was taken away from any other national forest. Burke and Kennedy were still looking for a free lunch. Reimers carefully explained that the only difference was that all national forests were giving up resources to fund these projects, and not just forests in California.

This sleight of hand has even more serious political ramifications. This method shifts even more funds from "poor" regions to the "wealthiest" in the Northwest and makes an already-lopsided funding situation worse. It also shifts funds from regions with lower timber sale preparation costs to regions with higher costs. The result: there will be less timber produced with the same amount of money.

The bill that authorizes AmeriCorps specifies that no regular jobs will be sacrificed or forgone to finance it. It is a very difficult task during a period of laying off regular workers because of budget constraints to maintain that off-the-top funding of AmeriCorps personnel without displacing additional regular personnel. I fear that we are in violation of the law.

And I likewise fear that the Forest Service is flirting with violation of the Federal Advisory Committee Act in the working arrangements with the Quincy Library Group. However, the secretary has decided that he has the authority to order these actions, and it is the duty of the Forest Service to institute those actions. The Forest Service has expressed reservations up to the point of being openly labeled out of control and accused of following our own agenda, and other such terms.

In my gut, I know both of these actions are likely to violate the law, and are certainly less than forthright. If things go wrong (and that is a distinct possibility), we will testify that in spite of expressed reservations, the Forest Service carried out its orders.

The politicals (Burke and Kennedy) seem to have a better grip on the circumstances and difficulties, but they do not seem at all troubled with the questions of budget line authority integrity and flirting with exceeding legal authority. For them, receiving orders, or even their interpretations of desires of higher-ups, is adequate to justify action. My admonition to the Forest Service workforce to tell the truth and obey the law is easier said than done.

15 March 1996, Washington, D.C.

There were two press conferences on the Nye County decision—the Nevada case involving the so-called County Rights Movement—via conference calls. In the first, Lois Schiffer, Department of Justice; Mike Dombeck, director of the Bureau of Land Management; and I represented the government.

We followed the press plan to the letter and there were no surprises. In midmorning, Dombeck and I gave television interviews to CNN. Again, we stuck to the plan as close as the questioning allowed.

In the next conference call, Peter Coppelman of DOJ replaced his boss, Lois Schiffer. Coppelman did not stick to the plan and was decidedly aggressive toward the Nye County commissioners in a way that could be described as in your face.

This was not the approach that Dombeck and I had taken in the earlier press conference and in the TV interviews. After the conference, we were both seriously concerned that our efforts to tone things down had been negated. We could feel hackles coming up across the West, and we had aimed for exactly the opposite result.

19 March 1996, Washington, D.C.

This was the day of the weekly meeting of the "Tuesday Afternoon Club" at the Council on Environmental Quality. This proved a contentious meeting, as time has run out on delaying the cutting of the so-called First and Last timber sales on the Umpqua National Forest. These sales were original 318 sales that had been withdrawn by the Forest Service for environmental reasons. They were ordered released by the now-infamous salvage rider. The sales were released by my order pursuant to court instruction. The purchaser has announced that cutting will begin tomorrow morning.

Without consulting or informing me, Undersecretary Lyons opened face-to-face negotiations with the purchaser to forgo the cutting of the First and Last timber sales in lieu of being provided alternative timber. Lyons conducted these ad hoc negotiations without a full understanding of the surrounding circumstances, including the law, applicable regulations, relationship to the President's Forest Plan for the Pacific Northwest, and the situation on the ground.

The Forest Service line officers and I were significantly irritated by Lyons's actions, which were variously interpreted as grandstanding, political meddling, or a direct insult to the agency. I was sufficiently irritated to tell Lyons and Deputy Assistant Secretary Gaede that if such interference (i.e., micromanagement) in Forest Service activities did not stop, I was resigning as chief.

The meeting at CEQ was dominated by lawyers (twenty-two of twenty-six persons present were lawyers). Discussions of legal strategy, often with several persons speaking at once, went on for several hours. I was amazed that during all of the discussions of legal strategy, not one person ever bothered to address the operative question: Do the Forest Service and Bureau of Land Management have the legal authority to do the job, and if so, do they have the promise of additional personnel and financial resources? If the answer to either question is no, the debate over legal tactics is a waste of time. The answers at this moment are that the land management agencies lack both legal authority and resources to prepare the necessary alternative timber volumes to replace 318 sales.

The only feasible course of action is dependent on quick formulation of an emergency rule under the National Forest Management Act to allow offering timber volume without full compliance with the normal procedures. This cannot be accomplished any more quickly than the middle of next week. The entire premise of the emergency rule is legally debatable, at the very best.

My recommendation is to proceed with the cutting of the First and Last sales—which is likely to be mandatory in the end, anyway—and let the proponents take the credit and the political heat. Lois Schiffer of the Department of Justice concurred with that recommendation, with the caveat that "someone high in the administration," preferably the president or the vice-president, make the announcement that all efforts to stop the two sales had failed and that the responsibility was now on the timber industry. The intent of such action would be a political ploy to focus displeasure on the timber industry and their Republican allies.

I objected strongly to that suggestion on two grounds. First, the purchasers of these timber sales have operated in good faith and do not deserve, individually at least, to take the political heat on the issue. Second, it is irresponsible for the political games to focus wrath on the Forest Service and BLM field people. It is one thing to sit in a conference room discussing strategy and quite another to be doing work on the ground with the only thing between your hide and a bullet being a shirt.

It was quickly clear that this course of action was not yet politically tenable. My second suggested course of action was to offer timber volume being prepared under the President's Forest Plan, on the theory that it would be better to risk violating the provisions of the salvage rider than to risk violating the plan, which would lead to a renewed injunction by Judge William Dwyer. That seemed to be the consensus. However, CEQ General Counsel Dinah Bear, who chaired the meeting, thought the final decision had to be made by Katie McGinty, CEQ director. Arrangements were made to meet with her at 6:00 P.M. in the Old Executive Building.

Katie McGinty acceded to the strategy of offering the purchaser timber volume being prepared under the President's Forest Plan as a substitute for forgoing cutting the First and Last timber sales. She fully understood the drawbacks.

During that discussion, Deputy Assistant Secretary Mark Gaede was called out of the room to take a call from Jim Lyons. He returned to say that Jim had just finished a conference call with the purchaser, who had offered to delay beginning cutting operations for one additional day (i.e., cutting will begin the day after tomorrow). He also sent word that he would bow out of discussions and let the Forest Service deal with the matter, which may be a little late at this point.

And so the discussion ended with a firm decision from Katie McGinty as to the course of action to be pursued. This is the first time in one of these sessions that I have seen a clean, clear decision made by a responsible individual.

26 March 1996, Washington, D.C.

Late in the afternoon, I had an hour-long conversation with CEQ General Counsel Dinah Bear. Several interesting points emerged. The first was that the political appointees dealing with timber issues "do not consider you one of us." She said that I made my dislike for the "necessary political decision" obvious and that "the Forest Service pursued its own agenda," which frustrates the politicals.

I told Dinah that in my opinion, the entire approach to dealing with the salvage rider was chaotic and getting worse because of an unclear and fractured decision-making process and the absence of a table of organization and clear lines of authority and responsibility. The conversation was frank but friendly.

I suggested to Dinah that if I were in a position of authority and actually felt the way the political appointees feel about the chief and the Forest

Service, I would ask for the chief's resignation. And if that happened, I would be more than happy to resign, having done my duty and finding myself in increasing disagreement with what is going on. Since I can visualize that the situation will only worsen as the election approaches, I offered one more time to guide an effort to develop a table of organization and a campaign plan to steer us through the deteriorating situation in the Northwest. She was noncommittal and really only has the power to pass on what I say to others. No one so lowly as an agency head ever speaks to real power, and rarely in a straightforward manner.

10 April 1996, Washington, D.C.

Brian Burke, USDA assistant undersecretary for Natural Resources and the Environment, and I met for nearly three hours. The purpose was simply to talk matters over. Burke was transferred from the White House to the assistant undersecretary's job to "repair the disconnect" between the Forest Service and NRE (i.e., the administration). He frankly admitted that he came into his present position firmly convinced that the problem lay with the Forest Service, which is seen as intransigent, resistant to change, and dominated by white male foresters (the timber beasts) all raised in the Forest Service. However, Burke has come to realize that those perceptions are for the most part incorrect and that the facts simply don't fit the perceptions. Now, how do we deal with the misperceptions?

Burke listened to a number of the aggravations and frustrations of mine and of others in the Forest Service. I was forthright in saying directly that in terms of natural resources management, this is the most confused administration in my memory. It is simply impossible to know, from hour to hour, who is in charge. Orders can and do come from anywhere in USDA or the Council on Environmental Quality, the Office of Management and Budget, or the Department of Justice. These orders or inquiries may be inserted into the Forest Service at any level. This leads to considerable confusion and is demoralizing to the workforce.

The next pressing problem is the increasing inability to get any speedy decisions out of our political bosses. Whether these delays are related primarily to unwillingness to make decisions, tough or otherwise, to sheer incompetence, or to decision overload, I don't know. These folks, having little trust in the agency's personnel and no real experience with running a large, complex organization with a high political profile, bring more and more decisions to their own level. This lack of experience and trust in the agency is coupled with the fact that the politicals don't trust each other, and therefore no single person is able to make a decision without extensive consultation with others spread across the government. As a result, decisions are simply stacking up in USDA, and the situation is getting worse.

There are problems associated with selection of personnel. The chief has been delegated authority for personnel actions up through GS- or ST-14. In practice, more and more of these personnel actions are being reviewed by the undersecretary. It had been routine for the chief to recommend personnel actions for Senior Executive Service positions and for GS- and ST-15 or above positions and have those suggestions routinely approved. This is no longer true, and these positions are scrutinized over and over by the undersecretary. I have not been able to discern whether this is a strategy to wear down the agency until it complies with selected political suggestions in order to get the major chunk of actions approved.

It is clear that there is increasing involvement in the personal selections of Undersecretary Lyons for these positions. And these choices are based strictly on his opinion and preference. In many instances it is evident that the person selected, according to Forest Service teams, was clearly not the best or most deserving candidate. This has two demoralizing aspects on the process of personnel selection. It destroys confidence in the system and over the long term will put the onus on personnel selected for promotion by an undersecretary. They will also be marked worthy of suspicion by the Forest Service's political overseers when the political party in control of the White House changes. This is, in my opinion, bad business for the Forest Service, the people we serve, and the individuals involved.

17 April 1996, Washington, D.C.

There was a hearing today in front of the House Committee on Resources and chaired, initially, by Don Young of Alaska. The witnesses were agency heads of the Forest Service, Fish and Wildlife Service, National Marine Fisheries Service, Army Corps of Engineers, and Bonneville Power Administration. Mr. Young described the purpose of the hearing as "proving that the secretary of the Interior is a liar" and to "reveal to the American public exactly how much it costs to carry out the Endangered Species Act."

It seems that Secretary Babbitt had said that the Fish and Wildlife Service's budget for administration of the Endangered Species Act was a paltry $44 million or so, and that sum was suggested for costs by the Congress. Congressman Young chose, for obvious political purposes, to say that Babbitt indicated that $44 million was the total cost to the federal government for executing the act.

Young railed at length about Babbitt being a liar and how he (Young) intended to expose him as such. Young promised to haul the secretary before the committee, put him under oath, and show him up for a liar.

Mr. Young told the witnesses that he ordinarily swore in any witness from the administration. As he knew each of us, he would forgo swearing us in. However, we were warned, if any of us lied to him, we would surely be sworn the next time we were brought before his committee.

Young's behavior was an embarrassment for Congress. All but the most hardcore right-wing Republicans on the committee were obviously chagrined. When Young left the room after the opening statements and his histrionics, he put Congressman Pombo of California in the chair. Pombo is a first-term hard-right Republican with a real hatred for the Endangered Species Act. In fact, he is the primary author and sponsor of a bill to reform (actually gut) ESA. Even Pombo was obviously embarrassed by Mr. Young's behavior and made a number of, for him, conciliatory statements toward the witnesses and Secretary Babbitt.

If the purpose of the hearing was to point out that federal agencies spend significant resources on ESA conformance, the committee succeeded. The Forest Service, for example, spends at least $40 million to $50 million per year in direct costs of management and research. However, as I pointed out in my testimony, the Forest Service was obligated by the National Forest Management Act, the National Environmental Policy Act, the Multiple-Use Sustained Yield Act, and the regulations issued pursuant to those laws to carry out, in essence, the basic requirements of ESA.

It was quite clear that the agencies represented at the table collectively spent well over a billion dollars in direct and indirect costs to carry out ESA. The intent of the hearing was to demonstrate that dramatic revision of the act was needed and that the bill offered by Mr. Pombo was just the ticket to make that happen.

As is usual in these hearings, there were only brief appearances by a couple of Democrats, and their performances were lackluster, to say the very least. We, as usual, were left to the tender mercies of the Republicans, primarily the freshmen—Cubin, Chenoweth, Pombo, et al. Overall, the hearing was poorly attended by committee members. I would guess that well less than a sixth of the committee showed up at all.

After Chairman Young left, the hearing was actually quite mild compared with the usual turkey-shoot atmosphere that prevails in the hearings of the natural resources committees in both House and Senate. It is a sad state of affairs in that ESA does need to be revised. But the issue is such a hot potato in an election year that the Republicans must attempt to assuage their constituents by gutting the act, and the Democrats see the most political mileage in standing with the greens to protect the status quo. The result is that there will be much sound and fury but nothing will come out the Congress this year. If something does come out, the president will relish the chance to cast an advantageous veto just prior to the election.

The questions to witnesses were directed primarily to Mollie Beattie, director of the Fish and Wildlife Service. Mollie is a formidable person and has just returned to duty following a second operation to remove a brain tumor. She was obviously suffering the effects of chemotherapy but still full of spunk. The combination made it very difficult for the Republicans to really pursue her with their usual vigor, or just maybe, the issue of ESA has not proven the political winner that the Republicans—particularly the freshmen—initially thought it would be. The room was packed with people wishing to observe the proceedings but the dais was noticeably sparse in congressmen. It was, all in all, a sorry show.

This is the week for hearings. Also today the Senate Natural Resources Committee, chaired by Frank Murkowski of Alaska, held hearings on the Tongass land management planning operations. The draft plan was being released as the hearing proceeded. This was Murkowski's show. The supporting cast was composed of Senator Ben Nighthorse Campbell of Colorado, Larry Craig of Idaho, and Ted Stevens of Alaska. Campbell said something to the effect that he loved animals but he really cared more about jobs. Craig in essence said "Howdy" and left. Stevens was his usual vitriolic self and said that a General Accounting Office report on expenditures in Alaska had shown that money had been misused (it actually said just the opposite) and that the responsible Forest Service officials should be punished. And as he left, he said that by this time next year, the Tongass National Forest would be turned over to the state of Alaska.

Murkowski started off by swearing in all the witnesses "so that all could tell the truth without fear of recrimination." The purpose of that maneuver was soon very clear. Murkowski had insisted that Steve Brink, former coleader of the planning team, who had been relieved and assigned to other duties because of the stresses he produced on the planning team, be a witness.

The first hour's questioning was directed to Brink. It was clear from Brink's actions that he knew what questions were coming from the committee and that he was giving prepared answers, while appearing (or trying to appear) to speak off the cuff. There, likely, is the direct pipeline from Janik's office to Murkowski. The situation was so contrived that I was not even offended—just disgusted.

At one point, while Murkowski was deriding the economic and social assessment, committee staffer Mark Rey held up a Ouija board. Even Rey seemed a bit embarrassed. Murkowski plowed ahead, on his own, for nearly seven hours of testimony. I left after two hours when I saw what lay ahead. This was strictly showboating for Alaska radio and television, and there was no sense in making the chief of the Forest Service into a foil and straight man for The Frank Murkowski Comedy Hour.

Dealing with Murkowski, Stevens, and Young would almost be entertaining in terms of the vaudevillian theatrics they employ if so much were not at stake for the long-term management of natural resources in Alaska. I am not angry or appalled sitting in front of these men. Rather, I feel sad and embarrassed for my country—and concerned for the long-term welfare of Alaska.

I caught up with Regional Forester Phil Janik and his entire crew at supper. They were in good spirits and believed they had performed well and that the report and the selected alternative would stand up.

18 April 1996, Washington, D.C.

I gave the briefing on replacement timber to Katie McGinty, director of the Council on Environmental Quality, and the usual cast of characters. There was much discussion and much replowing of old ground as to possible ways to dodge the bullet of cutting the old-growth 318 sales. Finally, the inevitable was made clear—there is no way out. It might be possible to buy out or trade out of the worst of the worst of the 318 sales, but in the end, many (perhaps most) of those sales will be cut. As that cold fact came to be accepted, the discussion turned to the politics of how to put the onus on the Republican Congress and give the president some cover. I did not participate in that discussion except to point out that the timber industry did not consist solely of timber barons or chief executive officers. For every one of those people, there are hundreds more who carry a lunch box to work each day, whose jobs are at risk.

McGinty eventually looked at me and asked for my recommendation-finally. I said that the first thing to do was recognize that the President's Forest Plan was already compromised and would be further compromised. That means that the plan would have to be revised—and such work would have to begin soon.

In the meantime, efforts should focus on saving as much of the basic structure of the plan as possible. This means that in conjunction with the Bureau of Land Management, Fish and Wildlife Service, National Marine Fisheries Service, and the Environmental Protection Agency, the most damaging of the 318 sales must be identified. Every effort should then be made to prevent the cutting of those sales through any or several of the following actions: substitute volume prepared in compliance with the plan; substitute volume not in compliance with the plan but significantly less damaging than cutting the 318 sales; buyout after contract cancellation; sale modification as negotiated with the purchaser; etc.

Simultaneously, an emergency edict should be issued to do two things: (1) stop all further inclusion of any old-growth forests in the matrix, and (2) begin revisions of the forest plans. Then see what adjustments should be made to compensate for the cutting of some 318 sales. I believe this approach is in complete obedience to the law, which allows a response to an emergency situation while following the instructions in the timber rider to the 1995 Rescission Act.

After another half-hour or so of the same old discussions, McGinty said, "These meetings are a waste of time. Secretaries Glickman and Babbitt are going to have to take more responsibility for making decisions." I have no idea what finally brought her to that conclusion, but I could not possibly agree more with her. Maybe now some order can be brought to the situation and the issue assigned to the field troops for execution.

It seems most likely that there will be no easy or completely acceptable political way out of this deepening quagmire. So it is time to move the responsibility out of the White House to somewhere else so that there is potential for the blame (if "failure" occurs, and it will, to some extent) to fall somewhere other than the White House. My cynicism runneth over.

23 April 1996, Washington, D.C.

There was a meeting today with a contingent from Nye County, Nevada, which included two county commissioners, a "consultant" (i.e., lobbyist), and the county planner. Their leader was Dick Carver, who has achieved his fifteen minutes of fame as the highly visible leader of the County Rights Movement. He carries a copy of the U.S. Constitution in his shirt pocket and has espoused the idea that the states are the owners of the public lands. Carver also defied the Forest Service by mounting a bulldozer and opening a closed road, almost driving over a Forest Service enforcement officer who was holding a sign telling Carver to cease and desist.

The Department of Justice took Nye County to court to settle the issue. The county, and the County Rights Movement, lost totally and accumulated some big legal bills in the process. Last week, Nye County offered to settle the case by agreeing not to appeal the decision, if the government would agree not to prosecute any county officials and to drop the remainder of the counts. They further suggested a formal agreement in which we (the feds) stipulate that the county had rights to special consideration in the management of federal lands within the county, plus a formal memorandum of agreement as to how the county government and the federal land management agencies would deal with one another.

The Nye County contingent was no longer defiant or even antagonistic. Carver, in particular, was polite to a fault and advocated cooperation and coordination. I probably should not look with disbelief at someone who has just undergone a "come to Jesus experience" and has therefore seen the light. They lost, and Carver in particular bears the responsibility for running up big legal bills on what has now been clearly revealed as political bunk. It is likely that the citizens of Nye County are upset about this expenditure of their tax dollars and the notoriety that has come with the entire episode.

After listening to their offers of cooperation, I made several points. I was not in charge of the ongoing litigation and could not comment on their settlement offer. There will not be any concession or granting of comanagement over the federal lands. However, the Forest Service would recognize the arguments of the past and the court actions as symptomatic of a serious problem that must be addressed. I quoted Pinchot's observation that for the forest reserves to be maintained, it is essential to get along with the people who live within or near those forests. I said that the Forest Service had been a part of western communities for nearly a hundred years and that we intended to continue for another one hundred years.

I turned to Dick Carver, who was seated on my left, and said that our hand was always out to our neighbors, in spite of differences that might arise from time to time. With that, I put out my hand to Carver and he took it. Then, looking him straight in the eye, I said, "There are a number of permittees that, in the spirit of the moment, violated their grazing permits or in some other way willfully violated law or regulations. We have held off on action until we had a clear decision in the Nye County case. Now that there is no doubt as to Forest Service authority over the national forests, we will move forward to enforce the applicable laws and regulations. That is a fact, not a threat. Therefore, I suggest that it might be a smart move for those folks that are in willful violation to meet with the appropriate Forest Service people and arrange to get right with the law, regulations, and agreements. We don't want to inflict punishment and are willing to go some distance to reach a solution, but the laws and regulations will be obeyed."

Still concentrating on Carver, I encouraged the county commissioners to use their influence to encourage people in Nye County to get any problems straightened out—and soon—to prevent confrontations that they will lose. Carver, without hesitation, agreed to play that role, if we would provide him with the names and circumstances of Nye County residents who are in violation.

I further promised that the regional forester (Dale Bosworth) and the forest supervisor (Jim Nelson) would come to Nye County to visit with the county commissioners and see what can be done to ensure a better Forest Service–Nye County relationship.

This seems a bit too good to be true; however, there is no real alternative to making a good-faith effort to make a new start. And in justice, I think that these folks have some legitimate concerns. Where such concerns exist, they should be clearly identified and quickly addressed. These folks are our constituents and, in many ways, our partners. Our mission is caring for the land and serving people. These people are a significant part of the constituency we serve.

24 April 1996, Washington, D.C.

This was the day of a hearing before the Appropriations Subcommittee for Interior and Related Agencies, chaired by Senator Slade Gorton. Undersecretary Lyons was the lead administration witness. He did not follow his script and early on quoted figures from the draft Resources Planning Act report that mentioned the contributions to the gross domestic product that come from various activities that take place on national forests. The contribution from recreation was hugely greater than that from timber production and harvesting. I flinched when I heard Lyons read the figures, as I know, without a shadow of a doubt, how the timber industry and Senator Ted Stevens of Alaska, who was, as usual, glowering from the dais, would react. Stevens pounced almost immediately. He did not comprehend that the disparity came from the dramatically increasing amount of recreational use that is occurring on national forests almost independent of any Forest Service action. It simply does not matter if the recreation contribution to the gross domestic product is 100 percent too high and the contributions from timber are 100 percent too low; the differences in favor of recreation are enormous and the disparity is growing rapidly.

Senator Stevens did not, as is his custom, ask any questions except in a rhetorical manner. He lectured us about the terrible way Alaska is being treated and chastised us for the decline in the timber program. He ended with the rhetorical question, "If recreation is so important and the timber program so much less significant in terms of the gross domestic product, what do we need with a Forest Service?" He announced in a press release later in the day that he would hold a series of hearings to determine how best to break up the Forest Service and divide its functions among other agencies.

As usual, the hearing was a series of questions and expressions of concern by various senators regarding quite parochial concerns. There was essentially no evidence of any senator's having any view of or concern with broader natural resources issues. I did try to introduce some such issues. I raised the issue of how firefighting costs are addressed and made the plea to cut Forest Service management some slack in closing locations and passing back administrative overhead in response to budget cuts and the desire to shift as many resources to on-the-ground management as possible.

Chairman Gorton ended the hearing by noting that our requested increase of 3 percent over 1996 was modest and would not cover inflation. However, he warned us that such an increase was not likely to be granted and told us to begin work to identify priorities for further cuts.

With the exception of Senator Stevens's standard performance, the hearings were friendly enough but, typically, poorly focused and parochial. There simply are no senators who seem to know or care much about natural resources management. Senator Gorton is the best informed, but he has little real interest outside the Pacific Northwest.

28 April 1996, airborne between Atlanta and Washington, D.C.

I spent the past three days in the Forest Service's southern region, enjoying a pleasant respite from the Washington, D.C., scene. This trip to visit the regional headquarters in Atlanta has been much delayed. At least three times I have had a visit scheduled and had to cancel at the last minute because of the press of urgent business of one kind or another. This visit completed my pledge to visit every regional headquarters. My consolation is that I have visited varying locations in Region 8, including North Carolina, Georgia, Florida, and Tennessee, before I got around to the visit to Atlanta.

Regional Forester Robert Joslin met me at the airport and took me to his home for the night. That gave us a chance to visit about the state of Region 8 and to discuss personnel matters.

The next morning began with a one-hour meeting with the Region 8 leadership team. The most glaring problem seems to be the pending forced move of the regional offices from the present location to a new federal building in downtown Atlanta. The General Services Administration is compelling the Forest Service to move to these less convenient and much more expensive quarters in order to fill up the space. This building, representing political pork at its best, was provided by Congressman Lewis of Atlanta. The building is in the downtown area, which, it seems, is of greater significance than either convenience to the public or employees or the significantly increased costs to the taxpayers.

However, times have changed since Mr. Lewis obtained this prime piece of pork. The Democrats are no longer in power in the legislative branch of government and dollars are short. The owner of the building in which the Forest Service's offices are currently located has begun to raise questions—and a bit of hell—with the secretary's office and with Speaker of the House Newt Gingrich. I suspect some Forest Service employees are putting in whistleblower complaints on the matter. Just maybe this fiasco can be headed off.

The middle of the day was spent at the Conference of Black Mayors in downtown Atlanta. The Forest Service had some sixteen or so people staffing booths at the convention. I don't think my presence did much to impress the black mayors (politicians are politicians). But I do think my attendance made a significant difference to black Forest Service employees. If that indeed occurred, it was well worth my time and effort. Pride is a significant possession and this was about pride.

By midafternoon, we were on the road to Coker Creek, Tennessee. The trip had two purposes. First, to check out the progress on the construction of the Olympic whitewater venue on the Ocoee River. Second, to participate in the dedication of the reconstructed Tellico Ranger District and attend the associated reunion of members of the Civilian Conservation Corps who served at that location in the late 1930s. These men were the "tree army" who toiled for a dollar a day to bring back the forests of the southern Appalachians.

The visit to the Olympic whitewater river venue was satisfying as it is obvious that everything will be ready in time and it will be a top-rate facility that will serve for many decades to come. Again, I could see pride in the faces of the Forest Service people who were making it happen.

As the plane nears Washington and I finish these notes, I realize that perhaps I overestimate what my visits to the field accomplish for Forest Service people. But there is no doubt what contact with the field people means to me. They persevere. They achieve. They are proud. They inspire me. The struggle is worthwhile, it truly is.

30 April 1996, Washington, D.C.

Today was the House Subcommittee on Forests hearing (Congressman Hansen's subcommittee) on proposed Forest Service regulations on use of the Snake River where it flows through the Hell's Canyon National Recreation Area. Hansen was in the chair and, as is not uncommon, the only representatives present were Helen Chenoweth of Idaho and Wes Cooley of Oregon: the increasingly famous Wes and Helen Show was on stage.

This is an ongoing saga of how to ration use and protect the resources of this desert river. A decade or so ago, the Forest Service limited the number of launches of both recreational and commercial floaters. This decision was necessitated by the limited number of campsites and the limited capability for launching rafts at a single site below Hell's Canyon Dam. Conversely, a simultaneous effort was made to limit jet boats, which was delayed for ten years or so by the decision of then–Undersecretary of Agriculture John Crowell. During those ten years, the number of jet boats and commercial operators continued to increase steadily, and problems of conflict went up proportionally, or more.

If there was ever a piece of ground that has been abused by overgrazing, it is the Snake River canyon. When I first rode the trails of the canyon some twenty-two years ago, I was appalled at the range conditions and the domination of exotic vegetation, particularly cheatgrass. Over the past twenty-two years, things have slowly changed for the better as domestic sheep numbers were reduced dramatically and the remaining bands more carefully controlled and monitored. The area reacted much more rapidly than I would have dreamed possible with a resurgence of bunch grasses, dominated by blue bunch wheatgrass and Idaho fescue.

Then efforts were made to reintroduce bighorn sheep to Hell's Canyon, from which they had been extirpated sometime early in the twentieth century. Bighorns would have done well if they hadn't repeatedly contracted diseases from domestic sheep. In 1994–1995, the supervisor of the Wallowa-Whitman National Forest, Robert Richmond, made the tough decision to eliminate domestic sheep grazing from the Hell's Canyon National Recreation Area. That action did not go down well with county governments with lands in the canyon or with livestock interest groups and their congressional representatives.

That set the stage for the politics that surround the issue of regulating jet boat use. Two years ago, Richmond put forward a plan to limit the number of commercial jet boat launches and to reserve three days per week for the "wild" segment of the river for float boaters. The decision was appealed to the regional forester by a coalition of commercial jet boat operators. John Lowe upheld the forest supervisor on all counts, except that he directed the supervisor to do a further assessment of the economic impact of the decision on jet boat commercial use, both as a group and individually. That assessment has now been completed, and Richmond is ready to move ahead with his decision. The game has been played out so far as the administrative procedures of the Forest Service are concerned. It is now time to move the game into the political arena. Hence, this hearing.

The hearing was of the oversight variety—the purpose is to allow some members of the subcommittee to posture for the home folks and to pay off political contributors. The members in attendance were testimony to that observation.

The "questions" and statements by Chenoweth and Cooley centered on two points. The first was that the Forest Service's actions were illegal and unconstitutional and contrary to the desire of Congress as expressed in legislation. Chenoweth and Cooley read to the witnesses from various laws, oddly not including the primary operative legislation, which is the act designating the Hell's Canyon National Recreation Area.

The second point was that there was no conflict between jet boaters and any other users of the Snake River canyon.

Chenoweth and Cooley got their posturing done but in my opinion took a real whipping on the facts and in the arena of public opinion. The final punch was reserved until the last moments of the hearing, when I read the operative sections of the Hell's Canyon Act that specifically, and quite clearly, allow the regulation of both jet boating and float boating in the canyon. From the look on their faces, Chenoweth and Cooley seemed shocked that their staffs had missed the salient point of law, or perhaps they assumed that we did not know the law.

And so yet another hearing has been endured and some of the members have satisfied their political constituencies. The Forest Service will now, unless our political overseers jerk our chain, proceed with the regulation of jet boat use. And then the appeals and lawsuits will inevitably follow.

2 May 1996, Washington, D.C.

Senator Larry Craig of Idaho has introduced a bill concerning the management of the Snake River within the Hell's Canyon National Recreation Area. The bill is solely concerned with preventing the Forest Service from instituting any regulation concerning boat traffic on the river. This is accomplished by declaring that by law, no conflict exists between float boats and motorized boats. Therefore, by law, there is no problem. Further, the conditions that now exist in terms of boat traffic are, by law, the perfect situation and will be maintained into perpetuity. And that is that.

There were some ten to fifteen witnesses waiting to testify when the hearings began. Most of these witnesses had flown in at considerable cost in time and money to testify. Senator Nighthorse Campbell (Colorado) came in to chair the subcommittee session and announced that there were a number of votes lined up on the floor of the Senate that would occupy the senators' time. Therefore, the senator gave the witnesses the choice of staying around to testify before a subcommittee staffer or going home and coming back another day. The testimony proceeded. Senator Campbell left. I read my testimony in opposition to the bill and left.

My only conclusion was that this was just a drill to mollify Senator Larry Craig, and to let him have his moment to mollify in turn his constituency. Another day in Fun City during the silly season of an election year has passed.

7 May 1996, Washington, D.C.

The long-standing conflict in Congress over funding of the Columbia basin assessment by the Forest Service and the Bureau of Land Management came to an end with the passage of the 1996 Appropriations Act, albeit seven months late. Republican Congressman George Nethercutt of Washington and Republican allies such as Helen Chenoweth of Idaho and Wes Cooley of Oregon had attempted to limit the funding to complete the project to $600,000, rather than the $5,700,000 actually required, and to impose a time limit. They further wanted to prohibit any publication of an environmental impact statement.

These efforts failed as President Clinton insisted that he would veto any appropriations bill that contained this action and won the point. This hearing was Chairman Ralph Regula's sop to Mr. Nethercutt, so that the latter could posture for his constituents and put a good face on the loss.

The hearings did not go well for Mr. Nethercutt as his witnesses were made up of Helen Chenoweth, a county commissioner that was opposed, Jim Geisinger (a flack for the timber industry), and a labor union representative. The more telling witnesses were John Howard, a county commissioner from Union County, Oregon, who represented some 200 counties within the assessment area and testified in support of completion of the assessment and planning efforts. Paul Brouha, executive director of the American Fisheries Society, testified on behalf of a number of natural resources management professional societies and in full support of completing the effort.

The last panel was made up of six agency professionals, including the chief of the Forest Service and director of BLM. When this group came on, Chairman Regula turned over the chair to Mr. Nethercutt for Nethercutt's time in the sun to posture for his constituencies.

Nethercutt displayed a totally different demeanor than in previous hearings. He had visited Walla Walla, Washington, for a briefing on the operations, particularly those of the science team. He professed to be impressed with their talent, dedication, and hard work. His questioning was fair and intelligent. Having lost the battle to close down the operation, he seemed cognizant of the desirability of working within the Forest Service's operations to satisfy the concerns of his constituents. These concerns seemed to be centered on encroachment of the federal government on private land rights and ensuring that local people have a significant chance for input into individual national forests or BLM districts. There was a real concern that came through the questioning over a one-size-fits-all approach to standards and guides that will emerge from land use plans.

The witnesses were unanimous in saying that the Forest Service and BLM had no intention of seeking any authority over private lands and claimed no such authority, that standards and guides should be derived to fit localized conditions, and that extensive outreach and public participation would be sought at both macro and micro scale.

The oversight hearing seemed to serve its overall purpose, which was to allow the congressman to save some face and to posture for his constituents as a subcommittee member who failed to deliver the goods but was still fighting the good fight. Using Congresswoman Helen Chenoweth gave her a chance to mollify her constituents as well.

The witnesses, particularly the Forest Service witnesses, used the opportunity to make short speeches in response to Nethercutt's questions to refute testimony by Chenoweth, Geisinger, and the odd-man-out county commissioner. It was a good political day's work with both sides given a chance to say what they needed to say for their constituents and for the record.

8 May 1996, Washington, D.C.

The administration has grabbed on to the big fires in Arizona and New Mexico as an opportunity for political attention. As a result, Assistant Undersecretary Brian Burke has requested increasing amounts of information dealing with fires. Secretary Babbitt has seized this political opportunity. Babbitt has been on the firelines as a red-carded firefighter. Such has gotten the secretary's picture in the newspapers and mentions on the evening news. I suppose that such is also expected to have a positive influence on firefighter morale, and maybe it does. But among the seasoned firefighters I talked to, the reaction is bit more cynical or perhaps amused.

As a result of the administration's growing interest, we presented a briefing at the Old Executive Building to the White House "crisis team" on the current fire season and what potentially lies ahead for the rest of the year and beyond. Brian Burke, Mary Jo Lavin (associate deputy chief for Fire and Aviation), and representatives from the Park Service and the Federal Emergency Management Agency accompanied me.

I spoke frankly and told them that the Forest Service was seriously underfunded for firefighting because of personnel downsizing, budget restrictions, personnel rules limiting the number of hours that can be worked by temporary employees, aging air tankers, ownership of air tankers, and inadequate attention to reduction of fire danger.

I clearly stated that both the administration and the Congress had played an increasingly ridiculous game of pretending that the cost of fighting fire was unexpected, and that when fires did occur and the political pressure picked up, it was necessary to react to this "unusual" and "unexpected" phenomenon by borrowing from the Knutson-Vandenberg trust fund. These borrowed dollars were then repaid by means of a supplemental appropriation. This elaborate charade continues because it makes the Forest Service budget appear to be hundreds of millions of dollars less than it really is.

I said it was time to cease this charade and fully fund a fire management organization as opposed to a firefighting organization. This fire management organization would be funded at the level of the highest-cost fire year anticipated. In years when fire occurrence is below that maximum level, the fire management crews would be used to conduct controlled burns, mechanical fuel reduction, and stand thinning to reduce danger of catastrophic fire occurrences, particularly in the wildland-urban interface. Over time, this level of expenditures should be cost efficient as fuel reduction efforts reduce firefighting costs. And stand thinning should also pay dividends in increased timber yields.

The audience quickly grasped the fiscal, management, and political ramifications of this pitch. That was encouraging, but I suspect their attention span for the issue will be short and, ultimately, limited to the politics of firefighting.

However, I think it is worthwhile to continue staff work to package all the facts and figures to achieve the highly desirable goal of the creation of a true fire management–forest health organization within the Forest Service. It is time to seize the moment that is developing in regard to wildland fire, forest and rangeland health, and the problems of the urban-suburban-forest-rangeland interface. Sooner or later politicians will want something to grab on to, after or during a catastrophic fire season, to mollify public demands to "do something." Whenever that happens, it will pay to be ready.

13 May 1996, Washington, D.C.

I received several phone calls from the field and from other agencies in Washington to the effect that the decision had been reached that Gray Reynolds (deputy chief for National Forest System) and Mark Reimers (deputy chief for Budget and Congressional Affairs) are to be replaced and that Undersecretary Lyons is contacting potential replacements. The prime scuttlebutt was that the desired replacement for Reimers is Phil Janik (regional forester in Alaska) and that Mike Dombeck (director of the Bureau of Land Management) is to take Reynolds's slot. And Secretary Glickman is to present this to me on the morning of May 18.

I am constantly amazed at the duplicity and ineptitude of some of the folks associated with the administration. After three and a half years, amateur hour continues unabated.

My first reaction was to dismiss the information as pure rumor. But after about the sixth call, and considering this information in conjunction with the long delays in approving some pending Senior Executive Service positions, the plot—or parts of it—began to crystallize in my mind. Lyons has consistently resisted Bob Williams's appointment as regional forester for the Pacific Northwest, in response to "concerns" raised by "the regulatory agencies" and by Governor Kitzhaber of Oregon (one wonders how the governor reached that conclusion). Williams had been deputy regional forester under Mike Barton in Alaska and then acting regional forester there, until Phil Janik arrived. Likely he is "perceived" (a favorite word used frequently to justify action against someone, since perception for these folks is reality) as being close to the Alaska delegation.

By this maneuver, Lyons ("they") could knock out Gray Reynolds, vacate the Pacific Northwest region job for a player to be named later, and placate the Alaska delegation by sending Williams back to Alaska. That's not too hard to figure.

Now the question becomes, do "they" want me out as well? The information coming in is that they do not. How could that be? How can they believe that I would abandon these people and accept the insult of their having attempted to arrange this without my knowledge? The only thing that makes any sense is that based on previous experience, they believe they can prevail upon me to "put the agency's welfare first" and acquiesce in the actions. Or that they believe the chief's job means so much to me that I would accept in order to keep the job. I am sorely disappointed that after two and one-half years on the job, they know me no better than that.

Assistant Undersecretary Brian Burke called to inform me of two additional political decisions. "They" have decided to delay the issuance of the new planning regulations until after the elections in November. The environmentalists have applied considerable pressure on the White House over the issue of a new and more practical approach to a viability regulation. The environmentalists like the existing regulation because it is essentially impossible to comply with and provides a handy weapon that can be used to stop any proposed action, or at least significantly delay and increase the cost thereof.

These new regulations are sorely needed as many new national forest plans are due for revision. They would dramatically streamline the process and replace items that have proven impossible to implement with new approaches. Ironically, these regulations are a prime example of what the administration's National Program Review was all about: more effectiveness and efficiency, doing more with less, etc. Such concerns go out the window in an election year when the people opposing change are recognized as representing special interest groups that are considered important to election victories.

Oddly, the same thing took place in the last year of the Bush administration, when the decision was made to delay release until after the election for fear of offending some group or another. There is another maxim operating in this arena that is becoming clear: "Any significant decision can have political consequences, particularly in an election year. Therefore, any decision that can be delayed should be delayed as long as possible. As there are several levels in the hierarchy that must sign off on any significant decision, and there are multiple chances for risk aversion through delay, almost infinite delay of any significant decision can and will occur."

The second political decision was to delay the release of the California spotted owl plan adjustments—CASPO—until after the release of the Sierra Nevada Evaluation Program later in the summer. Environmental groups are content with the current state of affairs in northern California, where the national forests are operating under intense protection guidelines that severely reduce timber cutting. The CASPO amendments to forest plans would change that situation.

The request (or demand) from the environmentalists to the political levels in the government (the White House) to delay the institution of CASPO until after the release of SNEP is based on the logic that the new information in SNEP must be considered. I am told that the CASPO planning team has been fully aware of what is emerging from the SNEP effort and has accounted for that information. Hence, there is no new information to consider.

That, obviously, does not matter. The environmentalists want a delay? They get a delay. Burke verbally ordered that the Forest Service release

CASPO immediately after the release of SNEP. I would give that scenario a very low probability of taking place. As the SNEP release should come in mid- or late June, CASPO institution should occur within a week of that time. I predict that immediately on the release of CASPO, there will be appeals and/or legal actions to delay institution of activities until all aspects of SNEP have been considered and incorporated in the CASPO environmental impact statement. At that point, it is likely that "they" will decide to acquiesce to that request, thereby delaying any on-the-ground action based on CASPO until after the November election.

It was not clear just who made these decisions, but they were made above the level of the Department of Agriculture. And again, the people who made the decisions did not afford themselves any opportunity to be briefed on the situation or to consider the consequences.

This was a real political day. Late in the afternoon, I spoke via conference call with Brian Burke and Ray Clark of the Council on Environmental Quality on the subject of the proposed New World Mine. CEQ wants to start up an effort to prepare an environmental impact statement to withdraw from surface occupancy some 1,900 acres in the basin where the New World Mine is to be located. This is in keeping with President Clinton's administrative action concerning the mine that was announced some months ago while he was on vacation in Jackson Hole, Wyoming.

The Forest Service and the state of Montana have coleadership on the preparation of the EIS for the proposed mine. It is obvious to me that all involved are doing, or attempting to do, a fully professional job in full compliance with law and regulation. On several occasions, I have had to stand tough in meetings at CEQ that the job would be done right and that the appointed decision maker would make the decision with no interference from any higher levels in the Forest Service. Furthermore, if the decision is appealed, the official charged with making the decision would similarly decide the appeal.

I think that stand and the stand of the forest supervisor in charge of the assessment has led some to conclude that the Forest Service is committed to going ahead with the mine. This has proceeded to the point that there has been insistence that an outside consulting firm be hired to review the work and attest to the adequacy of the effort. That has been done.

So in response to the president's gesture on surface occupancy withdrawal, the Department of the Interior requested the Forest Service to carry out the EIS for the withdrawal without offering any assistance whatsoever. As BLM has responsibility for approving such actions and Interior made the request, the Forest Service, with my approval, suggested that BLM do the work, or at least take on a coleadership role and provide one-half the resources of money and people to get the job done. This has led to a spitting match that has gone on for several months. The White House has become impatient and, of course, the Forest Service has been identified as the roadblock.

I think the primary problem is that Interior does not want to put any resources into the effort. And people in Interior realize—correctly in my opinion—that this is a political gambit that can create some economic liability. Further, given the landownership patterns in the drainage, the withdrawal seems unlikely to stop the mine in and of itself. The most the withdrawal could accomplish would be the political benefit of having done all that could be done to halt the operation and perhaps increasing the difficulty and the public pressure to persuade the mine proponents to just back off. Given the resources already spent and the potential of the ore body, it seems unlikely that the mine proponents will give up.

However, it is pointless to argue the matter any further, as no minds will be changed and the facts and costs don't matter. However, I believe, and Brian Burke concurred, that BLM should serve as coleader on the effort and provide one-half the resources. Ray Clark of CEQ concurred.

14 May 1996, Washington, D.C.

A civil rights forum occupied the entire day. Everyone from USDA who wished to attend was in the auditorium, and others were connected via satellite. There was a press conference at the National Press Club on 13 May held by several civil rights organizations. The Forest Service was a primary target, with emphasis on Colorado and Oregon. These are the same old tired targets that have been investigated time and again. There is a real problem in trying to respond to such charges: those making the charges can say whatever they wish and charge whomever they please. Those charged must be cleared, as the mere accusation creates the presumption of guilt. If there is no truth to the charges, there are no consequences for the accusers. The fact that agency officials are restricted from discussing any personnel case with the press and releasing any material from personnel files exacerbates the situation for the agencies.

The forum began with four political appointees whose speeches were clearly political. They unabashedly praised the president and his record on civil rights. Each of the speakers belittled claims by white males to have been discriminated against. It is clear to me that we are sometimes biased toward achieving goals for workforce diversity. The question is, can one be biased for something without being biased against something? Obviously not; anyone who can perform a chi-square test can look at the data for Forest Service employment in California and know that there was profound bias toward females in hiring, training, and promotion. Further, whether right or wrong, it was court-ordered.

It was not a comfortable day. The purpose was to renew attention on civil rights, to allow folks to vent, demonstrate concern, and increase managerial awareness. I suspect that the outcome is more likely to be a surge in complaints, cases filed, and increased militancy. Maybe that is good, but I don't think so. We do not need more acrimony and division at this extremely tense time of downsizing and declining budgets; fighting over sex and race divisions over pieces of a shrinking pie will divert time, money, and energy and that will make the pie even smaller, faster.

As expected, when I returned to the office, I was told that Secretary Glickman wanted to see me in his office at 0830.

16 May 1996, Washington, D.C.

The meeting with the secretary took place as scheduled, and the agenda was exactly as I thought it would be. Also present were Undersecretary Lyons and Chief of Staff Greg Frazier. As soon as the secretary had revealed enough of the purpose of the meeting to justify my speaking, I told him that I knew the details of what had happened and that the general facts were known in parts of the Forest Service, including the field, and in the Department of the Interior. The secretary seemed a bit taken aback by the fact that I already knew exactly what was in the offing, but he plowed ahead anyway. Deputy Chiefs Gray Reynolds and Mark Reimers were to be "moved to other jobs in the Forest Service." Mike Dombeck, director of BLM, was to fill the position of deputy chief for the National Forest System, and Phil Janik, regional forester in Alaska, was to become deputy chief for Programs and Legislation.

I told the secretary that Dombeck and Janik were top hands and would do well in those positions—that either man would a fine chief of the Forest Service—and nothing I said should be taken as a reflection on either man. Then I laid out my position. First, I was deeply offended that these decisions would be made without my knowledge, advice, and consent. There was no way I could interpret this action except as a slap in the face, at best, or an expression of either distrust or contempt for me and for the position of chief, at worst. That was bad enough.

Second, I had no reason to have anything but the utmost trust and respect for Reynolds and Reimers. I found it unconscionable to dismiss thirty-year veterans on the basis of rumor, hearsay, and innuendo without even affording them an opportunity to respond to the charges or the validity of the perceptions. Such was not in keeping either with Forest Service tradition or with standards of common decency. Therefore, I would resign as chief in protest of these actions. I recommended Mike Dombeck as the next chief of the Forest Service. He is a Forest Service exemployee (once Forest Service, always Forest Service) and therefore has proper breeding; he is career Senior Executive Service, and his appointment would return the chief's position to that status; he has experience and an outstanding record as an agency head; and he is a respected professional and a decent and principled man.

Secretary Glickman made it clear, several times, that he and "they" did not want me to leave the chief's job, particularly not prior to the November elections. He asked me to consider my decision overnight and to meet with him again tomorrow morning. I said that I would do as he asked, but there was no chance I would change my mind. This was a matter of honor and integrity. I would not be a party to what I consider a travesty. A lesson was imparted to me many years ago about loyalty: If leaders take care of their people, their people, in turn, will take care of their leaders.

I told the secretary that it would be wise in this process to replace Lyons at the same time that Reynolds, Reimers, and I are replaced. If, indeed, the problems with the Forest Service are as pronounced as the secretary and "they" perceive, it should be recognized that many of the problems have roots in the chaos and style of management in the undersecretary's shop.

Those problems include extreme micromanagement by the undersecretary and distrust of him within the agency.

There was no animosity, and we shook hands as I took my leave. Soon after I arrived in my office, Lyons was on the phone to tell me that he had been ordered to conduct himself as he had, and that he didn't agree with what was going on. I said nothing but thought plenty. I don't believe that I would have carried out such an order. However, that may be easier for me to say as I can retire and move on to another job and have no one to take care of but myself. Mr. Lyons, on the other hand, has a family to take care of and probably no immediate job prospects. Where you stand depends largely on where you sit and under what circumstances.

I went back across the street at midday to have my regular lunch with Jim Lyons and Assistant Undersecretary Brian Burke. Burke turned out to be unavailable, so Lyons and I visited for some two hours. Lyons did not seem aggravated by my statements to the secretary and accepted the situation of my departure as a done deal. He was considerate enough to offer to transfer me back to La Grande as a research scientist if that was my choice. I told him that my continued presence in the Forest Service would not be a good thing for either the agency or the administration and that it was simply time for me to leave.

He, too, is weary and used up by the circumstances that have gone on too long. We are still friends, I think.

In the evening, Mike Dombeck and I visited for well over an hour. Mike was incensed at what was going on and, in particular, at the way things were being handled. I briefed him thoroughly on what had happened in the morning meeting, particularly on my recommendation that he succeed me as chief, and assured him that he would get my full support.

He indicated that while he would like to be chief of the Forest Service, he wanted nothing to do with the situation as it was developing. He indicated that he would visit with Secretary Glickman in the morning before my meeting and make the following points: (1) the political fallout would be severe and confirm what certain members of Congress have been saying about politicization of the Forest Service; (2) the action would likely trigger congressional hearings; (3) the reasons that the chief resigned would be made clear in a resignation letter and a final message to Forest Service employees that would not reflect well on the administration; (4)

the reputation of the chief exceeds that of the people who are forcing the issue; (5) the actions being carried out are not well thought out, justified, or honorable; (6) the administration would be wise to keep the chief and search elsewhere for the root cause of the perceived problems; and (7) the chief should have direct and frequent contact with the secretaries of Agriculture and Interior and the vice-president and president. The punchline will be that the chief's departure would set back the conservation agenda significantly and that he, Dombeck, would have nothing to do with it under the circumstances.

17 May 1996, Washington, D.C.

I woke up feeling the best I have felt in months. The struggle is over and it is time to move on to other things. There are no regrets for I have done my best and have kept faith with the Forest Service, my colleagues, and myself. A huge burden has been lifted from me.

I was in Secretary Glickman's office at the appointed time of 8:30 A.M. Assistant Undersecretary Brian Burke was also waiting, to represent Undersecretary Jim Lyons at the meeting (Lyons was traveling). It was nearly 9:15 when Secretary Glickman came out and asked me to come into his office. He asked Burke to excuse himself from the meeting.

When I entered the office (actually the office of the chief of staff, Greg Frazier), Rich Rominger, deputy secretary of Agriculture, and Frazier were present. The secretary opened the meeting by going over the mission of the Forest Service and how important it is. He suddenly broke off that rather meandering discussion and came directly to the point. He wanted me to stay on as chief. There would be no interference with my staffing decisions: deputy chiefs Reynolds and Reimers would remain in their positions at my discretion. The secretary would spend more time with the Forest Service, and he wished to meet weekly with the chief, associate chief, and deputy chiefs for an hour or so. The chief would henceforth report directly to the secretary. Undersecretary Lyons would no longer have any direct contact with Forest Service operations. The details would be worked out later with Greg Frazier.

This was a complete 180-degree reversal of the situation that existed after yesterday's meeting with the secretary. Not only had the decision to remove Reynolds and Reimers been reversed (though it was clear that was still what they wanted), the problems caused by Lyons's management style had been recognized and had resulted in his being removed from the chain of command.

I was stunned as it became clear that all of my points in yesterday's meeting had been addressed, and on my terms. I did not feel victorious. Rather, I felt entrapped in a situation that is not likely to improve markedly, as the distrust of the agency by significant players in the administration has not changed. Yet given this reversal, there was little that I could do but agree to stay on.

There are details, myriad details, to be worked out, but there is little doubt that for the moment at least, the situation has changed dramatically. Will this last? I will be out of town for the next week, and that will give the political operatives time to do their thing, and who knows what will result.

Shortly after I got back to my office, Mike Dombeck called to find out the results of my meeting with the secretary. I laid it out for him. He told me that he had been on the phone with Secretary Glickman for nearly an hour before the meeting with me. In fact, that phone call had been what delayed the meeting. He told Glickman exactly what he told me yesterday, including the refusal to take the chief's job under those circumstances. Evidently Secretary Babbitt had weighed in as well, and the deal that Glickman had proposed to me had also been cleared by Vice-President Gore.

Dombeck and I discussed the likelihood, and mutual desirability, for him to replace me upon my retirement after the elections in November. I can and will fully support that move, as Mike would be a good chief, and perhaps most important, the move would return the chief's position to the career Senior Executive Service. That would lift a burden from my heart, because to know that I, however inadvertently, had been used to make the chief's position a political appointment would wear on me for the rest of my life—and affect the history of conservation.

I took pleasure in telling Reimers and Reynolds that their positions were no longer in jeopardy. That, at least, made me feel good, and they were obviously relieved. I also met with the associate chief and the deputy chiefs to tell them of the latest turn of events. I made it clear that there was no victory here, just another chapter in an ongoing saga. I certainly do not feel either victorious or vindicated. I just feel tired and, strangely, somehow disappointed. The decision to resign and to get on with a new life had been made, and I had quickly come to grips with that reality. I had, in my mind, done right. It was over. Now there is still more time to serve on my sentence. The challenge is to go back and give it everything I've got for another six months. The Forest Service and its people deserve no less.

23 May 1996, airborne between Sacramento and Washington, D.C.

I have been on the road the last six days with a Forest Service team composed of Robert Nelson, director of Fish and Wildlife; Chris Risbrudt, director of Ecosystem Management and Planning; David Hessel, director of Timber Management; and William Supulski, Timber Management staff. We have been looking at salvage sales being carried out under the auspices of the salvage rider. The political rhetoric surrounding salvage sales under this law is high pitched and incessant. And it is directed to all political levels in the administration, from the secretary of Agriculture to the White House.

As the whiners are a constituency that the White House wishes to cultivate as supporters in the 1996 elections, their accusations and complaints are given credibility. Among the most frequent charges are these: green sales that have been delayed by appeals and court challenges are being repackaged as salvage sales, roadless areas are being entered under the guise of salvage without proper analysis, green trees are being cut as salvage, and clearcutting disguised as salvage is rampant.

Our mission is to get a firsthand view of examples of each of these concerns and to get some feel as to the legitimacy of these concerns. The first site we visited was the Clearwater National Forest in Idaho. The salvage being conducted there involves sales that were once part of a larger overall project that had been appealed. The appeal had been upheld due to insufficient attention to some aspects of the environmental assessment. Some parts of that project clearly meet the definition in the salvage rider—in this case, the removal of dead and dying trees and removal of trees imminently in danger of fire and insect attack. Evidently Forest Supervisor James Caswell, along with other forest supervisors in the region, had received orders from the regional office to move ahead with such actions.

The next stop was the La Grande Ranger District on the Wallowa-Whitman National Forest in northeastern Oregon. The example case was the salvage of a large burn resulting from the so-called Boundary fire in 1994.

The issue here was the long delay in getting the salvage operation on line, the treatment of a roadless area that was committed to multiple-use management in the extant forest plan, and concerns with habitat for threatened anadromous fish. It seemed clear that appeals would have been inevitable had they not been excluded under the provisions of the salvage rider and that any further delays would have dramatically affected the financial viability of the sales. The reason for the delay in getting the sale out was reported to be the inability to work in or reach the area during the winter months. That did not seem a good enough reason to me.

Some of the forests are in pitiful condition, particularly those stands on relatively dry sites that were historically open ponderosa pine maintained by frequent fires. Fire exclusion has allowed shade-tolerant, fire-susceptible grand fir and Douglas-fir to dominate those sites. Severe outbreaks of spruce budworm that persisted far longer than what had been considered normal, coupled with drought, have caused severe mortality. These dead trees have produced fuel loadings far in excess of historic norms. So not only is fire danger (or probability) enhanced, fires that do occur will likely burn much hotter and over greater areas than has been the historic norm. And over the past several decades, more and more homes and associated buildings have been built in rural areas surrounding the national forests. The result is big problems that cannot be addressed quickly or cheaply, and there is no political consensus as to the proper approach.

The next day we were off to Sacramento for a visit to the Stanislaus National Forest. This forest was on the schedule as an example of the use of the controversial provision in the salvage rider that salvage would include stand treatments where there was an immediate susceptibility to fire or insect damage.

James Lawrence, deputy regional forester, and Johanna Wold, forest supervisor, and I flew to the site of some proposed thinning of fuel-ladder firs under an overstory of ponderosa pine and removal of some mature pine to increase spacing between the overstory tree crowns. Dead and down woody materials (the fuel loading) would be significantly reduced during the process. This forest is fortunate in having a biomass-fueled electricity generation plant nearby that can take such material.

When I asked the district ranger why he had moved forward under the provisions of the salvage rider, thereby precluding appeals, he replied that the situation obviously met the definitions in the law, there was a real and immediate danger of fires that would cause significant damage to homes and threaten human life, there was opportunity to use salvage trust fund dollars to deal with a forest health problem, and wood could be made available to mills in northern California that desperately needed raw materials. It was difficult to argue with his rationale.

So, all in all I agreed, but with full recognition that environmentalists also have a point in that these sales could have waited a year or so and gone through the appeals process. The real question was the risk associated with a delay of a year or so and the consequences of guessing wrong. These are management calls that should be made by district rangers and forest supervisors, not by the chief.

24 May 1996, Washington, D.C.

Today was the first meeting with Secretary Glickman since he told me that I (the Forest Service) would have direct access to him and that Undersecretary Lyons would have less to do with the agency's day-to-day operations. Deputy Secretary Rich Rominger, Chief of Staff Greg Frazier, Assistant to the Secretary Anne Kennedy, and Lyons accompanied Secretary Glickman. Forest Service attendees were limited to executive team (chief, associate chief, and deputy chiefs or acting deputy chiefs).

The secretary opened the meeting by announcing his intention to be more personally involved with Forest Service operations. However, he had no interest in being involved in micromanagement. His role, and that of his staff—including the undersecretary—was to be limited to dealing with policy. In that vein, weekly meetings would be held with the Forest Service's executive team to accomplish several things: assuring that the secretary was up to speed on significant Forest Service issues, making policy decisions in an open fashion and after full debate, and having action items brought to him for decision.

There was frank discussion on both sides as to perceived shortcomings and what should be done about them. Glickman expressed his aggravations with "Forest Service independence" and noted that his picture was not hanging beside the president's in the downstairs lobby. Upon later inquiry, I found that his picture had been hanging in the lobby but had fallen and broken and was being repaired. However, the point is that he did notice that his picture was missing and interpreted that omission as a statement of either defiance or indifference to being part of the Department of Agriculture.

I told him that the lack of connection between USDA and the Forest Service was a two-way street: USDA (the secretary and assistant and undersecretaries) traditionally paid very little attention to the Forest Service, and by his own admission, he had maintained that tradition. Deputy Chief Mark Reimers entered the conversation at that point and, more forcefully and intensely that I had ever heard him speak before, laid it out for the secretary. The secretary was uninformed about Forest Service affairs, listened too quickly to "perceptions" of others, was too quick to criticize, was parsimonious with praise, was not stating policy clearly, and had not addressed lack of organization and discipline within the USDA hierarchy above the chief's level. He said the Forest Service was always willing to work for those in legitimate power over the agency, but it was essential that the secretary be involved in understanding Forest Service activities and setting a course for action.

Gray Reynolds, deputy chief, chimed in on the same theme, but Reimers had said it all. And I believe the secretary listened with patience and without being visibly upset or taken aback.

The secretary returned to his pet grievances: the Forest Service had resisted his instructions to support the Quincy Library Group (not true, but he believes it), the Forest Service had "screwed up" the letter to the Senate Committee on Energy and Natural Resources concerning potential liabilities (though the letter contained nothing but facts that were already in hearing records), and some other mundane screwups. It is well to note that the secretary does not dismiss mistakes—he hangs on to them.

We were expecting the secretary to lay out some definitive guidelines as how we will operate under the new orders I had been led to expect in my meeting with him when he had rejected my resignation and said that I would deal directly with him. Late in the afternoon, Mr. Glickman's chief of staff, Greg Frazier, met with Associate Chief Dave Unger and me in my office to discuss what the new table of organization should be. Unger and I laid out several options of how to deal with the situation with Undersecretary Lyons still in place. Each time, we came back to the same point. With Lyons in place and with his personality of command-and-control and his propen-

sity to break lines of authority, it was difficult to see any of these options
working.

Frazier listened carefully and made suggestions of his own. He realized
that the situation is a shambles of disorganization from the White House
through the secretary's office through Lyons's shop to the Forest Service.
There simply is no coherent table of organization, no clear line of authority,
no accountability, no defined decision-making process, no accountability
for decisions, and no consistency of purpose.

Unger and I like and respect Greg Frazier. He listens. He will make decisions. He is a pragmatist, and he doesn't waste time. He is all business, and
we desperately need that.

28 May 1996, Washington, D.C.

In late morning, I met with Undersecretary Lyons at his request. He again
told me that he was ordered by the secretary to approach Regional Forester
Janik and BLM Director Dombeck as to their availability to replace Deputy
Chiefs Reimers and Reynolds. He said that he openly opposed the action
and warned the secretary that I would resign before acceding to such an
action. But he asked, "What could I do? I only followed orders."

I didn't cut him any slack and merely commented that he had a choice;
we always have a choice if we are willing to pay the consequences. I looked
him in the eye and said, "You could have refused."

He shrugged and told me that he did not appreciate my offering his resignation while I was offering mine. Again, I cut him no slack, saying that
I told the secretary straight out that whatever the perceived problems with
the Forest Service, those problems could not be fully addressed by removing the top manager in the agency without facing the mess that exists at
the undersecretary's level. I said that I meant it then, had not changed my
mind since, and made no apologies for saying what I thought needed to
be said.

The subject turned to the matter of the 1997 budget, now being put
together by the Forest Service. The president's budget has already been submitted to the subcommittees in the House and Senate. That budget called
for a 3 percent increase, which we were told we would not receive. And it
now appears as if even more severe budget cuts of an additional 6 to 7 percent, or more, may be in the offing. Work is under way to determine options

for reacting to such scenarios. Mr. Lyons said he wanted to participate "as part of the team" in working through these budget problems.

I told him that there was no problem with that provided that two conditions were met: He must spend the hours in meetings working on the issue as part of the team, and he must not behave in a dictatorial fashion, as is his custom. I told him that my estimate of the probability of his performing in such a manner was somewhere between slim and none. He still wants to participate and I agreed, again insisting that he follow the ground rules that I lay down. This was not a pleasant experience for either one of us, but things got said that needed to be said.

31 May 1996, airborne between San Juan, Puerto Rico, and Washington, D.C.

I have been in Puerto Rico for the dedication of the El Porta Visitors Center on the Caribbean National Forest, and to visit the only tropical rainforest in the National Forest System. The El Porta Center is, in my opinion, a remarkable architectural creation and shows that a government building can be a work of art.

The building has several purposes. The first is to provide a center for edification of visitors of all ages about the workings and values of the 2,800-acre rainforest. The second is to provide a facility for training and scientific gatherings to discuss management of tropical rainforests. The center will serve both the Caribbean National Forest and the Institute of Tropical Forestry.

Pablo Cruz, supervisor of the Caribbean National Forest, and Auriel Lugo, director of the Institute of Tropical Forestry, have done a marvelous job of creating public support and fostering partnerships that have made this effort successful in these times of controversy and tightening budgets.

We were able to spend a day touring the rainforest and benefited from a series of presentations from various experts from the national forest and Forest Service Research. We were able to visit the Fish and Wildlife Service aviary, where the severely endangered Puerto Rican parrot is bred in captivity to provide birds for release into the wild.

After listening to presentations by the Fish and Wildlife Service people and Ernesto Garcia, an old colleague and fellow "combat biologist" from earlier years in the western United States, it was clear that in spite of the skill and enthusiasm of all involved, the odds against long term success for the parrot are great. Not only is habitat severely limited and the population dangerously low both in numbers and in genetic material, the ecology of the rainforest as it affects the ecology of the Puerto Rican parrot has been dramatically altered by the introduction of an exotic bird that outcompetes the parrot for nest cavities.

This competition is so severe that it is essential for the wildlife crew to clean out nest cavities and construct or enlarge cavities and then predator-proof their handiwork. This all leads to inevitable question that arises from such efforts: When are the chances of success so low with a particular species as to make it wise to give up and turn the resources of people and money to circumstances where chances of success are more favorable?

I suspect that in cases such as the Puerto Rican parrot, the answer is that efforts will be curtailed only when the species is extinct in the wild and it is decided to keep the species extant in zoos. Why? Consider the name of the parrot. Then consider the beauty, size, and behavior of this bird. And then think of the symbolism of the continued existence of the bird in the remnants, the treasured remnants, of the only tropical rainforest in the trust and care of the United States and the U.S. Forest Service. Abandonment of the Puerto Rican parrot is not even a remote possibility.

The dedication of El Porta was well staged and well attended. The featured speakers, in order, were the chief of the Forest Service, the congressional representatives from Puerto Rico, the governor, and the regional forester for the Southern Region. The ebullient pride of the Forest Service employees and the citizenry was contagious. It was a magic moment and I was glad to be there.

Last evening I received a phone call from Dr. John McGee, forest supervisor on the Apache National Forest in Arizona regarding his pending decision on whether to let construction begin on the third telescope on Mount Graham.

Mount Graham is home to a subspecies of red squirrel (the Mount Graham red squirrel) that is listed as a threatened species under the Endangered Species Act. Some dedicated environmentalists have used the welfare of this creature as a ploy in the continuing fight to preclude construction of a third telescope by the University of Arizona on Mount Graham.

There was a rider attached to legislation signed by President Clinton that authorized the construction of a third telescope on Mount Graham. That seemed to end the argument. Then a forest fire swept part of Mount Graham and destroyed some squirrel habitat. This reignited the issue. McGee reported to me that he was being "worked over" by Department of Justice lawyers, who are conveying the message that the White House does not want the telescope constructed and, therefore, McGee should stop construction and reinitiate consultation with the Fish and Wildlife Service. The Fish and Wildlife Service will, presumably, make the correct decision or at least slow down construction until after the November elections.

McGee does not see any connection between the issue of consulting with the Fish and Wildlife Service, which he intends to do, over the consequences of the fire and the question of construction of the telescope. There has already been consultation over the telescope, and the consequences were judged insignificant to the viability of the squirrel. McGee, a wildlife biologist, does not believe that there is any reason to reconsider that decision.

In short, McGee intends to obey the law and tell the truth. This is a political issue in which the decision desired by the White House is not supported by either the law or the facts. He intends to authorize construction of the telescope. The purpose of the call was to clarify his thinking and give me a heads-up on a nasty developing situation.

From what I remembered of the situation from my visit in December 1995, I agreed with McGee's assessment, and I told him so. More important was my statement that (1) this was his decision to make; (2) the decision should be made on the facts; (3) it was not appropriate to be swayed by the politics of the situation; and (4) I would back his authority to make the decision. I asked him to give me an update early next week so that I can warn the politicals as to what is coming down.

11 June 1996, Washington, D.C.

The Forest Service top staff met with Secretary Glickman in the now-regular weekly meeting. The meeting took place in the chief's conference room. Staffers Greg Frazier, Charles Rawls, and Anne Kennedy accompanied the secretary.

The primary item covered was a letter of direction being prepared for the secretary to send to the chief of the Forest Service with emphasis on how the Forest Service actions will be carried out under the salvage rider until the expiration of that legislation.

The secretary had made a speech in front of an Audubon Society group concerning the Forest Service actions under the salvage rider. The Forest Service (at least the chief) had no chance to review the speech and thereby warn the secretary of the technical, legal, and political consequences of what he had to say. In this case, operating on perceptions in the White House, formed on what "they" are told by environmental activists, the secretary promised to do something about "abuses" by the Forest Service of its discretion granted under the salvage rider. This centered on cutting green trees under the guise of salvage, etc.

So Anne Kennedy began to prepare a memo for the secretary to send to me to respond to those concerns. This was done, again, without consulting me. Anne merely summoned folks from the timber staff over to help prepare a memo. Bill Supulski and Bob Lynn were the primary staffers involved. When Lynn and Supulski told me what was going on, I immediately asked for the consequences in terms of timber volumes to be cut, dollars lost to the Treasury, dollars lost to the states, consequences in terms of staffing, and political fallout.

The first cuts at assessment indicate that when the instructions are issued, the Forest Service will end up operating under more stringent rules than would have been the case without the salvage rider. The timber volumes will likely be 2.6 billion board feet (compared with the 4.5 bbf that was targeted in reports to Congress), about 44 percent below the targeted amount. Because salvage operations are conducted using salvage trust fund resources (funds outside the agency budget), it will be necessary to cut additional employees. That, coupled with the ongoing downsizing efforts, will obviously require the use of reduction-in-force procedures, which will likely produce costs that will require even further cuts in personnel to free up the funds.

As I brought these facts to the secretary's attention, his expression indicated that he suddenly understood there was much more to the issue than he had realized. He said that we had best slow this down until we fully understand the consequences.

Anne Kennedy then commented that this was all caused by the Forest

Service in California "going over the top" in using the salvage rider definition of salvage as including areas "imminently susceptible to fire." I could
feel the color rising in my face, but I calmly (I think) expressed the opinion
that none of the politicals making these decisions had viewed the situation
on the ground, had ever seen a forest fire, had any idea of the situation of
the urban-forest interface, had any experience in evaluating fire danger,
had ever stood in overstocked stands with heavy fuel loadings on the ground
with Santa Ana winds blowing and smelled resin in the air. I concluded,
"When you've done that and seen the consequences of past fires, then you
can make at least a semilegitimate judgment of the appropriate application of the 'imminently susceptible to fire' standard. Until that time, you
are making judgments you simply are not qualified to make."

I suggested an immediate trip to California by Burke, Kennedy, and
whoever "they" are in the White House for a firsthand look. Then they can
substitute their opinions for that of the professionals and be prepared to
take the consequences of their being wrong.

That must have rung a bell, because I received a call just as soon as they
had returned to their offices in the Agriculture Building. I issued the order
to staff the trip. We will see.[2]

12 June 1996, Salt Lake City

I am in Salt Lake to attend a meeting of supervisors in the Law Enforcement Investigation Division. But in keeping with a policy to make as much
as possible out of each trip, I met with Bernie Weingardt, supervisor of the
Wasatch-Cache National Forest and his district ranger. They are dealing with
the preparations of the Snow Basin ski area as the venue for the downhill
racing events in the 2002 Winter Olympics. This entire issue continues to
be politically sensitive, and it seems to become even more contentious with
time. The Forest Service, as usual, seems caught squarely in the middle of
the hassle. The United States has committed to the 2002 Winter Olympics.
Being prepared for the downhill events at Snow Basin using the standard
process of assessments, reviews, appeals, lawsuits, construction delays, etc.
seems virtually impossible. One possible solution to that problem is to make
the necessary arrangements of land exchanges, etc. a matter of law.

2. None of the political appointees ever accompanied me on a fact-finding trip.—*JWT*

That, however, is not as simple to achieve as it might seem on the sur-face. There is a long-standing issue of the intricacies of the "required" or "desired" land exchanges at the base of the ski area. The ski area owner-operator is Mr. Earl Holding, one of the most affluent and aggressive entre-preneurs in the Intermountain West. Many object to any more than the minimal land exchanges required to make the Olympic venue possible. Oth-ers look on this as an opportunity to clear up the issue once and for all. They hope to bring an end to the interminable hassle by providing the area needed and desired by the ski area developer and simultaneously creating a long-term management situation for the Forest Service that is as rational and efficient as possible under the circumstances. The key to achieving that is the amount and configuration of land to be exchanged. There is a pro-posal by the developer to exchange some 1,300 acres (which includes a "regular" configuration and all the land needed—or at least desired—by the developer for the Olympic venue and future development). There are other possibilities, which include an irregular boundary of wetland inter-spersed through the area. Some Forest Service folks have told me that the most businesslike thing to do is to exchange the 1,300 acres now, through legislation, and be done with it. Others believe that the minimum exchange necessary to accommodate the development of the Olympic venue should be made and the remainder considered through regular procedures.

Congressman James Hansen of Utah has introduced legislation to accomplish the 1,300-acre exchange. Further, at a recent Society of Amer-ican Foresters meeting in Utah, he reportedly said that he was only doing what he was asked to do by the chief and Deputy Chief Gray Reynolds.

Reynolds and I had in fact met with Mr. Hansen on this matter and on the grazing bill proposed by Senator Peter Domenici. My recollection of the meeting was that we discussed the problems of meeting the timelines to be ready for the Olympics under regular processes. And we did discuss the various proposals for the amount and configurations of the land to be exchanged. We made no requests of Mr. Hansen and told him that we did not know the position of the administration on the matter. We suggested that he and Senator Orrin Hatch meet with the president or vice-president and reach an agreement with them as to the best legislative solution.

However, Mr. Hansen's statements have caused some consternation among those Forest Service employees who, for whatever reasons, oppose the exchange, some aspect of the exchange, or handling such an exchange by means of legislation. That puts me in an awkward position. I do not want to start an argument with Mr. Hansen, which would solve nothing and would almost assuredly embarrass and aggravate the congressman who is chair of one of the key subcommittees with which we deal routinely. On the other hand, letting Mr. Hansen's statement stand uncorrected aggravates the administration, which suspects we are conspiring with the Republicans and straying from administration policy. And obviously, some Forest Service employees disagree or at least are concerned by the chief's support of this action.

I was waiting in Weingardt's office when he returned from a negotiating session with Earl Holding. Bernie said that Mr. Holding wanted to come over and meet me. I told Bernie that I would be pleased to meet and talk with Holding. But it should be made very clear that he understand that I would not enter into any agreements as the entire matter will be handled at the appropriate level by the appropriate people, that being at the forest level with the forest supervisor.

Mr. Holding, after talking with Weingardt, showed up after about a half-hour and we talked for some forty-five minutes or so. Most of that time was spent with Mr. Holding talking and me listening. His purpose was obviously to convince me that (1) he is a reputable and honest businessman; (2) he wants to do a first-class job with the Olympic venue; (3) he has the financial resources and the entrepreneurial skills to achieve the objective; and (4) he has no hidden agendas or ulterior motives.

The conversation was pleasant enough, and I assured him that I questioned none of the points he was making. I reiterated that Bernie Weingardt is the person charged with getting the job done on the Forest Service side—not the chief—and that, so far as the land exchange via legislation is concerned, the matter lies in the political arena between Mr. Hansen, Congress, and the administration. I enjoyed my visit with Mr. Holding and think we parted friends.

After Holding left, I was informed that there had been a whistleblower complaint on the district ranger, whose district includes the Snow Basin ski area, for allegedly taking bribes and gratuities from Mr. Holding. I was assured that the charges are unfounded. The Office of the Inspector General is investigating with the full cooperation of all involved. Obviously, this will be one of those issues that will take on a life of its own and exact a cost in terms of suspicion and the diminution of the reputations of all involved.

The meeting with the staff of the Wasatch-Cache National Forest went as most such family meetings go. They are frustrated by the constant instability caused by downsizing and declining budgets, policy shifts, and slow decision making. They are desperate for things to settle down.

Yet in spite of their concerns, morale is obviously better than it has any right to be. Such is a common trait of Forest Service employees wherever I meet them. I attribute this to the kind of people who are attracted to work in the natural resources field: they—most of them, anyway—have a vocation, and that spirit is infectious.

In late afternoon, I joined the ongoing meeting of the supervisors of the Law Enforcement and Investigation group as they were winding down the day's activities.

I had a chance to visit with Billy Ball, special agent, who is stationed in east Texas. He told me that he is continually receiving phone calls and written information concerning the proposed land exchange in Arkansas and Oklahoma between Weyerhaeuser Company and the Forest Service. There is a contention that the land exchange has been improperly put together as regards appraisals of the values of the lands being exchanged, and that as a result, the exchange is heavily in favor of Weyerhaeuser. He is turning the information over to the Office of the Inspector General to be pursued.

I did not ask for any details, as I did not want enough information that could compromise an investigation. I encouraged Ball to continue to serve in any way that would facilitate the Office of the Inspector General's review of the matter.

This matter has come up before, and the Lands staff has assured me that all is in order. As I recall, some congressman had also been concerned and had asked for an independent review by some assortment of academic types. That review confirmed that not only was the exchange fair but, in their opinion, the government had come out ahead in the deal.

The source of the concern is likely that the Forest Service is trading relatively high-value timberland for many more acres that are valued primarily for recreational purposes, which complicates the computation of values. Further, some Forest Service employees who dedicated their careers to the reestablishment of forests that are now being exchanged are concerned. Those feelings are understandable enough. And some appraisers, given the legislation to achieve the exchange, had opportunity to make some significant fees, or so the staff tells me.

And so here is yet another case where any significant action that involves a land exchange will raise concerns and prompt whistleblower complaints and investigative reviews. That seems to be the way of things.

13 June 1996, Boise, Idaho

In the afternoon, I met with some eighty employees of the regional office and the Intermountain Experiment Station. Questions centered on the proposed combination of the Rocky Mountain and Intermountain experiment stations, which I indicated was a done deal so far as I was concerned. People were obviously irritated that this decision has been dragged out over several years and they could see no excuse for that. Frankly, I agree with them and could only say that the decision had been delayed at levels above the Forest Service, a true but lame statement.

As many of these employees are administrative types, there was much concern over potential combination of administrative units. I assured them that every effort would be made to continue to wring every ounce of efficiency out of the system—and that probably would include additional combinations of units, some units providing services for several regions and stations, etc. However, I made it clear that there is receptivity to other ways of doing business including, but not limited to, telecommuting, job sharing, and entrepreneurial activity, among others.

Beyond those concerns, there were the standard questions about budget, downsizing, surplus lists, buyouts, early retirement, political atmosphere, doing more with less, burnout, and so on. Again, I was impressed that morale was better than we had any right to expect. I am convinced that if we could simply provide some stability and predictability, morale and efficiency would surge. However, that is not likely to happen in an election year. And perhaps the Forest Service merely mirrors what is going on across western society. This, I believe, is what is causing such unrest and volatility in the body politic in the midst of a period of remarkable prosperity, slow but steady growth, and low inflation.

However, it is hard to be satisfied as the titular head of a huge agency without acceding to the observation that we are caught up in a much larger surge of events. To do so is to accept impotence in having any real effect on the course of events. The only way to break this spiral is to resist the trend, to refuse to feel and act out powerlessness, and not to allow expediency to become a philosophy in and of itself. The nation deserves better.

14 June 1996, Prineville, Oregon

I spent the afternoon touring, via helicopter and on foot, the ongoing Thunderbolt timber sale. This salvage of a very small portion of a large 1994 burn in the South Fork of the Salmon River drainage has been a cause célèbre for environmental groups and the regulatory agencies (National Marine Fisheries Service, Fish and Wildlife Service, and EPA).

The flap comes from a decision made more than a decade ago. A rain-on-snow event coupled with too many clearcuts and a poorly engineered road system on granitic soils produced landslides that severely damaged salmon spawning grounds in the South Fork of the Salmon River. It was promised that no further timber removals would occur until salmon spawning populations improved. Those populations have continued to decline.

In retrospect, there were two aspects to that promise that were naïve at best and remarkably uninformed at worst. The first was that it would make a difference, when in fact the primary controls on the numbers of returning spawners lay elsewhere: offshore fishing, power dams on the Columbia, Indian fishing nets, etc. The second was that habitat conditions on the uplands would improve with time—that there would be no significant losses of trees to fire, insects, or disease. The 1994 fires put an end to that illusion.

In light of those changed conditions and the high volume of large ponderosa pine and Douglas-fir killed by the fire, the Forest Service proposed a very carefully designed salvage program to remove a small portion of the volume of dead trees via helicopter logging. When environmentalists and the regulatory agencies attacked the original plan, the Forest Service asked for a review by highly qualified scientists. That report largely supported the work done, with some recommendations for improvement. Those recommendations were followed.

Still, the regulatory agencies continued to express disapproval. As

months went by, Regional Forester Dale Bosworth bucked the decision up
to the Washington level. The regulatory agencies (at least some of their
regional people) continued to denigrate the Forest Service and disclose
material that could be useful to plaintiffs in a court case.

I told the regulatory agencies to issue a jeopardy call and I would pull
the sale. Of course, if that was done, the regulatory agencies would bear
part of the responsibility, and such is to be avoided. So those agencies
deferred to my authority but not without writing a cover-your-ass letter
stating that they had grave reservations about my decision.

My review left me with the distinct feeling that the only problem with
the Thunderbolt timber sale was that the salvage approach was far too cautious and far too few trees had been removed. The crews had set out to produce a benign salvage sale and they succeeded. Environmentalists and some
fellow fish and wildlife professionals vilified the responsible Forest Service
people, individually and collectively. But our people did what they viewed
as professionally correct and reasonable considering all factors, and they
behaved professionally throughout.

By late summer, it would be well to conduct a field review by a team
from the Forest Service, National Marine Fisheries Service, Fish and Wildlife Service, and EPA. However, I don't think that will happen, as it would
be very difficult to find much wrong, and that would be embarrassing.

18 June 1996, Arlington, Virginia

There was a briefing today on the New World Mine with Secretary Dan
Glickman and his staff. Regional Forester Hal Salwasser, Forest Supervisor Dave Garber, and associates came in for the briefing. I wanted the secretary to hear directly from the responsible people about the status of the
environmental impact statement and his realistic options for a decision,
now that he has decided to elevate the matter to his level.

My desire was for him to see the quality and caliber of those Forest
Service people responsible for the EIS. I wanted to disabuse him of the
mythology and wishful thinking in the Council on Environmental Quality and the Department of the Interior that the secretary can, with the stroke
of a pen, do away with the issue of the New World Mine.

It was a surprise to the secretary to find that, in accordance with the 1872
Mining Law, he cannot deny the New World Mine unless there are viola-

tions of other federal laws that cannot be mitigated. Further, the process cannot be manipulated, even if that is desired, as the state of Montana is a full partner in the EIS process. And Montana state law makes it mandatory that all documents be open to scrutiny at the time they are completed.

The secretary ordered that he does not wish the Forest Service to put forward a preferred alternative. He did understand that the state of Montana would have such an alternative and that the "no action" (i.e., no mine) alternative was, in this case, included only because regulations require such and it provides only a baseline for comparative purposes.

The most difficult of the applicable laws for the New World Mine to meet is the Clean Water Act, and compliance standards for that act lie with Montana and Wyoming, not the federal government. I believe it is likely that these states will want the mine.

This is another example of getting the political cart in front of the reality of the situation. The Forest Service and the state of Montana, which were preparing the environmental assessment documents and knew most about the matters at hand, were not even consulted prior to politicization of the issue. And once an issue is politicized and the politically correct action is obvious, it is very difficult to bring it into compliance with law, custom, and rationality.

19 June 1996, Arlington, Virginia

I was summoned to a meeting with Senator Frank Murkowski of Alaska, chairman of the Energy and Natural Resources Committee. The subject was announced as a discussion of the Tongass Land Use Management Plan. Dr. Chris Risbrudt, director of Ecosystem Management and Planning, accompanied me. Murkowski's aide, Gregg Renkes, also attended.

The first subject was indeed the Tongass Land Use Management Plan. In contrast to his persona when conducting hearings—bullying, abrasive, aggressive, combative, and arrogant—Murkowski and his aide were calm, rational, and even friendly. The senator's pitch was that there should be a six-month extension of the comment period on the draft plan. That would allow for a more detailed study of the economic and social impact on individual communities in southeast Alaska. He started to run down the quality of the plan. But perceiving that we were not receptive to that pitch, he switched back to his request for a delay.

I told the senator that so far as I was concerned, Regional Forester Phil Janik in Alaska will make the decision as to an extension of the comment period: a decision made in Alaska by Alaskans, which the senator had previously expressed as the desired state of affairs. I believe the problem is that Murkowski simply does not believe that decisions concerning Alaska are not being made at the White House. I told him that I could see his point. When individual timber sales are canceled or delayed by orders from the White House, it is difficult to believe that something as high profile as the Tongass land-use plan will be left to the discretion of the regional forester in Alaska.

At that point, Murkowski shifted to a discussion of the desirability of a fifteen-year extension of the fifty-year timber sale agreement with Ketchikan Pulp Company, a subsidiary of Louisiana-Pacific Corporation. KPC is in deep trouble with EPA for environmental transgressions. It is estimated that KPC must invest some $150 million to upgrade the facility to meet environmental laws and regulations. The company makes the pitch that it cannot make such an investment without an assurance of a set level of wood supply from the Tongass National Forest.

The senator made the point that it is essentially impossible to have a viable timber program on the Tongass without a facility that can utilize the significant amount of timber that cannot be used for sawlogs. He is correct, in my estimation, and I told him so.

I also told him that, realistically, the Forest Service would not make the decision. In this administration, Vice-President Al Gore holds the environmental and public land organization portfolio. I advised him to seek an audience with the vice-president. Murkowski indicated that such a meeting was scheduled within the next few days.

20 June 1996, Arlington, Virginia

The day was quiet enough until I dropped by the Timber Management staff director's office on some routine matter. My office assistant, Sue Addington, found me and told me Congressman Norm Dicks was on the phone. The Forest Service budget was being debated on the House floor, and Dicks was dealing with two issues that he wanted to head off. One was an amendment by Congresswoman Elizabeth Furse of Oregon to prevent any expenditure of funds to carry out the salvage rider. The amendment, as usual, was so poorly constructed as to preclude an exact determination of what it meant (will they never learn to ask for drafting assistance?). Dicks was looking for quick answers, and the Forest Service was the place he and other representatives were used to getting that information from—quick and accurate information.

Earlier, the congressmen from western states with significant timber programs had been shocked by the passage of an amendment—inspired by something Ted Kennedy of Massachusetts had introduced on the Senate side—removing the ability of the Forest Service to use road credits to timber purchasers to build needed logging roads and cutting the road budget by 30 percent. In combination, these two aspects of the amendment would have stifled the Forest Service timber program rather decisively. The amendment initially passed 212 to 211. Upon reconsideration, it failed on a 211–211 tie vote. The additional backlash in the House against the salvage rider, in the form of Furse's amendment, is now spilling over into the entire (and already much reduced) timber program. This situation should not go unnoticed by political observers.

Furse's amendment was another approach to killing the salvage rider but was so carelessly contrived that there would likely be huge liabilities incurred, and a chain of unforeseen consequences that would produce literal chaos in the timber industry—particularly in the Northwest. This amendment was eventually defeated.

Late in the day, I had a conversation with Secretary Glickman's assistant, Anne Kennedy, who blithely said that we should be referring any calls such as the one from Norm Dicks, and other congressmen, to Legislative Affairs in the Department of Agriculture. Then she said that the administration was "generally supportive of the Furse Amendment" and thought it was understood that although the amendment was "terribly flawed," it was all O.K. as problems would be worked out in conference with the Senate. I nearly exploded but kept it inside.

Here we go again. The administration decided it was in "general support" of a piece of legislation without consulting with the Forest Service or having any understanding of the consequences of its passage. Then, having made that decision, the administration did not inform the agency head so that actions could be guided accordingly. How many times will that same pattern of behavior repeat itself?

Urban tree house dedication, Washington, D.C., June 21, 1996. Chief Thomas
and Secretary Glickman are at the center. Photo courtesy of Jack Ward
Thomas.

Can they imagine the response of a congressman calling the chief of
the Forest Service if I said, "Congressman, I am unable to respond to your
request for information. Let me give you the address of the Legislative
Affairs director in the Department of Agriculture"? I don't think I want
to find out.

24 June 1996, Rosslyn, Virginia

Over the past several weeks, one of the infamous "group grope" letter-
writing efforts has been going on, involving the secretary's office, the Council on Environmental Quality, and the Forest Service. The letter being
prepared for Secretary Glickman to send to me will spell out constraints
to be applied to the execution of the provisions of the hotly debated salvage rider.

The administration can't stand the heat from the environmental community over the salvage rider. Thus, the secretary's letter, which will
severely curtail Forest Service efforts in salvage logging until after the provisions of the rider expire at the end of 1996.

The last version of the draft letter that I saw had a number of phrases inserted that the field troops and the press would seize on as highly critical of the Forest Service. I kept striking those phrases yet they continued to reappear, I suspect after more circuits through CEQ. This political spin is sure to have a good political effect on the environmental constituencies that have become so important and has the added impetus of some payback to the Forest Service for not being immediately compliant to changing administration positions.

However, these spins will have an immediate chilling effect on Forest Service line officers, who are finally showing an aggressive posture toward salvage and forest health problems. Besides, these gratuitous insults were absolutely unjustified. The Forest Service had been encouraged to aggressively use the salvage rider to accelerate salvage and forest health activities. And it was made clear that the agency was not to fail in meeting the target of 4.5 billion board feet.

I felt it was appropriate for me to meet personally with Secretary Glickman and his chief of staff, Greg Frazier, to discuss the matter. After being presented with the facts, Secretary Glickman agreed with my points, particularly about unjustified and unfair chastisement of the Forest Service for carrying out past administration policy.

He simply said that the administration's policy had changed, that he had the authority and responsibility to change policy for what he considered appropriate reasons, and that the policy was being changed to ensure a much more conservative and more politically acceptable approach to salvage under the salvage rider. I said that I had no disagreement with those prerogatives or with his exercise of that authority. I did, however, object to using Forest Service personnel as scapegoats that had somehow misused the authority granted them under the salvage rider as a ruse or subterfuge to cut green timber under the guise of salvage.

I said that if it were truly necessary or appropriate in the secretary's or the administration's opinion to chastise someone, such chastisement should be directed at top Forest Service management and not at field personnel. He indicated that he saw no reason to chastise anyone. He had changed policy, he would accept responsibility for that decision, and he would state that in the letter he would send me.

I told him that in my opinion sending this letter was a mistake, politically and technically. The social and economic effects in local communities would be more serious than anticipated and the administration's read on the politics was off base. The environmentalists involved in this effort cannot be satisfied, no matter what the concessions. And public opinion in rural areas is strongly in favor of an even more aggressive salvage and forest health program.

I told him that no matter how gently the policy reversal of the salvage program was sent out, this switch would have a demoralizing and chilling effect on Forest Service personnel. The result would likely be a failure to reach the 4.5 bbf target far worse than is being estimated. My best guess was approximately 3.0 bbf.

Mr. Glickman heard me out and acknowledged that the Forest Service had done exactly as it was instructed. But the decision to concede to the environmentalists' pressure had been made, and he wants us to execute the new policy faithfully. I agreed, with the proviso that this change is clearly set forth in writing. He agreed.

25 June 1996, Arlington, Virginia

The weekly meeting with Secretary Glickman took place in midafternoon. He started with a discourse on the letter he was about to send to me concerning changes he wants in operations under the salvage rider. He laid it out for the Forest Service top executives exactly as we had worked through the issue yesterday. He complimented the Forest Service's efforts to carry out an aggressive salvage program as directed under the rider. He said that policy had now changed and that he would assume responsibility for that and would spell it out in his letter of instruction.

Mr. Glickman made it clear that the establishment of policy was his responsibility. However, he also clearly said that he had no wish to micromanage Forest Service operations, and he was certain that I would tell him the difference. At that point, I said that in my opinion, the administration's and the secretary's dealing with individual timber sales was micromanagement. He nodded but said nothing.

Secretary Glickman then made some observations concerning Senator Larry Craig's letter to him of some weeks ago. Senator Craig had made it an open letter, in that his committee staff director, Mark Rey, had spread copies all over the bureaucracy and to selected newspapers. The secretary allowed that he did not have any considerable argument with much of the letter. But he preferred that the letter not be distributed via the Forest Service's electronic mail system (as requested by Craig) until his (Glickman's) answer was ready to go back to the senator. At that time, both letters could be distributed.

The secretary was bemused by one of the senator's points: the senator chastised him for interference in Forest Service affairs. The secretary recalled that in his confirmation hearings, Senators Craig and Murkowski had intensively browbeaten him to agree that he (Glickman) would involve himself to a much greater degree in Forest Service affairs than had past secretaries of Agriculture. Now, the secretary was criticized for doing just that. Perhaps there is something to the old adage, "Be careful what you wish for. You may get it."

Secretary Glickman also made it clear that he would be less inclined in the future to passively accept suggested actions from CEQ or the secretary of the Interior. He made it clear that when he signs a letter, it will be his letter and he will not be rushed to sign something that he is not ready to sign. That's progress.

Mr. Glickman is gaining respect from the Forest Service brass. He has one attribute of a leader that I find sorely lacking in most political appointees. He is willing to make decisions and publicly take responsibility for them. Others play politics and then try to camouflage the fingerprints by forcing the agency head to "make" the decision. If things go right, the politicals take credit. If things go wrong, the agency gets the blame. This is done so routinely that I sometimes wonder if the players even think about what they are doing.

The rest of the meeting dealt with routine matters.

27 June 1996, Washington, D.C.

In midafternoon, Nancy Green of the Wildlife Staff dropped by to give me a heads-up on the issue of proceeding with the construction of the University of Arizona's third telescope on Mount Graham. Interior Secretary Babbitt and Peter Coppelman of the Department of Justice want to satisfy the demands of environmental groups and further delay the initiation of construction of the telescope. Though not said in the conversation, I would suspect that a delay until after the November elections would be about right.

I was given the heads-up on the likelihood that Secretary Babbitt might call to use his powers of persuasion on me to produce a decision to delay construction pending renewed consultation. I called Mark Gaede from the Natural Resources and Environment staff, who has been attending the Tuesday Afternoon Club meetings at CEQ to get his view of the situation.

His take was that it was obvious that I would not go along with reinitiating consultation and delaying construction, as that would override the Forest Service people responsible, and it was very clear that my personal opinion was that such an action is not warranted. And there is no stomach in Interior or the Fish and Wildlife Service to take the heat. Therefore, he believed that the University of Arizona would begin tomorrow to finish clearing the site and move in equipment. I hope so. This whole issue has dragged on too long.

3 July 1996, Washington, D.C.

There was a meeting today in the chief's conference room of representatives of all the cooperating federal agencies (now there is a real oxymoron) to come to terms on the draft environmental impact statement on the New World Mine. The meeting had been arranged on very short notice, over Forest Service objections, at the insistence of Janet Potts, counsel to the secretary of Agriculture, and Brian Burke, assistant undersecretary of Agriculture, to bring the "federal family" together on the issues. We told them repeatedly that it would take more than one day, that these were the wrong people, that Washington was the wrong place (Denver was better for the people who would have to do the real work), and that the attendees would not have time to review the massive document.

And sure enough, the meeting began with all the representatives from the various agencies making all of the above points and being a bit put out with the Forest Service for trying to railroad them to some hasty conclusion. Janet Potts was on holiday and Brian Burke did not speak up to claim credit for calling the meeting.

As is usual in these events, there was no one present in the room who could make decisions for the group as a whole. That, as usual, drove me to the point of extreme frustration. I left the room at noon and did not return. I was merely an observer, anyway.

It is obvious, at least to me, that the players from the Department of the

Interior, the Council on Environmental Quality, and EPA do not want the New World Mine to start up under any circumstances, and certainly not before the November elections. The hassles over the EIS are, again in my mind, merely a ploy to delay action. I have always suspected that there is some behind-the-scenes deal being negotiated with the mining company. If this were not so, the company would, quite legitimately, be raising unmitigated hell over the interminable delays and the deep involvement of the White House in this matter.

At the break, Ray Clark of CEQ pulled me aside and said that this effort would ultimately be unnecessary as "a deal" was close to being "a done deal." The EIS process will likely become OBE—"overcome by events." He said he would like to meet with me early next week to let me in on what is coming down. I knew it—I could smell it—but I told him that I didn't think I wanted to know.

9 July 1996, Arlington, Virginia

Tom Lyons, regional special agent in charge for the Pacific Northwest Region, called to tell me that the law enforcement people were ready to move in to clear the demonstrators blocking the access road into the Warner Creek timber sale. All was considered routine except that because an informer had told our officers about one person in possession of what could be a machine pistol, the officers would go in full tactical gear (helmets, flak jackets, etc.). Lyons also told me that four Forest Service agents had infiltrated the group of protestors, local law enforcement officials and prosecutors will assist, the U.S. attorney for Oregon is in agreement, the FBI has been consulted, and the U.S. Marshals Service has agreed to transport prisoners.

As a matter of courtesy, I made an appointment to give Greg Frazier, chief of staff for Secretary Glickman, a heads-up on the situation. After listening to my report, he suggested that Secretary Glickman receive the same briefing.

Upon hearing the report, Secretary Glickman expressed great concern and asked a number of questions. Had the FBI been informed? Answer: Yes. Was the U.S. attorney in Oregon in agreement? Answer: Yes. Were local and state law enforcement personnel involved? Answer: Yes. Were the marshals involved? Answer: Yes. The more questions he asked, the more agitated he became. He invoked images of Waco and Ruby Ridge.

He wanted to know if, given the danger, we could simply back away from the sale. Answer: If we back off, we have merely capitulated to illegal action. Such will damage the timber purchaser and will almost certainly encourage more such activities in the future. He asked me to ensure that all precautions had been taken.

We got Special Agent Lyons on the phone, and he responded to all the secretary's questions and concerns. To my mind, Lyons (who has handled numerous demonstrations more dangerous than this one without a hitch) had everything down pat and should be given the go-ahead.

To my dismay, the secretary ordered the operation put on hold until he checked things out with people "above his pay grade." I tried again to convince him that such delay imperiled the element of surprise and would likely compromise the operation, and this delay would increase the danger to the officers involved. He repeated his order.

I called Lyons back and told him to put the operation on hold until he received clearance from me to proceed. He was professional in his response, but he was obviously upset and concerned that a carefully conceived strategy was being upset by political interference.[3]

19 July 1996, Bozeman, Montana

Today was spent in taking a look at the proposed site of the New World Mine that has become such a controversial issue. I was flown to the site above Cooke City, Montana, via helicopter, which afforded me a chance to see the entire layout and get a feel for the situation that is impossible to get from maps and photographs. We landed near Cooke City and drove to the head of the drainage that contains the proposed mine site.

The drainage has been mined for some hundred years. The creek is sterile from acid mine drainage and tailing deposits. Old mining equipment and deteriorating buildings dominate the upper drainage and would be considered junk if not defined as "historically significant artifacts." There is a switchback road at the head of the drainage that leads to a partially reclaimed open-pit copper mine that is one ridge closer to Yellowstone Park.

Much of the drainage is private land, and the public land has been almost

---

3. The Warner Creek salvage sale was eventually bought back from the purchaser by order of the secretary of Agriculture. The Forest Service was not asked for an opinion.—JWT

entirely covered with patented mining claims. The alternatives (options) that have been developed in the environmental impact statement to date were explained to me once again. Again, the effects of the alternatives are much easier to visualize standing at the site.

As we returned to the helicopter, we met with a group of eight citizens of Cooke City and environmental activists, two of them hired guns. The discussion was cordial enough. Their bottom line was that they oppose the mine, period. Obviously, no amount of analysis, assessment, or mitigation is going to change their minds. And that, I think, clearly spells out the situation. The New World Mine has become a symbol, a cause, and a rallying point for the environmental community. Arguing about technical details is merely a ploy and a delaying tactic. Facts are merely tools to be debated, twisted, and elaborated.

It is time to cut to the chase and quit playing as if the technical assessment will have any influence on the outcome. The mining company will mine. I have absolutely no doubt of that. The only question is whether it will mine New World or the United States Treasury. I bet on the latter. The environmentalists win a victory. The administration gets credit, and just before the election. The company gets to "mine" and the taxpayers get the shaft.

29 July 1996, Cleveland, Tennessee

July 26, 27, and 28 were the days of the Olympic Whitewater events at the Whitewater Venue on the Ocoee River. The Forest Service, primarily the Cherokee National Forest, was the lead agency in putting the venue together. By necessity, this was a partnership effort that included the state of Tennessee and the Tennessee Valley Authority. Holding the coalition together required a masterly job of leadership by John Ramey, supervisor of the Cherokee.

The Forest Service was given a clear mission—a rarity in these times—and the troops came through brilliantly. I visited the site three times over the past two years. Each time I came away convinced that the people in charge would get the job done, although on the surface completion by the time of the event seemed unlikely.

As I toured the venue the day before the event, I never felt more pride in my life in the Forest Service. The venue was ready. Nothing was left undone. As I was briefed on the operation to come, I was even further impressed. The battle plan was inclusive of every contingency of which I could conceive. Given the threat of terrorism, security measures were intense and involved officers from the Forest Service, TVA, FBI, Secret Service, Sheriff's Department, Tennessee State Police, and Tennessee Park Police. The Forest Service handled the overall coordination. SWAT teams were assigned to patrol the adjacent hillsides, snipers occupied key positions, SWAT teams were in reserve on site, uniformed officers were stationed at every key point, other uniformed officers moved constantly through the crowds, plainclothes officers were strategically located, and TV surveillance cameras constantly swept the crowd.

The Olympic Committee was effusive in its praise for the Forest Service and told me over and over that this was simply the finest whitewater venue in the world. Hundreds of officials from all levels of government and, more importantly, hundreds of citizens expressed appreciation for a job well done.

And the games went off without a significant hitch. The weather cooperated. The stands were filled. All the technology worked. Everyone did his or her job. It was a pleasure to watch. I reveled in the atmosphere of total, complete success.

At the closing ceremony, at which the winning athletes were awarded their Olympic medals, an unprecedented event occurred. Representatives of the Forest Service (the chief), TVA, and the state of Tennessee were called to the awards platform. For the first time in Olympic history, the people who staged an event were honored with Olympic medals, a gold for the Forest Service and bronzes for the other two partners. It was a good feeling to have the Forest Service recognized for excellence in performance. Such recognition has been scarce these past few years.

Today was the day after the closing ceremony and time to honor the hundreds of Forest Service employees who, in a synergistic combination, produced the agency's part of the 1996 Olympics. It was my pleasure and honor to be present, along with Regional Forester Robert Joslin and Forest Supervisor John Ramey, to shake hands with and say thank you to each awardee.

But the high point for me came when I presented the Olympic gold medal for achievement to all those gathered at the picnic, and handed the medal to John Ramey. The crowd reaction clearly showed their pride in achievement, and their pride in the Forest Service.

Later, thunderstorms rolled across the sky and the rains came. A Forest Service plane flew me from Cleveland, Tennessee, to Knoxville to catch the last flight to Washington, D.C., so that I could be the witness at another hearing tomorrow. As the pilot twisted the plane through the thunderheads, I could see the clouds light up with lightning flashes and hear the rumble of thunder. I fancied that as applause from the heavens as I stared out in reverie. My God, I felt good!

2 August 1996, Washington, D.C.

There have been two hearings during the past week in front of the Senate Energy and Natural Resources Committee. The first was a courtesy to Senator Kyl of Arizona on the subject of forest health and fire in the Southwest. Kyl was in the chair, and the hearing was low profile and obviously intended to give him a platform useful to the politics he faces at home. It is clear that Kyl is frustrated with the situation on public lands in the Southwest, and I don't blame him.

The timber program was already in free fall in the Southwest with the exhaustion of old-growth ponderosa pine and the rising concern over threatened species. This ongoing trend has been exacerbated by a series of federal court injunctions that have arisen out of arguments over appropriate levels of consultation with the Fish and Wildlife Service over effects of proposed management activities. These process problems have temporarily shut down all timber harvesting in the national forests, and mill after mill has closed.

This loss of infrastructure has now gone to the point that, even if there is a wood supply of smaller-diameter trees from thinning operations associated with management aimed toward forest health, there will be too few mills to process the wood. That, in turn, deprives the Forest Service of revenue to offset, at least partially, the cost of such operations. When the jobs and wood products forgone are considered, it is difficult to see the situation as anything but a growing problem.

Senator Kyl wants the Forest Service to put on the table a plan, with associated funding and manpower requirements, to address the building forest health and fire danger problems over some period of time. That seems reasonable. But the administration, through the Office of Management and Budget, does not want any proposals—certainly not before the election—

that will increase spending or require more government workers. And of course, there is the risk of offending environmentalists, who argue that there is no forest health or fire problem and disagree with cutting any trees. Since environmentalists are considered a key to the election, there should be no move that would offend them.

Kyl knows that. I know that. He wants that plan. I would love to give it to him. I can't do it. He knows that. So now what?

I took the opportunity to give a speech about the increasing problems in public land management caused by the interactions of the crazy quilt of law and regulation, overlapping agency responsibilities, lack of clear policy direction, ever-increasing litigation, and paralysis in political will. I pointed out that we are involved in partisan political wrangling over trivia such as the salvage rider to such an extent that there is no will nor capability to face up to spiraling problems.

The Forest Service could lead in laying the groundwork to face those problems. But we will not be able, or allowed, to do so in the current political situation. I know that. Kyl knows that. So now what?

As I was leaving the hearing room, I heard Mark Rey, chief of staff to the committee, say to the press, "Don't miss the hearing later this week with Secretary Glickman. That will be the mother of all hearings."

That hearing was designed to castigate Secretary Glickman over his recent direction to the Forest Service to pull back on some portions of the efforts to put up 4.5 billion board feet of wood under the salvage rider. The Forest Service had the secretary well briefed and prepared for the hearing. Undersecretary Lyons and I were at the witness table with the secretary.

The walls of the hearing room were covered with blown-up photographs of stands of dead and dying timber from a wide array of national forests. The aim of the committee majority was twofold. First, they wanted to show that Mr. Glickman did not have any concern over the fate of men and women in resource-dependent communities, had capitulated to pressure from environmentalists, and had knuckled under to orders from the White House (i.e., the Council on Environmental Quality). Second, they intended to drive a wedge between the Forest Service and the administration by portraying the Forest Service as striving to carry out the law and good forest management but being thwarted by the politicians in charge at the White House.

This change in tactics—bragging on the Forest Service as the professionals—is a recent shift. Oddly, this belated appreciation of the professionalism of the Forest Service vis-à-vis other agencies only damages the Forest Service's reputation in the administration.

The attempt to drive a wedge did not work. And for a change, a Democrat showed up at the hearing, Senator Dale Bumpers of Arkansas. What a difference that made. Someone had prepared Bumpers well and he single-handedly took Senators Murkowski and Craig and associates apart. The hearing dragged out far too long and ended with a rambling discourse by Senator Murkowski, interspersed by questions that made little sense. Like most hearings, this hearing produced more heat than light and was the same hearing held for the tenth time.

Mark Rey has badly handled the opportunity afforded him and the committee he serves. He simply had too many hearings with no focus. The press quickly became bored with the show. The window for radical change, or even significant change, in the laws driving and constraining national forest management has in my opinion closed. The carefully constructed drive to devolve public land ownership or management has failed.

But maybe something good can come of all this. I have tried to foil Rey's effort but at the same time lay the groundwork for a serious bipartisan effort to revamp the laws and regulations related to public land management.

Perhaps that effort can be instituted in a bipartisan manner after the elections in November. One can only hope. The time is right.

6 August 1996, Washington, D.C.

I met today with Kathy Maloney, who is in charge of the staff preparing the Resources Planning Act report. This report, required at five-year intervals, outlines long-range options for management of the Forest Service. I had not been present for the last review of the RPA by the national leadership team. Therefore, I was taken aback by the proposed RPA program, which is based on the reality of a flat budget situation over the next decade, with essentially the same program distributions.

I asked questions, many questions, but my primary thrust was that the document was devoid of vision, devoid of new ideas, and devoid of hope. This was hardly what one would expect from an agency that bills itself as a conservation leader.

I told Kathy that this draft was not acceptable. The scenario of the flat-line option should be presented exactly as put forward with full suggestion of the modest results, including deterioration of infrastructure and a declining ability to produce goods and services projected in the forest plans. Maybe, as the senior staff thinks, that is the likely reality.

Maybe, but maybe not. Without consulting staff, I ordered that an additional scenario be prepared based on the assumption that Congress can be sold on the idea of allowing the Forest Service permanent authority to collect and retain fees. Those fees would be used to finance a much more aggressive program in areas of recreation and fish and wildlife along with an aggressive forest health program.

Kathy explained that staff had heartburn with that idea and fear that any such additional revenues would simply result in further reductions in appropriated funds. Maybe, but maybe not. That will depend on how good a job Forest Service people do in cultivating support among constituency groups who are paying fees for enhanced services. And even if that does occur, it may simply represent a new way of doing business for the Forest Service in those areas of interest that are in keeping with the idea that users should pay for those services, as opposed to the taxpayers at large. Besides, nothing ventured, nothing gained.

I was somewhat surprised when Kathy did not argue against my instructions. This was, after all, an individual order from the chief that essentially countermanded a decision by the national leadership team. And it was certainly an order that would cause significant additional work for her and her staff. I suspect she was relieved that somebody was willing to think outside the box.

Why can't we put forward a new vision of how the Forest Service can accomplish its mission? As an agency that wants to be considered conservation leaders, do we have any choice? I think not.

Mike Dombeck met me for lunch at the Cosmos Club. We began to lay plans for what we both consider the high probability that he will succeed me as chief, assuming that President Clinton wins reelection. If Bob Dole becomes president, all bets are off. I believe that I could remain as chief under such circumstances, if that were my desire. However, in either case, it is my desire to retire as soon after November 15, 1996, as possible.

On the other hand, I do not believe Dombeck would be an acceptable chief to a new Republican administration, as he is too closely identified with leading the resistance to the grazing bill championed by a segment of the livestock industry and sponsored by Senator Pete Domenici of New Mexico. But even so, Dombeck would enter the chief's job as a career senior executive, which means that he must be placed in some equivalent-level position. So there is no real risk to his career if he assumes the chief's job before the end of the year and then is replaced by a Republican administration early in 1997.

Our main objective is to see to it that the transition is orderly, professional, and carried out with dignity. That is essential to the morale of the Forest Service. It is essential that the fiasco that surrounded the departure of my predecessor, Dale Robertson, not be repeated. Such a moment can be a proud one for the Forest Service. Dombeck and I are determined that this passing of the mantle of Gifford Pinchot be just that. We have decided to include Max Peterson, chief emeritus of the Forest Service, in our planning sessions, at least some of them. He is a wise man and likely more politically astute than either Dombeck or I.

8 August 1996, Washington, D.C.

There was a meeting in the early afternoon with Greg Frazier, chief of staff to Secretary Glickman; Gray Reynolds, deputy chief for National Forest System; and Tom Lyons, special agent in charge for the Pacific Northwest region. The meeting concerned the still unresolved issue of how the Forest Service is to deal with illegal demonstrations on timber sales. Things have been at a standstill since the panic reaction to the removal of demonstrators that blocked the access road into the Warner Creek salvage sale. The secretary, fearing violence, called off pending actions to remove the demonstrators. Then, confabs were held with the Council on Environmental Quality, Justice Department, FBI, and others that just produced confusion and more fear and ended up compromising the operation.

Then, that error was compounded by the decision to avoid any potential violence by buying out the timber sale purchaser. This handed a clear victory to the protestors, who had illegally destroyed government property, blocked a road, and committed other acts in defiance of law.

So, thoroughly confused and considerably demoralized, we have been waiting for the administration to figure out what they want to do. The best move is for them to simply get out of the way. Our strategy has just been to pull back and wait for the public backlash to what amounts to highly selective application of law enforcement.

That decision has in fact produced a backlash from the timber industry, local government, state government, and individuals. Pressure is obviously building on politicians, and they are bringing pressure on the administration. Even liberal, proenvironmental politicians are feeling the heat and pushing for us to enforce the law. Ron Wyden, the newly elected senator from Oregon, has called to complain.

Greg Frazier feels highly embarrassed that interference ever took place at all and wants to figure a way out of the mess. This meeting was designed to give Tom Lyons a face-to-face opportunity to vent frustration to Frazier and to get Frazier's support to turn our job back to us to execute.

That may be difficult because the hardcore environmentalists have learned to go to the top to get what they want, and they get what they want most of the time. Worst of all, the people at the top have a tendency to believe what they are told and make no effort to discern the truth, or at least the truth as perceived by the Forest Service.

Frazier listened to us very carefully and agreed that the current situation was becoming increasingly embarrassing and untenable. He plans to visit the Pacific Northwest region next week for some firsthand knowledge of the situation. He clearly understands that day-to-day operations cannot be directed by means of interagency group gropes in Washington.

12 August 1996, Bozeman, Montana

Late last week I received a pointed request from the secretary's office via Janet Potts, the counsel to the secretary. She requested that I attend the ceremony in Yellowstone Park at which President Clinton was to announce a deal with the mining company to assure that the New World Mine on the Gallatin National Forest would not proceed. It had become clear to Potts and to Ray Clark of the Council on Environmental Quality that the Forest Service had been scapegoated in the political drama surrounding the New World Mine.

They have arranged for the entire Forest Service crew involved in the long, ongoing preparation of the associated environmental impact statement to be present at the ceremony. The plan is to have the president visit briefly with the group for a group picture.

Ceremony marking agreement with New World Mine, Yellowstone National
Park, August 1996: President Clinton and Chief Thomas with Brian Burke,
deputy undersecretary of Agriculture; Richard Bacon, deputy regional forester;
and Dave Garber, supervisor of Gallatin National Forest. White House photo,
courtesy of Jack Ward Thomas.

The ceremony was a well-staged political event with the target being the
voters of green tint. The master of ceremonies was from the Greater Yellowstone Coalition. His accolades, exclusive of the lexicon of environmental
activists, were directed to the mining company (which had agreed to a better
deal than the New World Mine) and to Mike Finley, the superintendent of
Yellowstone Park. Finley has become a new hero to the environmentalists
based on his carefully conceived untrue and slanderous assaults on the Forest Service and state of Montana professionals charged with the preparation of the EIS. Where are we when a "professional" can so blatantly use
propaganda to achieve a political objective and become a "hero" to politicians and a significant segment of the American public?

Actually, no deal was announced, only the agreement to make a deal. I
could not help wondering about the wisdom of announcing, for political
purposes, a deal to make a deal with details to follow at a later date. Does

President Clinton at Yellowstone, August 1996, with Dave Garber, supervisor, Gallatin National Forest; and Katie McGinty, chair, Council on Environmental Quality. Photo courtesy of Jack Ward Thomas.

this not hand the mining company a decided advantage in negotiations that are ongoing? This deal is a long way from completion.

The agreement makes it clear that the EIS process will be terminated as soon as possible. The document, which would indicate a very, very low risk to Yellowstone Park from the New World Mine, would never emerge to reveal the true situation.

Most, if not all, Forest Service employees and I could not care less if there ever was a New World Mine. We cared deeply that the process was carried out correctly. It was not.

The president's speech was a good one, and the punch line was, "Yellowstone is more precious than gold." The signatories to the agreement were gathered on the stage for a formal signing: Katie McGinty for CEQ, the head of the Greater Yellowstone Coalition, and the chief executive of the mining company. The president and Mike Finley looked on. Odd, truly odd, that the proposed mine is in the Gallatin National Forest.

The president is a masterly politician. When he walked up to the Forest Service group, who were not happy with how the Forest Service has been treated in this entire episode, he completely disarmed the group by saying, "You folks really got the short end of the stick in this whole deal. I know that. But I appreciate your dedication to a good professional job." It worked, at least to some extent, as the entire group seemed to relax a bit.

15 August 1996, Washington, D.C.

During the chief and staff meeting, one of the Office of General Counsel's attorneys announced that the Forest Service had prevailed in the suit called The Friends of Animals v. Thomas, or more commonly, the bear baiting case. That was a significant win that required two distinct victories. The first was to finally persuade the Department of Justice to defend the Forest Service's position that the setting of any hunting regulation of resident animals was the prerogative of the state within whose boundaries a national forest is located. The Justice Department initially refused to back that position. They simply failed to see the longer-term consequences of setting the precedent of having the Forest Service execute an environmental impact statement on a hunting regulation set by a state.

Such would impose a requirement, at least implicitly, that anyone could use to disrupt a hunting season of any kind. Worse yet, it is likely that the EIS would be subject to reconsideration every time "new information" became available. Such new information could include new census data, changes in habitat conditions, etc.

I felt certain that we could prevail in court if I could persuade the Justice Department to defend our position. So once again, it is obvious that Justice has tremendous influence on natural resources policy simply on the basis of what legal actions they choose to pursue.

In the afternoon, there was a conference call between Mark Schaefer, assistant secretary of the Interior; Brian Burke, assistant undersecretary of Agriculture; Chris Risbrudt, director of Ecosystem Management for the Forest Service; and me concerning data collection and vegetation mapping in the Greater Yellowstone ecosystem. Interior has charged ahead in this area while, I think, pointedly excluding the Forest Service from planning and strategy sessions.

The initial purpose of the call was to determine where were the Forest Service funds to carry out the data collection and vegetation mapping exercises contracted for by the Park Service. I said, rather frankly, that no such money was forthcoming. The Forest Service was eager to proceed in the Yellowstone ecosystem in a cooperative fashion and in partnership—equal partnership—with Interior agencies. The Forest Service has most of the federal land in that ecosystem, a research division larger and more diverse than the Biological Services outfit of Interior, and considerably more experience in leading ecosystem management assessments. The Forest Service, however, was not willing to be part of a tail-wagging-the-dog scenario.

I agreed that Chris Risbrudt would serve as Forest Service liaison with Interior agencies in Washington to see what could be worked out. I also pointed out that Forest Service–Park Service relationships in the Greater Yellowstone area were greatly strained by the continuing insults from Superintendent Mike Finley. Those problems would need to be worked out before full cooperation would be resumed.

26 August 1996, Washington, D.C.

During my absence, the Pacific Southwest regional forester, Lynn Sprague, prepared to release the draft environmental impact statement for the Sierras. From what we could read in the papers (the Forest Service was not consulted), White House Chief of Staff Leon Panetta and Secretary Glickman decided to hold back issuance of the EIS as they had determined that it "was not based on the best science." The press release (we think prepared by the White House) was spun to make it appear that the White House had saved the Sierras from Forest Service incompetence. I was furious.

Glickman's assistant, Anne Kennedy, had dressed down Sprague and Associate Chief Dave Unger based on information that "outside scientists" and "Forest Service scientists" operating through back channels had informed the White House about the situation. None of those informers, of course, were identified. If we don't know they are, how can we deal with the allegations, and with the problems, if indeed there are any?

Barry Noon, along with Jared Verner and Kevin McKelvey, all top-notch Forest Service research scientists, led the California spotted owl assessment effort. I called Noon to see if he, Verner, or McKelvey was involved. Barry said they still had some concerns but thought the draft should be released. Most of their concerns had been met and the remainder would be appropriately handled as comments to the draft. They had nothing to do with the present situation.

Anne Kennedy had told Unger that we should start over, using "the A-Team." Bingo. That was Jerry Franklin of the University of Washington talking. I called Franklin and sure enough, he was the trigger. Franklin has entrée to Katie McGinty, chair of the Council on Environmental Quality, who bought his story without consulting the Forest Service. At that point, it seems likely that she gave Franklin access to Panetta or appropriate aides.

Franklin has been a friend and colleague for over twenty years. That he would do this without even talking to me pains me. That the administration would act so precipitously without talking to me pains me even more.

Now we have to try to manage the situation. There are three options that I can see. First, reconsider the decision and go ahead and release the EIS but without a preferred alternative—enough of a change to save face. Second, announce a hold until the Sierra Nevada Ecosystem Project publications are complete and have been in the hands of the public for thirty days, prepare an additional SNEP alternative (a reserve system similar to that in the Pacific Northwest), and release after the presidential election. Third, start over with "the A-Team," which would include the critics.

Option 1 is best to maintain faith with the process but is not politically correct so far as the administration is concerned—it is not likely to fly. Number 2 is the best compromise and has a certain plausible logic that can be presented while passing (maybe) the laugh test. However, it will not satisfy those with the ear of the White House, who will not be satisfied until they set the course, so it is not likely to be adopted. Number 3 will infuriate many people inside and outside the Forest Service, mess up the process, require the chartering of a "balanced" Federal Advisory Committee Act group, and be expensive (and perhaps impossible) to fund. But it will be popular with the environmentalists and their pet scientists, so it is very likely to be adopted.

This is but one more clear sign that the administration (at least the White House) does not trust the Forest Service, or is overly concerned about the political ramifications of offending the environmentalists in a presidential election year, or is simply inept, or some combination of the above.

28 August 1996, Washington, D.C.

Joan Comanor (deputy chief for State and Private Forestry), Bill McCleese (associate deputy chief for State and Private Forestry), and Mary Jo Lavin

Undersecretary of Agriculture Jim Lyons and Chief Thomas, North Minam River, Eagle Cap Wilderness, Oregon, August 1996. Photo courtesy of Jack Ward Thomas.

(director of Fire and Aviation Management) met with me in my office to discuss the current and projected fire situation. In terms of national forest lands burned, we are headed for a record. And the long-term weather analysis does not offer much hope for a quick end to the fire season.

All qualified resources (both people and equipment) from all federal and state agencies are committed to fires. Canada has provided assistance in the form of fire crews and air tankers. Two battalions of military have been committed. Mary Jo (or Joan) then said, "We could activate as many as four more battalions of military but we can't do so due to lack of available strike teams to provide overhead."

Upon my probing, she said that perhaps there were some additional strike force people around but that they were being held back by some line officers. I asked my secretary to have Gray Reynolds (deputy chief for National Forest Systems) come to my office. While I was waiting for Gray to arrive, I called BLM Director Mike Dombeck and said that I was about to issue an order establishing a telephone tree (Washington to regions to forests to districts) to shake out every last person we could field as strike force to military battalions. I asked him to do the same. He said he would.

The orders were given and executed. In an unprecedented move, I also made available all qualified persons in the Washington office. By close of business, between the departments of Agriculture and Interior, we had found enough strike force personnel to put two additional battalions on standby.

I was encouraged to go to the fires personally to keep morale up, get a firsthand feel for the situation, and bear down on the absolute necessity of placing safety first on every fire, every time. I told staff to prepare a schedule for me stretching from tomorrow afternoon through the Labor Day weekend. That will take up only one workday (Friday) and will make the most of the three-day weekend, and get me back to Washington at 11:00 P.M. on Labor Day. It is important for the troops on the fires to see their leaders, particularly when they are going twelve to fourteen hours a day, seven days a week, holidays notwithstanding. And it is important to stay in the fire camps, eating and sleeping in the same manner as the firefighters. What the chief does and where the chief goes mean more than what the chief says.

30 August 1996, Fire Camp, Umpqua National Forest, Oregon

The day started early with a tour of the National Interagency Fire Center at Boise. This is the nerve center that deals with wildland firefighting across the nation. No one could fail to be impressed with this operation and the state-of-the-art nature of the facility. We visited every section of the operation. The visit started with the 0800 national situation briefing. That briefing gave me my first opportunity to reemphasize my desire for increased attention to safety and to express confidence in and appreciation for the efforts expended so far this fire season. That message was repeated to each group I met with during the day.

Stan Hamilton, Idaho state forester and president of the National Association of State Foresters, was with our group during the entire visit. I took the opportunity to discuss with him my ideas about the creation of a national fire management cadre. This would be a new mechanism for funding fire management (as opposed to firefighting), whereby we can put a stop to the collusion between Congress, the administration, and the land management agencies to hide the actual cost in dollars and manpower of dealing with fire.

There is no logic in funding a low level of presuppression activity, and then spending whatever proves necessary to fight wildfires above an artificially (and arbitrarily set) low level by borrowing from trust funds, while promising that Congress will pay back the funds through emergency appropriations.

We must collectively begin to deal with the conditions that contribute to large, hot, and uncontrollable wildfire. Our firefighting capabilities are better than ever. However, it is clear that current conditions caused by a combination of fire exclusion, insect and disease control, insect and disease outbreaks, and drought (perhaps due to changes in weather regimes) are producing more and more fires. In spite of ever-increasing success in initial attacks on fires, more and more acres are burning on public lands.

By noon, we were on the way to Portland on an aircraft lent by the Bureau of Reclamation (all Forest Service aircraft are tied up on fires). We landed at the Portland International Airport for a working lunch with the Multi-Agency Command located there. MAC is established when the fire situation in a regional area reaches the point that available resources to respond to the fires are being strained.

We discussed the current situation and five-day outlook with MAC. James Brown, state forester for Oregon, was invited and was present at the meeting. The Oregon Department of Forestry had sent Regional Forester Robert Williams a rather pointed letter last week chastising the feds for what it considers a misguided new fire policy. Brown particularly dislikes the change in priorities in the new policy. Under the old policy, priorities were (1) protection of human life; (2) protection of property (structures); then (3) protection of forest resources. The change in policy allows incident commanders, after protecting human life, to evaluate the situation and decide on the most cost-efficient basis how to fight the fire. In other words, the decision might be to not protect a house in order to protect forest and rangeland, if the cost-benefit ratio favors that course of action. That differs from the Oregon policy that gives private property higher priority.

Brown also had nothing good to say about the new prescribed natural burn policy. Under that policy, a fire in a wilderness area that starts naturally from a lightning strike is allowed to burn as long as it is within "prescription"—in accordance with that area's fire management plan. When the fire moves outside the prescription limits, efforts are made to bring it within prescription or suppress it altogether. There have been at least four large fires, where full suppression efforts are being applied, that resulted from naturally ignited fires in wilderness. I told Brown that an investigation team had been named and a review was forthcoming. Beyond that, I made no commitment.

31 August 1996, Fire Camp, Umatilla National Forest, Oregon

By midafternoon, our aircraft had delivered us to Roseburg, Oregon, to visit firefighters battling a large fire on the Umpqua National Forest. We went first to the headquarters of the Umpqua and were joined by Forest Supervisor Dan Ostby and Deputy Supervisor Bernie Rios, a friend from our former days as wildlife biologists. Though it was a weekend, the office was filled with folks serving as the central dispatch and purchasing crew.

We drove to the fire camp, which was at one end of a long dirt airstrip, and split up in order to visit as many folks as possible. The young lady handling ground operations for the six or so helicopters arrayed at the other end of the airstrip asked if I would visit her crew. I replied that I would be pleased to do so but didn't have time to walk the half-mile or so to the other end of the airstrip and back.

Later, she approached me again; Could I ride a four-wheeler? She showed me to my assigned mount, and we crept along at less than five miles per hour. Once we had cleared the part of the airstrip filled with people, I picked up the speed to some twenty miles per hour and passed my companion. She yelled something that I didn't understand; I assumed it was some comradely joking comment.

I soon came upon a law enforcement officer—I think a deputy sheriff—who put up his hand. I waved back as I passed him. It turns out he wasn't waving—he was commanding me to stop. He yelled, "You a—, STOP!" I circled back and dismounted. He dressed me down in a forceful manner about exceeding the five-mile-per-hour speed limit and was not impressed when I said no one had told me about any speed limit. His castigation complete, I put out my hand and said, "I'm Jack Thomas, chief of the Forest Service."

He did not shake my hand. He said, "Yeah, right! And I'm f—ing Santa Claus. Do you think I give a damn who you are?"

I drove away toward the helicopter operations. When my companion caught up with me, she tried to apologize. She seemed surprised when I told her to forget it and tell the officer to relax. I was speeding, it was late, and the officer had been on duty for several days, twelve to fourteen hours a day. He had likely endured a number of smart remarks and probably thought my claim to be chief of the Forest Service was just one more. I found the incident funny. Besides, it is well to have one's ego deflated from time to time.

The officer later found me in the chow line and began to apologize. It took a bit of effort to convince him that I was not offended or angry. He joined me for supper and finally seemed at ease after I told the story to our dinner companions, who seemed to enjoy it. The instance will serve as a good addition to my catalogue of self-deprecating humor—the only sort that is politically correct.

We left the Umpqua by van shortly after breakfast in Bend, Oregon, to look over the results of a fire that swept in from open juniper and sagebrush through a subdivision three days ago. Nine houses were lost and it was a miracle that many more were not destroyed.

Some neighborhoods that were not burned shared the same characteristics of the ones that had burned, making the potential danger glaringly obvious. How many times does this same thing have to happen before the lessons become apparent and somebody—government or individuals—does something about the situation?

Listening to the reports of the incident commander, I realized that in the effort to save people's homes, the firefighters took some chances that I don't consider acceptable. And it was clear that in the heat of battle (no pun intended), danger to lives and property leads firefighters to be much less cautious than they should be. That, sooner or later, will produce a catastrophe. But it is heady stuff to drive through such a neighborhood and see signs leaning against trees or mailboxes containing such messages as "God bless you firefighters, you saved our home," "Thank you, firefighters," "There are still heroes—our firefighters," and "Bless the U.S. Forest Service."

All this leads to a question that begs an answer, and very soon, because the situation worsens each day with the inexorable expansion of the urban-forest interface. How much does society as a whole owe to people who deliberately put themselves in harm's way?

After meeting with the firefighters who fought this fire to hear their accounts, observations, and concerns, and conveying appreciation to them for a job well done, we were off via charter aircraft to Pendleton. At Pendleton, the deputy forest supervisor for the Umatilla met us and took us to forest headquarters to visit the folks manning general dispatch and purchasing for the firefighting efforts to the south. Such office workers are as essential to the firefighting effort as the folks on the firelines. They are the people behind the scenes who are seldom recognized for their services. I took no more of their time than to introduce myself to each person and say thanks.

From Pendleton, we drove to the district office at Ukiah. Smoke from the Tower fire could be seen for thirty miles. We arrived in time for the afternoon briefing on the status of the firefighting efforts. The incident commander was from the Oregon Department of Forestry. I was surprised to see Gordon Smith, current Oregon state senator and Republican candidate for the U.S. Senate, at the briefing. Smith was narrowly defeated by Democrat Ron Wyden in a special election last year to fill the seat vacated by the resignation of Senator Bob Packwood. He is favored to win the Senate seat being vacated by the retiring Mark Hatfield.

When I addressed the group briefly, I spoke primarily about the necessity to begin to take a different tack to combating wildfire. We need to assume a posture of fire management, as opposed to continuing emphasis on firefighting. Smith seemed interested and asked some good questions. I don't think he was there to politic but to learn something. That's encouraging.

1 September 1996, Fire Camp, Summit fire, Malheur National Forest, Oregon

Every time I view a fire like this from the air, I am struck by the consistent inaccuracies in the media reporting on the fires. Part of that is our fault. The media report that the number of acres burned is the entire area within the perimeter of the fire. In reality, it is common that many acres within the perimeter did not burn at all or were relatively undamaged. To compound the inaccuracy, the acres burned are described as "acres destroyed," or more melodramatically, "acres devastated" or "acres ravaged." No acres were destroyed. The vegetation burned but the acres are quite intact.

This fire has burned into some older burns, and that has yielded the firefighters some advantage. There is probably something to learn by examining the timing and accumulation of these burns over the past several decades. The acres are beginning to mount up to impressive totals, giving some credence to the old saw, now heard more and more often, "It is not a question of *if* it will burn; it is a question of *when*."

We arrived in the fire camp in the late afternoon and repeated the ritual of visits, handshakes, and briefings. Each camp is very different in the composition of agencies and personnel but all operate under the same tried and tested Incident Command System. In one camp, there were personnel from four federal agencies, the Army, the Marines, the National Guard, thirty-seven states, Oregon prison inmate crews, contract crews, and myriad contractors. It worked, and it worked beautifully.

I ate supper with a lieutenant colonel commanding a battalion of engineers. He and his troops, all with close-cropped hair, clean shaven and in clean fatigues, stood out in stark contrast to the civilian firefighters in dirty Nomex, many with beards and long hair, and nearly all unshaven. Most looked the kind of tired that comes from day after day after day of weariness.

Yet that young colonel said that the military could learn much about how to bring quick coordination between "mixed troops" by participating in such firefighting exercises. And by damn, he meant it.

As I sat in the mess tent after supper watching the late-arriving fire crews in the snaking chow line pass my seat, I was entranced by the faces and the snatches of conversation. They were grimy and tired after a twelve-hour day. There was no joking or horseplay. The nights are turning cold and they shuffle along in the line with their hands in their pockets and shoulders hunched under their light jackets. Their breathing leaves a cloud of moisture in the air that reflects the light from the bare bulbs in the tent.

I can hear different accents as the groups pass: soft southern drawls, Canadian, Maine, Yankee, and Spanish—lots of Spanish. They come from far-flung locales, yet they are all here for one purpose. The mission is clear and there is no time or energy for anything else. And it pays the fire crew members the princely sum of $8 an hour and three meals a day and a place to sleep on the ground in a tent. How can you beat that?

Just now the media are making heroes of these young people, though they are unlikely looking heroes. And when the fires are out and the snows come, many of them will be without work and they will be heroes no longer. These heroes get no hospitalization, no pension credit, no sick leave, and no annual leave. They get their $8 an hour and $12 an hour for overtime and that's it. And when the fire season is over and the woods and structures are safe for another year, we owe them nothing. That doesn't seem right. Somehow, it just doesn't seem right.

2 September 1996, airborne,
en route from Portland, Oregon, to Washington, D.C.

Carl Pence, supervisor of the Malheur National Forest, picked us up just after breakfast. We stopped by the forest headquarters in John Day for a few minutes to visit with the dispatch and purchasing crews on duty this Labor Day holiday. There were Christmas decorations all over the office and each of the crew wore a T-shirt with a Christmas tree on the front. I think the message was that they had been there so long that surely it must be Christmas.

Morale was good and we all took pleasure in the "extra tall and fat" size Christmas tree T-shirt that they presented to me. I must have a hundred special T-shirts and a hundred special ball caps with various logos and messages. They are my trophies, given me by friends and "family." The question is, what does one do with so many T-shirts and ball caps?

After visiting at the John Day airport with the air operations folks, we went by charter to La Grande for a visit at the fire center and a briefing on the fires in the Hell's Canyon National Recreation Area, on the Snake River. The same crews were on duty that had been there when I was here over two weeks ago. These fires resulted from prescribed natural fires that were started by lightning and were at first allowed to burn but then escaped the prescription limits of the area's fire management plan about the first of August and had to be controlled.

These fires have been "out of prescription" for six weeks and suppression costs have been high, which has provoked some severe criticism and some justified questions. No doubt we will be called to answer for this at some hearing or another. A full-scale review headed by Jerry Williams of the National Interagency Fire Center in Boise has begun.

I met with Bob Richmond, supervisor of the Wallowa-Whitman National Forest, at the fire center. We discussed the Snake River fire situation and the pending decision on powerboat use in the recreation area. The decision will be to limit jet boat use to four days per week. That will be contrary to the desires of the entire Idaho congressional delegation and the governor of Idaho. Tough call politically, but in my mind and in Bob's, the only right one.

Now on the way to Washington from Portland, I have time to reflect. To put it succinctly, we have a mess on our hands and we need a bipartisan agreement about what to do about it. That is it in a nutshell.

The Forest Service simply no longer has a clear mission. These outstanding firefighting efforts, the progress on timber salvage (until the rules were changed), and the preparation of the venue for the whitewater Olympic events demonstrate what we can do with a clear mission. On the other hand, the Forest Service muddles or mills very poorly. And that is what the agency has been reduced to by circumstances of interacting laws, downsizing in both budget and personnel without real consideration as to effects, and increasing politicization of Forest Service operations. And I see no light at the end of the tunnel.

3 September 1996, Washington, D.C.

In the letter of instruction sent by Secretary Glickman concerning salvage sales, there is a provision for the chief of the Forest Service to review sales submitted for exemption from the secretary's instructions and grant such exemptions where justified. This afternoon, I met with Gray Reynolds (deputy chief for National Forest Systems), Robert Nelson (staff director for Fish and Wildlife), David Hessel (staff director for Timber), and several other staffers to consider exemptions.

Regional foresters have submitted a list of sales for exemption. There were two sales in the Southwest that were under injunction in the ongoing legal argument over protection of the Mexican spotted owl. Decisions on those two sales were deferred until such time as Judge Muenke lifts the injunction. Seven sales were submitted for exemption because they fell into "identified roadless areas" under the Carter administration's roadless area review exercises. Under RARE II, areas not identified as wilderness study areas were allocated to multiple-use (i.e., timber harvesting was anticipated).

These seven timber salvage sales were designed for multiple-use management and have since been roaded and logged. As the areas involved are not roadless at this time, it was my decision that the constraint imposed by the secretary's directive was moot. The exemptions were granted.

Only six other sales were granted exemptions, for varying reasons. One was "imminently susceptible" to insect attack—it's a pocket of southern pine beetle infestation, which is growing rapidly in the southern region. Another was a large area of blowdown on the Six Rivers National Forest that was cleared because it was imminently susceptible to fire and rapidly deteriorating in value. These were clear-cut cases; the others were a bit less cut and dried but were well justified, in my opinion.

No more had the decisions been made than I received a call from Undersecretary Lyon's office with instructions not to give field notification until my "recommendations" (suddenly no longer "decisions") have been cleared by the secretary's office and the White House. That level of political involvement in such a minor matter is difficult for Forest Service staff to comprehend. It is hard to know whether to laugh or cry. Laughing is probably healthier.

4 September 1996, Washington, D.C.

There was a lengthy meeting with Janet Potts (counsel to Secretary Glickman) and Brian Burke (assistant undersecretary) on the New World Mine. Ray Clark of the Council on Environmental Quality has been heading up the effort to carry out the political directions (the agreement to make an agreement) concerning the mine. The first cut at a "team" to execute this effort did a pretty good job of excluding the Department of Agriculture, which strikes us as strange, given the mine is on the Gallatin National Forest.

We came to the conclusion that Janet Potts should be on the executive committee, and Brian Burke should be cochair of a subcommittee on identification of assets to be offered for exchange. I frankly stated that I could never be party to offering National Forest System lands for exchange, first as a matter of principle, and second as not being permissible under the law. I had something more of an open mind about trading timber rights, although I consider that a miserable idea. Such would be a precedent that should be avoided and, whether intended or not, would be akin to a long-term contract with a single source. Every long-term contract regarding timber supply has turned out to present severe problems over time.

Potts and Burke understood my position, agreed with it, and wanted me in a situation to make my case. I agreed but cautioned that Ray Clark, speaking on behalf of CEQ, would resist that idea. Clark knows that I do not agree with the approach to both the New World Mine and the Headwaters Redwood deals: I am "parochial" in my protection of the national forests and not much interested in the politics of the situation. The meeting ended on that note.[4]

13 September 1996, Washington, D.C.

There was a meeting in the afternoon at the Council on Environmental Quality concerning the thirteen sales, prepared under the auspices of the salvage rider but withdrawn pursuant to Secretary Glickman's letter of instruction, that I had determined to exempt. As I suspected, the secretary's delegation to me of authority to exempt sales meant nothing; here we are at CEQ to second-guess my decisions. Present were Dinah Bear of CEQ; Anne Kennedy, Jim Lyons, and Mark Gaede of USDA, a representative from the White House Domestic Policy staff, and a representative from the Office of Management and Budget.

It was obvious from the beginning that my decisions in the matter were no more than advisory. The discussion was *all* political and had nothing to do with technical considerations or logic related to the directive. It was further interesting that after some initial efforts to discuss technical matters and logic, Undersecretary Lyons and I were discounted and the discussion took place primarily between Anne Kennedy, Dinah Bear, and the young lady from the Domestic Policy staff. Their entire concern was, so far as I could tell, related to the potential reaction of the environmental community. They wanted to be certain no exemption of any sale was announced that was likely to cause any flak for the president during his planned political campaign trips to key western locales in the next two weeks.

The staffer from OMB wanted to know if the Fish and Wildlife Service

---

4. In June 1998, Crown Butte Mining, Inc., and other interested parties signed a consent decree, approved by the U.S. District Court, that contained the final terms of agreement on property transfer and the amount of funds available for cleanup of the area around the New World Mine.—*Ed.*

and the Environmental Protection Agency had been consulted on these exemptions. I said that other agencies had been consulted in the preparation of the individual sales, and since the sales had been cleared and appropriate assessments done, I did not consider it necessary to repeat the process. She begged to differ and insisted that no such exemptions be granted until approved by all the regulatory agencies, whether approval is required or not.

At that point, I saw that it was no use to say anything further. I doubt that any sale will be exempted, at least not in a timely manner. And if any sales are released, it will be done on a carefully reasoned political basis. So I simply went quiet and listened and watched the exercise of government power. The question is whether this is an abuse of power. If it is, what should I do about it? I am becoming more and more disgusted with the situation by the day.

14 September 1996, Rosslyn, Virginia

On 10–11 September, I attended the national reunion of Forest Service retirees in Utah. It was good to see so many old friends and colleagues. There is life after the Forest Service.

During that period, I had a chance to visit with Deputy Chief Gray Reynolds, Regional Forester Dale Bosworth, and Steve Mealey, who is heading up the team preparing the environmental impact statement for the Columbia basin assessment. They were quite concerned about events that had taken place during the most recent meeting of the assessment's executive committee.

After expressions of concern by National Marine Fisheries Service and the Fish and Wildlife Service over the preferred alternative selected by the executive team, they pulled the draft EIS from the printers in order to further analyze the alternatives under a new set of questions posed by the regulatory agencies.

I told the three regional foresters involved that they must make what they consider the appropriate decision, even if the regulatory agencies disagree. The two BLM state directors feel much more pressure to toe the line with what the regulatory agencies want (i.e., demand). From the beginning, the regulatory agencies have insisted that PACFISH buffers plus all roadless areas over 1,000 acres be placed in reserve status. They have never altered that position and therefore have insisted on a process that is contrary to the spirit and regulations of the planning process.

We have had an extremely difficult time getting through this process in light of the extreme political concern shown by a number of national political figures, including Senator Larry Craig of Idaho and Representatives George Nethercutt of Washington, Helen Chenoweth of Idaho, and Wes Cooley of Oregon. In addition, many county commissioners have expressed sincere concerns and were kept on board only through my personal influence and Steve Mealey's.

With the decision to pull back the EIS from the printers and to make one more attempt to mollify the regulatory agencies, the political situation is now about to come unhinged. And I can see why.

I have told the political players, over and over, that this would be a clean process run by the local managers without interference from politicals or Washington Office personnel. As a result, I have remained aloof from the process except to insist that those responsible get on with the job, and to meet with concerned politicians to insist that the process would be a straight deal with decision authority resting in the local managers' hands.

To my dismay, Undersecretary Lyons sat in on the last meeting of the executive committee. He called me ahead of time to ask me to call the regional foresters and tell them that it was imperative that the National Marine Fisheries Service and Fish and Wildlife Service be "on board" with the preferred alternative. I said that I was dedicated to a clean process and would not attempt to influence their decision. That decision may come to me on appeal, and it is essential that I be able to make an independent judgment. I advised him not to attend the meeting. As usual, he disregarded my advice.

He told the regional foresters that it was essential that they settle this at their level and not let it be elevated to the Washington level. That clearly meant that they must find some way to satisfy Will Stelle, regional director of the National Marine Fisheries Service, and John Rogers, regional director of the Fish and Wildlife Service. Stelle and Rogers remain adamant that regardless of any assessment, the result will include the position they dictated from the very start. That amounts to making an offer the regional forester cannot refuse and guarantees a single outcome. In the words of a regional forester, this is blackmail pure and simple. And it is a perversion of the prescribed process.

Lyons told me that he merely observed the executive team meeting and said nothing. Therefore, there had been no political interference and I could rest easy on that score. I couldn't accept that: the political pressure had occurred in his discussion with the regional foresters.

On Friday, the word was out to the concerned Republicans and the county commissioners. As I had warned Secretary Glickman last week, when the administration encourages back channels that go around proper channels, it should be expected that back channels will also open up to the concerned Republicans in Congress.

Senator Larry Craig of Idaho called me at home to discern what was going on. I told him that I was not sure, but I clearly remembered my pledge of a by-the-book process and I meant it. I told him that if I find otherwise after discussion with the regional foresters, I would come and tell him.

The next day I had a call from Anne Heissenbuttel, chief of staff to Congressman James Hansen. She wanted to know if the administration was interfering in what was supposed to be a process carried out by the professionals as opposed to a politically motivated exercise. I said I would get back to her. By midnight Washington time, I must have heard from a dozen county commissioners about their concerns.

This is a serious business. My sense of alarm is heightened by the recent interference in a similar process for the Sierra Nevada by Leon Panetta, White House chief of staff. Panetta called back a draft EIS because the "best science" had not been used. Panetta had no idea of the situation. He simply read the political tea leaves, obeyed orders from the environmentalists, and allowed his press secretary to scapegoat the Forest Service. This looks like the same thing all over again. I hope we can turn this one around because in this case, I personally promised an honest process and a decision, free of coercion, by the designated decision makers. If I can't keep my word, I will either have to resign or blow the whistle and then likely be fired.

23 September 1996, airborne, en route from Springfield, Missouri, to Washington, D.C.

Today was the opening day of the annual meeting of the National Association of State Foresters. At this same meeting last year in Louisiana, I managed to irritate a number of state foresters by stating that we needed to work together to resist further cuts in the Forest Service budget. I also told them that if additional severe cuts continued, the Forest Service would be forced to cut back on State and Private programs to ensure adequate funding to care for the 191-million-acre National Forest System. I thought that statement was simply telling like it is, but some state foresters interpreted it as an abandonment by the Forest Service.

After the uproar subsided, both sides increased efforts to heal the breach and strengthen relationships. It worked; budgets essentially stabilized and state-federal partnerships are now enhanced.

However, there are a few state foresters (Oregon, Idaho, and California, in particular) who are a bit seduced by the talk of devolution of federal land management, or even ownership, to the states. There have been several working groups looking at the issue in some degree of collusion, I believe, with Senate staffer Mark Rey's efforts to reform the National Forest Management Act. Title VI of the draft revision contains a provision for the states to petition Congress to take over management of selected national forests for management under state law, and maybe transfer of ownership to the states at some later date. I suspect that Title VI is a throwaway to be abandoned when negotiations start.

So in my talk I took pains to address the issue forcefully. I said the following:

> You know where I stand on the issue of devolution of our national forests. For me it is a *nonnegotiable* issue! While there may be individual parcels of land that would be better in either state or private land, those can be handled through systematic public procedures. But that is not what is driving this issue.
>
> In my opinion, if this issue is actively pursued with the serious intent of accomplishing it, it will set off the most divisive, most destructive political conflict over natural resources management in a century. In my opinion, it would be most wise for the state foresters to step back from this issue. In the end, there will be—again in my opinion—no devolvement of either land or management responsibility.
>
> The debate and discussion on private property rights and responsibilities are part of this larger discussion on the rebalancing of power and of roles. A positive outcome would better be an understanding of the costs for pub-

lic benefits derived from private lands and appropriate mechanisms to equitably share these costs. Equally important would be better recognition of the responsibilities for stewardship as a part of all landownership. Our challenge is to use this call for a rebalancing of power to stimulate responsible leadership action on the land and in communities. In the end, we may well gain more public support for better protection and management of all our nation's forests.

I knew when I addressed the group that Rey would address them later and that devolution would be part of his pitch. There was also a place on the program for a discussion led by Jim Brown, state forester from Oregon, to seek passage of a resolution on devolution.

As I expected, after lunch Mark Rey made his pitch on the revisions to the NFMA, including Title VI on devolution. It is too bad that he has included the devolution lightning rod; the rest has some merit for debate and potential resolution.

After Rey finished talking, I asked him a question that went something like this: If the issue is largely one of whether management should take place under state law or federal law and regulation, why not, instead of devolution, merely have the Forest Service manage the national forests in one state under trust doctrine and under state law? Wouldn't that be a much better test and one more likely to actually happen, given the personnel and infrastructure in place? It didn't matter what the answer was (which was evasive). Key state foresters understood the message in the question.

Jim Brown was immediately ready to follow Mark Rey's discussion with a discussion paper for a resolution by the group. The argument was that there are only two ways to address the present problems of conflicts in laws and resulting gridlock, the first being a simplication of laws and better federal-state cooperation and coordination. If that does not occur, then devolution to the states should come about. Oddly, in his presentation he clearly stated that if the states had to manage federal lands under current law, they could do no better than the federal land management agencies.

That, then, meant that devolution to the states would involve not only a shift in power and responsibility but also a simple cancellation of some federal laws and regulations. Now that is radical, and unlikely.

Oddly, it was some time into the discussion when the state forester from

Connecticut (where there are no national forests) said, in effect, "Those are the lands of the American people at large and not the property of the states." What he didn't say was, If the state wants the land, should it not be purchased from the United States? Or conversely, if federal lands are managed by the states, should not the federal government receive 25 percent of gross receipts, which the federal government paid the counties when the roles were reversed? Those are arguments for another day.

I left the meeting feeling fairly certain that a vote for devolution of management or ownership of the national forests would not occur. If so, my attendance was worthwhile.

This was my last meeting with the state foresters. In a way, that seems a shame, as I am just establishing rapport with the group. There are many outstanding men and women in their ranks.

26 September 1996, Rosslyn, Virginia

I met with Greg Frazier, chief of staff to Secretary Glickman, for an hour in the morning. The subject was my desire to announce my retirement at the upcoming meeting of the regional foresters and experiment station directors in Washington on 7–11 October. I told him that the rumor mill was operating overtime concerning my retirement. Further, several reporters are digging at the story that I have accepted an appointment as the Boone and Crockett Professor at the University of Montana and that I have purchased a house near Missoula, all true.

I promised Greg that I would provide the secretary with curricula vitae and one-page assessments of five candidates whom I would recommend and welcome as a successor. I promised to deliver that material on the morning of 30 September. I identified the five names I would submit: David Unger, associate chief, FS; Mike Dombeck, acting director, BLM; Phil Janik, regional forester, Alaska Region, FS; Pete Rusopolous, director, Southern Experiment Station, FS; and Elizabeth Estill, regional forester, Rocky Mountain Region, FS.

There was a very brief discussion of the pros and cons of each candidate for the position of chief. I was pleased with the exchange, and Frazier left my office with a full understanding of my position. I believe that he was in general agreement with my suggestions.

In midafternoon, there was a hearing before Wyoming Senator Craig

Thomas's subcommittee of the Committee on Energy and Natural Resources. The invited witnesses were Katie McGinty, director of the Council on Environmental Quality; Nancy Hays, undersecretary of the Interior with responsibility over BLM; and me. Ostensibly, the subject of the hearing was an examination of the workings of the National Environmental Policy Act, to determine whether improvements were needed, and what those changes might be.

The witnesses knew better. This hearing was to be a wide-ranging turkey shoot, and Katie McGinty, as surrogate for the Clinton administration, was the designated turkey. Interestingly enough, even with a senior White House official on the griddle and the elections only five weeks away, the Democratic senators did not even attempt to level the playing field by attending. Only Bill Bradley (in a cameo appearance and without enthusiasm) and Senator Daniel Akaka from Hawaii (who is totally oblivious to the issues covered by the committee) even showed up. They did not stick around when things began to turn nasty. The Republican members were out in full force. Senator Craig Thomas was in the chair. Senators Frank Murkowski of Alaska and chairman of the full committee, Larry Craig of Idaho and chairman of the Forests subcommittee, and Conrad Burns of Montana were present and loaded for bear.

Senator Thomas did his best to maintain some attention to the subject of the hearing—the functioning of NEPA—and to maintain some façade of a civil debate. But he simply couldn't hold the others to either the subject or the idea of a civil discourse.

The other senators one by one called Katie McGinty—personally and as surrogate for the administration—a liar (no mincing of words) and charged her, as part of a continuing pattern of behavior, with garnering political gain by bowing to the wishes of environmentalists. Senator Burns actually used the word bullshit to describe the course of events and the testimony at the hearing. The "high point" of the hearing was an attack on the just-announced decision to create a new national monument in Utah by presidential decree. The action was considered deceitful behavior on the part of Interior Secretary Bruce Babbitt, White House Chief of Staff Leon Panetta, and others. Though not mentioned by name, there were strong implications that the president and vice-president knew fully of the strategy that was executed.

The senators then went after McGinty about why the route of presidential decree, in circumvention of the entire NEPA procedure, was taken, rather than more usual prescribed NEPA process with full analysis of the economic, ecological, and social consequences of several approaches.

The give-and-take was entertaining, but it was clear that the decree was a political maneuver, pure and simple. This was a coup for the environment in the establishment of a national monument that will be a lasting legacy of President Clinton's. As there is no conceivable way that the president can carry Utah in the upcoming election, and there are so few electoral college votes in that state, the national political gain far outweighs any political drawbacks.

This likely is my last appearance before a congressional committee—at least as chief of the Forest Service. Oddly, over time I have come to enjoy the hearing process for a combination of reasons. First, preparing for a hearing requires some intense briefings by staff and review of pertinent documents, which in and of itself is educational and forces knowledge into some contextual arrangement. Second, it allows one to establish some rapport with committee members and their staffs. They are willing to accord respect, even if grudgingly, if the witness knows the issue, responds honestly, is forthright, unflappable, respectful of the committee, and uncowed by what is clearly designed to be an intimidating experience. Third, the adrenalin rush affords a brief period of maximum mental acuity and stage presence. The combination of knowing the facts, thinking three questions ahead of the inquisitors as in a chess game, and trying to contrive the best theatrics *ad hoc* is a real challenge. Strangely enough, these hearings are a part of the chief's job that I will miss. I don't know if I should admit that, as most folks seem to have a genuine aversion to being in that witness chair and in the glare of the lights.

4 October 1996, airborne between San Diego and Washington, D.C.

The past four days have been spent on an intensive review of circumstances surrounding the Forest Service's part in the intensive federal efforts to "stem the tide" of illegal immigrants entering the United States from Mexico. The background is that worsening economic conditions in Mexico coupled with the presence of low-wage jobs in the United States have set off increased levels of illegal immigration. The vast majority of those illegal entries routinely were taking place along a small section of the border near the California coast. That situation finally produced an intensifying political issue and led to the institution of an effort by the U.S. Border Patrol called Operation Gatekeeper. That effort, beefed up by new resources in both agents and equipment, proved successful in cutting down on the illegal entries near the California coast. Unfortunately, this merely shifted the problem to the east, and one of the primary routes of entry was suddenly the Cleveland National Forest.

Within one year of the institution of Operation Gatekeeper, hundreds of inappropriately located trails began to appear in the backcountry roadless areas of the Cleveland National Forest. Along with this increased human use came other problems, including litter and garbage; human fecal material along watercourses; increased danger to Forest Service personnel, recreational visitors, and the illegal immigrants themselves; and the increased danger from wildfire sparked by abandoned campfires.

The increased fire danger was of immediate concern due to the dramatic urban-forest interface problems in southern California. The fires are not extinguished, even when water is handy, because the guides (called coyotes) do not want to put up a plume of smoke and ash that might give away the camp's position. The fire crews stationed in the area had, like everywhere else, been pruned back significantly because of mandated downsizing and budget reductions.

Noting the increased (and increasing) fire danger, Forest Supervisor Ann Feege requested an emergency increase in presuppression firefighting funds. Upon requests from Regional Forester Lynn Sprague, funds adequate to come to full staffing were released from the chief's emergency funds. This situation is likely to worsen over the next six weeks, as the illegal immigrants are likely to build more fires as the nights are becoming cooler and fall comes on. And coupled with that, the season of the Santa Ana winds is upon us. Based on historical patterns, we can expect eight to fourteen days of strong Santa Ana winds over the next six weeks. These winds can produce very hot and fast-moving fires that are almost impossible to control in the heavy brush and rugged topography typical in the wilderness areas of the Cleveland.

Fire management people are very concerned with human safety—particularly that of illegal immigrants—as the situation worsens. If there are large and fast-moving fires, the fire organization is concerned about using backfiring and air tanker water or retardant drops when it is possible that bands of illegal immigrants or individual illegals are in the area trying to avoid detection and could be injured.

Local people, particularly those in rural areas, are also being adversely affected in terms of resource damage, shooting of watchdogs, poaching of livestock, destruction of gates, cutting of fences, threats to personal safety, trash, and the inconvenience of dealing with intensified law enforcement activities. There seems to be a split between Hispanic and Anglo members of the community. The Anglos, in general, are incensed with what they deem an invasion of illegal immigrants and demand that "the government" solve the problem, even if that means the use of military forces. They are angry that the government decided on a strategy (Operation Gatekeeper) without consulting the community and without considering the likely shift in the traffic of illegal immigrants from the west to the more rural areas. One of them referred to the situation as Paradise Lost, and they are angry and frustrated.

Representatives of the local Hispanic community are upset with the rhetoric surrounding the illegal immigration issue. These concerns include sensitivity to terms commonly used in the press and by government agencies—aliens, invaders, illegals, etc. In general, they are sympathetic to the illegal immigrants, resent both political parties for making points by scapegoating the illegal immigrants, and resent the government for playing games by concentrating on interception and deportation of illegal immigrants. They believe that the answer is to punish any citizen who hires illegal immigrants. That approach, of course, would deprive U.S. citizens of a source of cheap labor that can demand no fringe benefits.

These Hispanics have a generally good feeling toward the Forest Service and resent the Forest Service's involvement in Operation Gatekeeper, which they consider the business of the Border Patrol. Hispanics warn that Forest Service participation in this effort will cost the agency in terms of relationships with Hispanic citizens in local communities.

Discussions with Hispanic Forest Service employees revealed similar concerns. There were reports that Hispanic employees were being stopped and questioned by officers, including Forest Service officers, and are fearful of recreating on national forest lands. When I asked for specifics, none were forthcoming, and the reason given for not providing information was fear of retaliation. I am inclined to be suspicious of such cop-outs and tend to believe that this is merely an excuse to avoid facing up to an issue or that there are inadequate facts to back up the charges.

I continually pointed out that the assignment of twenty officers to the Cleveland National Forest to help with Operation Gatekeeper was in response to direct instructions from President Bill Clinton. These orders came down to the Forest Service from the president to the secretary of Agriculture to the chief on June 11, 1996. The twenty officers were in place by June 13, 1996. The Forest Service had no choice, in my opinion, but to execute the orders of the president.

It is easy to predict that the intensifying anger of the people around the Cleveland National Forest will produce a political reaction to stem the flow of illegal immigrants in this particular area. As the Border Patrol pours more resources into this area, it seems highly likely that the entry of illegal immigrants will be shifted, again, farther to the east. If so, this means that the Coronado National Forest is next. As a result, I have constantly emphasized to all those with whom I have visited the necessity for the Forest Service and BLM to be involved with the analysis and planning of such actions as Operation Gatekeeper. The view that predominated, and still seems dominant, in planning Operation Gatekeeper was focused on the simple measure of illegal immigrants intercepted, with very little consideration of the impact on land and on communities and other agencies.

10 October 1996, Washington, D.C.

This was the last day of the fall meeting of regional foresters and station directors. I had asked for a few minutes on the program just before noon. At that time I announced my intention to retire as chief effective in mid-November. And I announced that on that day I would become the Boone and Crockett Professor at the University of Montana in Missoula. It was an emotional moment for me, and I had difficulty getting through my short speech without breaking into tears.

Undersecretary Jim Lyons followed me at the dais and had some nice things to say. The thing that he said that I will remember was the story of his visit to La Grande to persuade me to take the chief's position. I had turned down the job in previous telephone conversations for two reasons. First, I simply had no ambition or desire to be chief of the Forest Service. Second, even when confronted with the argument that it was my duty to accept the challenge—particularly if the alternative was a political appointee from outside the Forest Service—my response was that my first duty was to Margaret, who was slowly dying of cancer.

Jim spent two nights at our home—he was an old friend and colleague from previous significant experiences. He had worked for the Society of American Foresters and was the staff assistant there who did the support work for a committee whose report was titled, "The Scheduling of the Harvest of Old-Growth Timber."[5] That title spoke volumes about the attitude of SAF toward old-growth at that moment in history. I suspect that Lyons, working with Warren Doolittle, associate deputy chief for Forest Service Research, was instrumental in appointing a couple of ringers to that committee that just might view old-growth forests through somewhat different lenses. Those ringers were Jerry Franklin and me. The report for the first time did take note of the value of old-growth forests as a special habitat, a unique attribute of diversity in the successional spectrum, and a thing of great beauty and even spiritual significance.

By the time the old-growth issue (disguised as an issue about the potentially threatened northern spotted owl) had exploded in the Pacific Northwest and the Interagency Scientific Committee was formed under my leadership, Lyons was chief of staff to the Democratic majority on the House Agriculture Committee.

He could see that the George H. W. Bush administration could never stand politically to allow implementation of the ISC strategy, and stalemate was inevitable. Lyons therefore persuaded Congressman Kiki de la Garza, Agriculture Committee chairman, and Harold Volkmer, chair of the key subcommittee, to convene a committee of scientists to produce an array of alternatives for Congress to consider as a solution to the growing problem. In doing so, Lyons reached back to players from the original SAF committee: John Gordon of Yale, Norman Johnson of Oregon State Uni-

---

5. Society of American Foresters. 1984. *Report of the SAF task force on scheduling the harvest of old-growth timber.* SAF Policy Series. Washington, DC.

versity, Jerry Franklin, formerly with the Forest Service and now at the University of Washington, and me. We became known as the Gang-of-Four.

When Congress failed to act on the alternatives laid out in that report, the issue continued to simmer until, just before the 1992 presidential election, Judge William Dwyer completely shut down timber operations on public lands within the range of the northern spotted owl. When newly elected Bill Clinton held his Forest Conference in Portland shortly after the election, one of the organizers was Jim Lyons (who knew he was to become undersecretary of Agriculture with authority over the U.S. Forest Service). Lyons approached me prior to the conference and asked me to organize a team to produce options to break the stalemate in the Pacific Northwest for the president to consider. We would have ninety days to do the job and the president would announce our existence at the close of the summit. This group became known as the Forest Ecosystem Management Assessment Team—FEMAT. We did our job.

Nine months later, Lyons was at my home in La Grande to persuade me to become the thirteenth chief of the Forest Service.

He made his pitch to me and to Margaret, and she and I talked until late into the night after he went to bed. He said he lay awake and listened to our discussion. I was reluctant, but she was insistent that duty prevail over our personal problems. At breakfast, we said we would go to Washington.

He said other nice things, but I most appreciated that he recognized Margaret for her courage and sense of duty, for we were a team. Those were kind words.

I was surprised at the reaction of many of the tough old-timers in the audience as they came up to me after the morning session. Several were openly weeping as they hugged me. Many others seemed obviously distressed. Of course, many said nothing at all.

In the afternoon, there was a press conference in the chief's conference room. I was a bit taken aback as the room was packed and there were TV cameras from ABC, NBC, CBS, and CNN. My last comment before leaving was something like this: "It is time for me to return to the West. My heart is there. I have been gone too long. I am sad to leave but eager to go."

A car was waiting for me. Thirty minutes after I arrived home, Bob Nelson picked me up to go to Dulles to catch a plane west. The wily elk wait for us in the high Wallowas. I am going home. And there are no regrets.

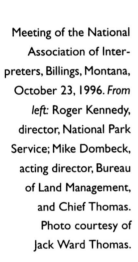

Meeting of the National Association of Interpreters, Billings, Montana, October 23, 1996. *From left:* Roger Kennedy, director, National Park Service; Mike Dombeck, acting director, Bureau of Land Management, and Chief Thomas. Photo courtesy of Jack Ward Thomas.

24 October 1996, Missoula, Montana

There were two purposes for my trip to Missoula: to make two speeches and to close on my new home. The first speech was to a workshop sponsored by the Law School on different approaches to the management of public lands. The usual cast of characters made varying pleas for schemes that would be "more efficient" or produce more revenue for counties or more raw materials at a lower cost. The options offered ranged from "disposal" of the lands to private ownership, to the transfer of management responsibility (and of course, negation of federal laws and regulations) to states or counties under some trust responsibility arrangement, to management whereby local managers are given full leeway to charge what the traffic will bear for all marketable products or opportunities.

The second speech was at the headquarters of the Boone and Crockett

Club on the subject of the place of the Forest Service in today's evolving resource management philosophies. This speech was part of a continuing series. The usual attendance was described as 30 to 40 people. Attendance this night was over 200. I think this was a result of two things—curiosity by some, and a warm welcome to Montana by others. It was an enjoyable evening, but it is easy to see that the split between preservationists and utilitarians is alive and well in this community.

During the social hour, the leader of the Missoula Smokejumpers called me aside and presented me with a Smokejumper's knife made especially for members of that elite corps of wildland firefighters. I was touched, and I guess it showed. He said, "We wanted you to have one of our knives. No other chief has ever cared about us the way you do. And so we wanted you to have a knife."

Little fanfare and not much flowery rhetoric was involved. They did not even gift-wrap the present. But the presentation was pure Smokejumper style—straightforward and to the point. It made my day.

25 October 1996, Missoula, Montana

Today I signed all the paperwork to become the owner of a home in Montana. A year ago I would have been incredulous if anyone had suggested that I would be moving to Montana to become a college professor—and bringing along a new bride, to boot.

The situation brought to mind a quote from John Chancellor, the TV news commentator: "You want to make God laugh? Tell Him your plans."

When it became clear that it was time for me to retire from the Forest Service, it became equally clear that I was not ready to assume the rocking chair. It was important to leave the stress of the chief's job behind, along with the frustrations of being micromanaged at every turn. Certainly, I received invitations to apply for other positions—in academia, state government, and conservation organizations—that would put me back in a role that I have come to despise in today's world. Though obviously essential to today's government administrators, I wanted no more of dealing with equal employment opportunity, civil rights, sexual harassment, employee complaints, whistleblowers, appeals, downsizing, rightsizing, reinvention, budgeting, micromanagement, political correctness, hearings, audits, investigations, reviews, etc.

When I was approached about an endowed chair at the University of Montana, I jumped at the chance to be active and to avoid, or at least significantly reduce, the trials and tribulations of administering a significant program. Teaching and doing research—and holding the bully pulpit for a while—in a wonderful part of the earth seem ideal for one last hurrah.

My heart may not be what it was, but there is nothing wrong with my mind and no dampening of the fire in my belly to care for the land and serve people. To help educate a couple of dozen bright and eager young people with a chance to be among the conservation leaders of the 21st century seems an attractive alternative to retirement. It would be a great legacy if some of these young people did great things and identified themselves as among Jack Ward Thomas's students.

I am truly eager to begin a new life with my bride and a new career as a teacher, and to meld into a new place.

4 November 1996, Washington, D.C.

Today was taken up with the performance review board on ratings, bonus awards, rank increases, and recommendations for Presidential Rank Awards. The board is composed of six senior executives from the Department of Agriculture, who based on a review of one-page writeups on each senior executive, comment on ratings, advancement in rank, and bonus recommendations of the respective agency heads. The two agencies covered by this review board are the Natural Resources Conservation Service and the Forest Service—the two agencies under the oversight of Undersecretary for Natural Resources and the Environment Lyons.

There was very little comment on anything to do with NRCS. I had little trouble with the concerns and recommendations of the review board itself. When it came to the Forest Service personnel, Lyons and Assistant Undersecretary Brian Burke began to exert pressure. Burke was adamant about increasing the ratings for two black female executives, from superior to outstanding. Clearly, he knew little of either individual but was insistent on upping their ratings. Clearly, they (or someone) had complained about the ratings. The board seemed a bit mystified and saw nothing in the writeups to justify such a change. As the change did not alter either advancement or bonus situations, I did not choose to spend any chips in backing Burke off from his recommendations. Burke's argument was the politically correct one, and it was not worth fighting.

Lyons then entered the fray and altered the ratings for several of his particular favorites. As these were white males, I could have resisted. But again, there was no change in status or bonus involved and it was expedient to offer only token resistance. However, other members of the board were becoming more and more puzzled as to what was going on. They were simply being passed over by the undersecretary's unilateral decisions unrelated to the established process.

Then the brawl began. Lyons wanted the ratings lowered for individuals he has been down on for at least three years, including the individuals targeted for removal on May 16 in the Thursday Night Massacre—Deputy Chiefs Gray Reynolds and Mark Reimers and Associate Deputy Chief Steve Satterfield. There was now no option but for me to stand and fight for these folks and for their ratings and recommendations for Presidential Rank Awards. The fight was between Lyons and me, and it got hot. Lyons became totally autocratic and dictatorial. I became adamant. And the board was somewhere, by my estimate, between mortification and anger at being bypassed and ignored in the process.

I said that I believed that this entire procedure—both as related to rewarding pets and punishing targeted individuals—was a blatant perversion of prescribed process. I was disgusted and my feelings were obvious. And so the farce of the performance review board ended.[6]

6 November 1996, Washington, D.C.

Bertha Gilliam, director of Range Management, came to my office with a letter that I had asked her to draft for my signature. The letter declares my resignation from long-standing membership in the Society for Range Management in protest of its board's endorsement of Senator Pete Domenici's grazing bill.

The current board has taken on the role of the guardian of and spokesman for livestock grazing on the public lands. That, in my mind, is

---

6. On January 25, 1997, two months after he stepped down as chief, Thomas recorded in his journal that Reynolds and Reimers had been relieved of their positions and had elected to retire rather than accept reassignments.—*Ed.*

not a proper role for the Society of Range Management, which should not take positions involving which of the multiple uses should be dominant on the public's lands and the transfer of property rights.

I told Bertha to make certain that the letter "gets around" within informal routes of distribution. This is accomplished when the first person leaks a copy or copies of the letter (both hard copy and e-mail) with an attached note that says something like, "Wow! Have you seen this?" or "Interesting reading," or "Your eyes only, keep this to yourself." That indicates that the information has escaped from tight security and ensures that people will read the material and pass it on.

This matter of resigning from an organization to which I have belonged for a number of years has not been taken lightly. My usual inclination is to stay involved and work from inside to change things. But this move by a professional technical society into the political realm of transfer and assignment of property rights on public lands and dictating of law that would set forth a dominant role for livestock grazing on public lands is simply too much and must be forcefully resisted. It seems possible that a visible resignation of the chief of the Forest Service from membership can bring some focus to the issue.

12 November 1996, Eugene, Oregon

I flew to Eugene, and the Willamette National Forest, from the Society of American Foresters' conference in Albuquerque to visit with Forest Service folks concerning the recent arson of the Oakridge Ranger District. My first meeting was with the people in the regional office. I told them that my primary purpose in the visit was to let them know that the entire agency—and particularly the chief—was concerned about them and stood with them in this most difficult time. I told them that we had among the most important jobs in the world and we simply could not and would not be deterred from carrying out our mission of caring for the land and serving people.

Only a fool would not be a bit frightened. But courage does not reside in not being afraid. Courage lies in being afraid and then conquering that fear by carrying on in spite of it.

As they warmed to conversation, it was obvious that they were dispirited and worn down, to say the very least. It was evident from the questions and statements that this was not the result only of the arson fires. It seemed to be cumulative reaction to the arson, the continual appeasement of confrontational environmentalists, downsizing, micromanagement by the administration, the flip-flops on the execution of the salvage rider, and the turmoil in the community in which they live. And these were mostly administrative personnel with little connection to the land and little understanding of what the battles over natural resources management are all about. In general, I think they were glad to see the chief, but the entire feel of the meeting left me uneasy.

I ate supper with the Willamette's leadership team and the district rangers. The leadership was much more positive and heavily engaged in how to react positively to the circumstances. However, they made it clear that they felt the administration's constant retreat from confronting illegal actions by environmentalists had set them up for increasing violence. Such "training" seems ordained to produce increasing levels of confrontation and violence. They felt abandoned by the administration, whose actions have contributed to the present circumstances.

13 November 1996, Eugene, Oregon

I visited two ranger districts on the Willamette National Forest—Detroit (where vandals set fire to a Forest Service truck and spray-painted the office walls with graffiti) and Oakridge (which was burned to the ground by arsonists). I met with the press standing in front of the burned-out hull of the Oakridge Ranger District office.

When asked my feelings upon seeing this mess, I said, "I can't decide whether I am sick to my stomach or just plain mad. Probably both." I went on to say, "This is not a Third World country. This is not acceptable. This is what people do who do not understand how to operate in a democracy. In this country, acts of terror nearly always produce the exact opposite results that the terrorists hope to attain. This is not going to deter us from carrying out our jobs as professionals."

A reporter asked me why I was so intense about the situation when I was retiring at the end of the month. I replied, "Until 1700 hours on the 30th of the month, I'm the chief of the by-God Forest Service—and then there will be someone else there to replace me. Until then, I'm the man."

Later, I met with employees of the Detroit, Oakridge, and Rigdon ranger districts at the Ridgon office. These folks were obviously shaken, more bewildered than frightened, and exhibited what might best be described as numbness.

We discussed just who we were and what our performance might mean in the history of natural resources management in this country and in the world. We talked for more than two hours. By the time the meetings came to an end, we had progressed from mourning to laughing and joking and looking to the future. It was time well spent.

These folks, who I think are typical of Forest Service people across the country, are good people and serve their nation well when properly led and properly appreciated, and when they have a clear mission. We have not done well by these people and they know it, yet they can come up laughing and willing to try to work through the mess that public land management has become.

On the short flight on the Forest Service plane, affectionately known as Smokey Bear Air, I contemplated the one thing that is troubling me about my retirement: walking out with acts such as these bearing down on the workforce. Down deep, I know that the troops don't feel that way, as many have encouraged me to go before something snaps in terms of health— and they are right.

But I have never walked away from a fight in my life and it pains me to step aside now. But down deep, I know that this fight never ends—it merely changes in form and intensity.

22 November 1996, Rosslyn, Virginia

The past five days have been spent in Asheville, North Carolina, at the biannual meeting of the North American Forest Commission. This year the Forest Service, representing the government of the United States, was host for the meeting. The leading representatives were Dr. Oscar Gonzalez of Mexico, Dr. Yvan Hardy of Canada, and me, representing the United States. Dr. David Harcharik of the UN's Food and Agriculture Organization in Rome also attended.

Staff work was excellent and all went off in good order. Primary business concerned questions of sustainable forestry—matters of sweeping international discussion. The debate seems to center on whether there

Meeting of the North American Forestry Commission, chaired by Jack Ward Thomas, Asheville, North Carolina, November 1996. Participants included members of La Secretaría de Medio Ambiente, Recursos Naturales y Pesca (SEMARNAP), the Canadian Forest Service (CFS), and the USDA Forest Service (FS).

Front row, from left: Pablo Navarro, SEMARNAP; Sheila Andrus, FS; Victor Sosa Cedillo, SEMARNAP; Laura Lara, SEMARNAP; Jose Cibrian Tovar, SEMARNAP; Bill Luppold, FS. *Second row, from left:* Rosalie McConnell, CFS; Oscar Gonzalez Rodriguez, Chief of Forestry, SEMARNAP; Bill Banzhaf, Society of American Foresters; Carlos Rodriguez Franco, Instituto Nacional de Investigaciones Forestales, Agricolas y Pecuarias; Peter Roussopoulos, FS; Allen Dunn, Clemson University. *Third row, from left:* Chief Thomas; David Harcharik, Food and Agriculture Organization; Bob Jones, CFS; Gordon Miller, CFS; Les Whitmore, FS; Yves Dube, Food and Agriculture Organization; Jan Engert, FS; Kurt Johnsen, CFS. *Back row, from left:* Pauline Myre, CFS; Yvan Hardy, CFS; Jerry Sesco, FS; Mike Lennartz, FS. Photo courtesy of Jack Ward Thomas.

should be a single set of criteria on which to judge sustainability on an international scale or whether these criteria should be more site-specific.

It seems likely that the international pressure for some formal approach to these concerns on an international basis will not abate. The position of the United States (which is determined by the State Department, with little consultation with the Forest Service) has been to oppose this movement. However, the movement continues, and the American Forest and Paper Association has led the way in the United States with a voluntary program, with the "teeth" in the program being dismissal from the organization for noncompliance.

The opinion of the commission, after hearing a report from a committee assigned to review the matter, is that some combination of criteria and indicators at an international level, coupled with more region-specific data, would yield the best approach. Pressure in this regard seems likely to increase and the United States would do well to get on the bandwagon. After all, we like to think of ourselves as international leaders in conservation. If so, we had best begin stepping up our efforts in this regard.

As the discussion went on, I could not help but be struck by the irony of the situation. The United States (at the direction of the congressional budget committees) is scaling back an already inadequate international forestry program at the same time that the rest of the world has awakened to the dramatic significance of sustainable forestry and the establishment of criteria and indicators for judging sustainability—and enforcement of those criteria—in international trade. Some 50 percent of wood commercially harvested in the world comes from North America, and the United States is both a major importer and a major exporter of wood.

At the very time the United States should be stepping out as an international giant in forestry and natural resources issues, the nation moves toward isolationism in such matters and leaves the field to others.

Canada, on the other hand, takes the situation much more seriously and, in spite of downsizing in their federal forestry programs that dwarfs the downsizing in the United States, has chosen to be a major player in the issues of sustainability. This interest is, no doubt, related to the critical importance of forest exports to the Canadian economy.

Mexico likewise takes the matter quite seriously but lacks the resources to be a major player. I think they hope that their role and interests can be magnified through the North America Forest Commission and the Forestry Division of the Food and Agriculture Organization.

The significance of this matter of sustainability, as related to exports and imports, has special significance when considered in light of the North American Free Trade Agreement that will become clear only over time. We heard numerous other committee reports, primarily dealing with research and technical assessments on matters of North American concern.

The social aspects of the meeting were most pleasant, as the participants have become friends over the course of the biannual meetings. There were two field trips. The first was to the Bent River Demonstration Forest to examine an array of regeneration cuts, ranging from clearcuts to shelterwood to seed-tree to single-tree selection. As most studies have indicated, the efficiency and effectiveness are best with clearcutting and diminish with intensity of tree removal. However, the visual effect is the dramatic and diametric opposite: acceptability diminishes with the intensity of the regeneration harvest. And there is the problem.

If it is recognized that in a democracy, both public and private forestry will be practiced only at the sufferance of the public, then this is significant. The day is past, particularly for forestry on public land, when land managers have a free hand in making silvicultural decisions. My predecessor as chief, Dale Robertson, fully recognized that and as a result, made it Forest Service policy to dramatically decrease the use of clearcutting as the primary regeneration harvest technique. I continued that policy, with very significant results in the percentage of timber sales using clearcutting.

However, decreasing the use of clearcutting has had significant consequences for whether sales are above or below costs and for attaining satisfactory regeneration of desired species. One of the troubles of being toward the end of a forty-year career is that I have been around long enough to watch fads in forestry come and go and come again.

I was pleased with the demonstration and research plots we were shown. But I was not pleased to find that few or no data were being collected beyond standard timber-type silvicultural data, such as regeneration and growth by species of both residual trees and reproduction. Missing were such things as aesthetic ratings by a sample of the public, browse produced per acre and the effects of browsing, response of birds (particularly neotropical migrants) and small mammals, etc.

I felt a sadness as the meeting ended. We exchanged gifts, shook hands and embraced, and then we were gone back to our individual jobs.

25 November 1996, Washington, D.C.

At the close of the day, Mark Rey, chief of staff to Senator Murkowski's Senate Committee on Energy and Natural Resources, dropped by the office with a bottle of champagne to say goodbye. He announced that his visit was a tradition, as he had done the same to bid farewell to my predecessors Max Peterson and Dale Robertson.

My relationship with Mark Rey over the years—both in his role as vice-president of the American Forest and Paper Association and in his present role—has been good. We were frequently on different sides of the fence on issues, but I believe we respected each other. He has always been a straight shooter and I believe he would accord me the same description.

I have never been able to discern his personal views and values concerning management of natural resources. He reminds me of a lawyer serving as counselor to a client, in that he does the best he can to advance the interests of his client under the rules of the game. And he certainly does that very well.

He and I do agree on one thing: the laws that govern public land management and their interpretation in regulations and by courts are a mess and need to be addressed. We disagree on what will be required to remedy that situation. He, pragmatically enough, and understanding the jealous guarding of prerogatives by various congressional committees, will take the route of dealing with individual pieces of legislation, starting with that legislation under the oversight of the committee he serves. He is ready to go with revisions to the National Forest Management Act, including a new title that allows for devolution of management of public lands to the states. I don't believe he seriously thinks that such will occur. He defines and defends the inclusion of that title in the bill as a put-up-or-shut-up provision.

It was a pleasant visit, and it was kind of Mark to drop by for a personal farewell. As he was leaving, he turned and commented that it was likely that I could be a better spokesman for conservation outside the government, "and we will see to it that you get that chance."

26 November 1996, Washington, D.C.

Today was the day of my farewell as chief to Forest Service employees in the Washington, D.C., offices. On the way to work I stopped by the office of the Law Enforcement and Investigations staff in Rosslyn, Virginia. They wanted time to say goodbye. I believe I have paid more personal attention to the law enforcement folks than any previous chief, and they know that and appreciate it.

After a bit of visiting, we retired to the break room for donuts and coffee and presentation of some gifts. First, I was presented my new credentials as a Forest Service law enforcement officer. Second, I was presented a stainless steel Ruger .357 revolver, appropriately engraved by the law enforcement folks. They said I could use a good sidearm in Montana, for protection from both grizzlies and militiamen. I replied that it would probably do a better job in a display case. We have made a lot of progress in the past few years with professionalizing the law enforcement division. There is still a ways to go, but we are going in the right direction.

At midmorning, I met in the training room of the Auditor's Building for a bit of roasting from the National Forest Systems staff. It was fun and it gave everyone a chance to say goodbye and laugh at the same time. Yet there were tears in a lot of eyes, including mine.

Just before noon, Mary Jo Lavin and her staff from Fire and Aviation dropped by to present me with my very own chrome-plated and appropriately inscribed pulaski, the frontline firefighter's primary tool. While the presentation was going on, I kept trying to visualize a chrome-plated pulaski on the living room wall. I had some difficulty with that; however, it will hang somewhere and I will display it with pride. The firefighting organization and I bonded tightly when we lost fourteen firefighters on the Storm King Mountain fire in Colorado in 1994. I have visited many fire camps over the past three years, and many frontline firefighters came to know me by sight. They appreciated the attention and felt a bit more appreciated and respected. Things improved so far as safety is concerned after 1994, to the point that we went through the worst fire season of the decade in 1996 without a single fatality. I'd like to think that I had something to do with that.

Then there was an open house for employees in the afternoon. Hun-

The view from the chief's office.
USDA Forest Service photo
by Karl Perry.

dreds of folks came by to bid me farewell. I had a photographer on hand
to take a photo of me with each employee. My intent is to inscribe each
and every one. As the hours went by and my hand began to ache from shaking hands, I appreciated the trouble people took to come by and share a
last moment with the chief. I kept thinking that the Forest Service is truly
a special organization and that I have been blessed to have worked thirty
satisfying years for "the outfit."

27 November 1996, Washington, D.C.
    I had one last visit with Secretary of Agriculture Dan Glickman and his
chief of staff, Greg Frazier, in the secretary's office.
    As the secretary began to speak, he laid out several newspaper clippings
in front of him. The headline on one, from a North Carolina newspaper
that I had given an interview, said something to the effect that "Forest Service Chief Blasts Politicization." He said I was soon to be a free agent and
could say what I wanted when I wanted—and he had no doubt that I would
do just that. However, he expressed hope that I fully understood the cred-

ibility I hold, generally, with the media and with Congress. Therefore, he said, he hoped I would be very careful and deliberate in my comments and observations.

With that, we stood and shook hands and I walked out of the secretary's office for the last time as chief of the Forest Service. I felt good about the meeting. I said what needed to be said "with the bark on," and the secretary was patient and attentive. An increasing sense of relief came over me as I walked back to my office.

29 November 1996, Rosslyn, Virginia

Today was my last day as chief of the Forest Service. Shortly after I arrived, the movers showed up and began packing up my office.

Over a two-hour period, I cleaned out my desk, and when it was as I found it three years ago—empty, with the top swept clean—I sat back to watch the packing. Soon the walls were bare and the memorabilia of awards and gifts that filled the shelves were packed away. That old familiar feeling of watching as something dies swept over me. Though I was eager to move on, I felt sad at the ending of this episode in life.

When the movers were gone, Jim Long from the Office of General Counsel came for my security debriefing. Then Sue Addington, my administrative assistant, checked in my property, took my Forest Service key, my office key, my identification card, my government credit card, and my telephone credit card, and removed my access to the computer.

My mind raced over the three years since I first sat down at Pinchot's empty desk and began to put *my* things in place. Was it all worthwhile? Only history will be able to tell that, I suppose. All I know is that I did the very best that I knew how to do. I worked as hard as I could work. I could not do more than that. In those respects, I am content.

INDEX

Accelerated harvest, 199

Adams, Brock, 21, 47–48, 162

Addington, Sue, 12, 98, 155, 340, 399

Agriculture, U.S. Department of: air tankers, 273; budget crisis shutdown, 275; CITES, 140; Legislative Affairs office, 341; New World Mine, 371; Office on Natural Resources and the Environment, 242; reinvention plan for, 134; Statistical Services Unit, 202; Thomas resignation offers, 240; *318* sales, 247–248; USFS relations with, 91, 245, 325–327

Air tankers, 235–238, 246–247, 271–274, 312

Akaka, Daniel, 79, 379

Alaska: Alaska Pulp Company timber contracts, 72–73, 80–82, 85–89; fish habitat in, 82; Ketchikan Pulp Company timber contracts, 72–73, 227, 340; subsistence hunting and fishing by natives, 280; timber harvest as USFS priority, 220; timber sales, 72–73, 80–83, 85–89; Tongass N.F. land management planning, 300–301, 339–340; Tongass timber targets, 227–228, 240–242

Alaska Pulp Company, 72–73, 80–82, 85–89

Aldo Leopold Wilderness Research Center, 130–131

Allard, Wayne, 151, 284–285

Allowable cut levels, 30–31

American Fisheries Society, 311

American Forest and Paper Association, 394

AmeriCorps, 280, 292

Anadromous fish. *See* Salmon

Andrus, Cecil, 69, 111

Andrus, Sheila, 393*fig*

Angeles National Forest, 112–114

Animal rights groups, 83–84

Annual sale quantity (ASQ), 41–42, 74

Anthony, Robert G., 53

Anti-Deficiency Act, 269

Antique Aircraft Act, 246, 271

Apache National Forest, 329

Applegate, Carol, 210

Applegate, James, 210, 211*fig*

Appropriations Act of *1990* for Interior and Related Agencies, 162, 289

Aquatic communities, 284

Armstrong, Bob, 130, 155

Army, U.S., 368

Arson, 390–391

Ashmore, Nick, 57

Assessments: of Columbia basin ecosystems, 205–208, 212, 245–246, 310–312, 373–375; lack of skilled personnel for, 166; of NEPA, 378–380; scale of, 174–176, 178

Association of Senior Executives, 257

AuCoin, Les, 41

Audubon Society, 331

Babbitt, Bruce: appointment as Interior Secretary, 9; ESA administrative costs, 298–299; fire fighting, 312; grazing reform, 106, 140, 209–210, 212–213, 225–226; Interior Department appropriations, 79; Kaibab Industries case, 141; Mt. Graham observatory expansion, 345; National Monument creation, 379; New World

Babbitt, Bruce*(continued)*
  Mine proposal, 234, 258; public land ownership and transfer, 276–277; Storm King
  Mountain disaster, 97; Thomas resigna-
  tion offers, 322; *318* sales, 248, 250, 302;
  timber industry lawsuit against, 62; "War
  on the West," 90
Backiel, Adela: APC timber contract, 85–87;
  CITES, 128, 140; Columbia River Basin
  ecosystem planning efforts, 212; grazing
  reform, 209; Oregon timber supply crisis,
  155; salvage logging, 190, 200–201; timber
  harvest targets, 147; viability regulations,
  110
Bacon, Dick, 120
Ball, Billy, 335
Bankhead-Jones Act, 133
Banzhaf, Bill, 393*fig*
Barry, Dan, 233
Barry, Don, 290
Barton, Mike, 72–73, 87–88, 314
Baucus, Max: grazing, 208; hearings on owl
  proposals, 47; roadless areas, 255; Thomas
  appointment as USFS chief, 69; USFS rein-
  vention, 134–135; wildfires, 117
Baum, Ray, 104
Bayh, Birch, 234–235
Bear, Dinah, 179, 296, 372
Bear baiting, 83–85, 359
Beasley, Lamar, 94, 97–98
Beattie, Mollie: endangered species listings,
  102, 139–140, 147; ESA administrative
  costs, 300; Oregon timber supply crisis,
  155; salvage logging, 153
Bent River Demonstration Forest, 395
Beuter, John, 43, 48
Bibles, Dean, 77
Bighorn sheep, 308
Bingaman, Jeff, 100, 198, 213, 277
Biodiversity, 7, 26, 54, 189
Blackwood, Jeff, 174
Block grants, 244–245
Blue Bunch Wheatgrass, 308
Blue Mountain Natural Resources Council,
  206
Board of State Foresters, 242–245, 265–266

Boise-Cascade Corporation, 251
Bombings, 172–174, 222–223
Bond, Kit, 264–265
Bonded debt, 80–81
Boomer, Bob, 100
Boone and Crockett Club Professorship, 256–
  257, 378, 383, 388
Border Patrol, 382
Bosworth, Dale: Columbia River Basin ecosys-
  tem planning efforts, 206, 373; grazing
  bill, 213; Nye County rebellion, 304; pub-
  lic land ownership and transfer, 277;
  salmon vs. rafting, 110–111
Bradley, Bill, 180–181, 198, 379
Bradley, James, 40, 64–65
Brink, Steve, 301
Brouha, Paul, 311
Brown, Bill, 40, 103
Brown, Hank, 283–285
Brown, James, 364–365, 377
Brown, Perry, 256
Buckman, Robert, 11
Bull Run Watershed, 148–150
Bull Trout, 101, 105–106, 207–208
Bumpers, Dale: forest service budget, 79; grazing reform, 213; public land ownership
  and transfer, 184, 277; Thomas Report,
  20–22; USFS actions under salvage rider,
  353
Bureau of Land Management (BLM): assess-
  ments, 175; bombing of office, 172;
  Columbia River Basin planning efforts,
  205–206, 208, 373; ecosystem management, 264; execution of Rescission Bill,
  233; God Squad actions on timber sales of,
  50–51; grazing, 197–199, 202–203; Jamison plan, 29; lobbying charges by legislators, 217, 226; national grasslands and,
  133; New World Mine, 317; overcutting
  of lands of, 46; PACFISH, 100, 102; salvage logging, 153–155, 199–200, 224, 295;
  Storm King Mountain disaster, 94–96,
  150, 171; *318* sales, 302; transfer to state
  ownership of land of, 202
Burke, Brian: bear baiting, 359; DOJ-USDA
  relations, 281–282; New World Mine,

316–317, 346, 371–372; performance reviews, 388–389; planning regulations, 314–315; Quincy Library Group, 292; spotted owl planning, 315; Thomas resignation offers, 320–321; USFS actions under salvage rider, 332; USFS budget, 286–287; USFS vs. politicians, 297–298; wildfires, 312

Burns, Conrad: assessments, 174; environmental laws and forest management, 179–180; grazing, 198; on International Forestry, 182–183; mining, 234; NEPA assessment, 379; private forest management, 183–184; public land ownership and transfer, 277; reinvention of USFS, 145–147; roadless areas, 255; salvage logging, 154, 230; testimony before Congress by, 152; wildfires, 117

Bush, George H.W., 54, 270, 384

Byrd, Robert, 196

C-*130* aircraft, 235–238

C-SPAN, 200

Cable News Network (CNN), 294

California spotted owl plan adjustments (CASPO), 315–316, 360

CalTrout, 89

Campbell, Ben: grazing reform, 198, 213; Interior Department budget, 79; jet boats in Hell's Canyon, 310; Tongass N.F. land management planning, 300

Canada, 124–126, 362, 394

Canyon Creek fire, 94–97, 115–116. *See also* Storm King Mountain disaster

Caplan, James, 190–191

Carhart (Arthur) Wilderness Training Center, 130–131

Caribbean National Forest, 328–329

Carpenter, Forrest, 174

Cartwright, Chip, 100

Carver, Dick, 303–304

Cassidy, Edward, 48

Caswell, James, 323

Cattle, 31

Central Intelligence Agency, 272–273

Chafee, John H., 47

Chancellor, John, 387

Cheatgrass, 308

Chenoweth, Helen: Columbia basin ecosystem planning, 310–312, 374; ecosystem management, 263–264; endangered species, 262–263; ESA administrative costs, 300; grazing, 204; jet boats in Hell's Canyon, 308–309; mining, 234; salvage logging, 260–261

Cherokee National Forest, 349

Chesapeake Bay, 196

Cibrian Tovar, Jose, 393*fig*

Civilian Conservation Corps (CCC), 307

Civil rights, 317–318

Clark, Ray, 316–317, 347, 356, 371–372

Clark, Roger, 55

Clean Water Act, 339

Clearcutting, 395

Clearwater National Forest, 323

Cleveland National Forest, 112, 381, 383

Clinton, William Jefferson: accountability for salvage targets, 222; AmeriCorps, 280; biodiversity, 7, 189; Columbia basin ecosystem planning, 310; election of 1994's effect on, 130; as environmentalist, 262; Forest Plan for PNW, 55–56, 58–59; grazing reform, 209; illegal immigration, 383; inauguration of, 6; National Monument creation, 379–380; New World Mine, 356, 358*fig*, 359; NFMA-driven planning regulations, 270; Oregon timber supply crisis, 155; owl as factor in election of, 54; Portland timber summit, 52*fig*, 54, 385; re-election of, 354; Rescission bill signed by, 225, 260; salvage logging, 189, 201, 251; Thomas appointment as USFS chief, 65; Thomas resignation offers, 257; Thomas visit with, 66, 67*fig*; water rights, 284; wildfires, 235–236

Coast Range, 46, 49

Collier, Tom, 58, 75, 166

Columbia River Basin: ecosystem planning efforts, 205–208, 212, 245–246, 310–312, 373–375; salmon habitat, 9–10, 100–103, 158–159, 208

Comanor, Joan, 194, 361–362

Commerce, U.S. Department of, 247

Committee of Scientists, 108

Conference of Black Mayors, 307

Conley, Jerry, 111

Connelly, Kathleen, 269

Conrad, Kent, 132, 213

Conservation Strategy for the Northern Spotted Owl (Thomas *et. al*). *See* Thomas Report

Convention on International Trade in Endangered Species of Wild Flora and Fauna (CITES), 126–129, 139–140, 183

Cook County, IL, forests, 195

Cooley, Wes: Columbia basin ecosystem planning, 310, 374; endangered species, 262–263; grazing, 204; as hearing participant, 264; jet boats in Hell's Canyon, 308–309; salvage logging, 260–261

Coppelman, Peter, 124, 249, 290, 294, 345

Coronado National Forest, 383

Council on Environmental Quality: environmental laws' effect on forests, 178, 181; implementation of Rescission Bill, 223; New World Mine, 258, 338, 347; salvage rider, 247, 294–296, 352; salvage sales exemptions, 372; timber sale protests, 355

County Rights Movement, 216, 293–294, 303

Court cases: Friends of Animals v. Thomas, 359; Pacific Rivers v. Thomas, 159. *See also* Dwyer, William

Craig, Larry: Columbia River Basin ecosystem planning efforts, 374–375; environmental laws and forest management, 178–181; forest health hearings, 117–119; Forest Plan for PNW, 176; grazing, 197–199, 213; jet boats in Hell's Canyon, 310; land-use planning, 174–175; mining, 234; national grasslands administration, 198; public land ownership and transfer, 276–277; salmon vs. rafting, 111; salvage logging, 137, 154, 230–231, 260; Thomas appointment as USFS chief, 69–70, 79; Tongass N.F. land management planning, 300; USFS actions under salvage rider, 344–345, 353

Crime, 114, 172–174, 390–391

Crowell, John, 308

Crown Butte Mining, Inc., 372

Cruise ships, 141

Cruz, Pablo, 328

Cubin, Barbara, 204, 300

Dahl, Bjorn, 70

Daley, Richard, 194

D'Amato, Alphonse, 235–236

Daschle, Thomas, 116, 118, 169

De la Garza, Kiki, 36, 41, 49, 384

DeCoster, Tim, 49

DeFazio, Peter: Forest Plan for PNW, 59; hearings on owl proposals, 46, 49; salvage logging, 200–201; Thomas Report, 26

Denver Post, 97

Departure, from sustained yield, 49–50

Devine, Dan, 282

Devolution of land management, 376–378, 396

Dicks, Norman D.: Furse Amendment, 340–341; salvage logging, 199–200; timber cut levels, 40–41, 49, 137, 161

Dole, Robert, 197, 213, 354

Dombeck, Mike: as acting BLM director, 9; Canyon Creek fire (CO), 94, 115; Congressional ire at, 198; county rights movement, 293–294; fire season projections, 362–363; grazing, 202–203, 209–210, 225–226; lobbying charges by legislators, 226; Oregon timber supply crisis, 155; PAC-FISH, 100; photo, 386*fig*; as replacement for Reynolds, 313, 318–319, 327; salmon habitat protection, 10; salvage logging, 153; Storm King Mountain disaster, 171; Thomas resignation offers, 320–322; *318* sales, 218; USFS chief appointment possible, 319–322, 354–355, 378

Domenici, Pete: grazing reform, 197, 213, 217, 225–226, 333, 355, 389; USFS budget, 79

Donations, 115

Doolittle, John T., 262

Doolittle, Warren, 384

Doppelt, Bob, 106

Dorgan, Byron, 132–133, 197–198, 204–205, 213, 217

Douglas-fir, 324, 337

Douglas-fir tussock moth, 171

Dube, Yves, 393*fig*

Dunn, Allen, 393*fig*

Dwyer, William: assessments, 175; ecosystem management test case, 264; Forest Plan for PNW, 137–138, 143–144, 176, 296; logging shutdown in owl habitat, 36, 45, 51, 385; Option 9, 55, 76, 162; train wrecks, 55–56

Eagle Cap Wilderness, 29–30, 32, 103–104

Earth First!, 118

Eco-corps, 114

Ecological Stewardship Workshop, 266

Ecosystem management: Chenoweth on, 263–264; legality of, 124, 175; Thomas on, 267–268; as USFS policy, 138, 141–144

El Porta Visitor Center, 328–329

Elections: *1992*, 6, 51, 54; *1994*, 123, 129–130, 139; *1996*, 345, 347, 353–354, 361, 380

Elephants, African, 126, 128–129

Elk, 31, 385

Emergency Salvage Timber Sale Program. *See* Rescission Act of 1995

Endangered Species Act (ESA): administrative costs of, 298–300; broadleaf mahogany listing, 126–128, 139–140, 183; calls for changes in, 47–48, 104, 107, 270; EIS requirements, 208; Forest Plan for PNW and, 248; listing of salmon, 159; old growth conservation and, 17; problems with, 188; public lands as focus for compliance with, 188–189; reauthorization of, 130, 137, 192; repeal proposals, 76, 262; requirements of, 108, 123; salmon habitat protection, 10, 101–102; Scientific Assessment Team report, 53; as superseding the Organic Act, 185; USFS regulations and, 14

Engert, Jan, 393*fig*

Eno, Amos, 266

Environmental Impact Statements (EIS): for the Endangered Species Act, 208; for grazing, 204; for the New World Mine, 233–235, 258, 338–339, 346–347; on the Sierra Nevada, 360–361, 375; on state hunting regulations, 359–360

Environmental laws, 178–181, 191–193

Environmental Protection Agency (EPA), 75, 144; environmental laws and forest management, 179, 181; KPC pulp mill violations, 340; New World Mine, 347; salvage logging, 191, 224, 337, 373; *318* sales, 302

Environmentalists: and the election of *1994*, 129–130; Forest Plan for PNW, 75–76; lawsuit over ISC plan, 45–46; New World Mine, 349; NFMA-driven planning regulations, 270; political support for, 122–123; political weaknesses of, 139; Republicans' view of, 153; salvage logging, 118–120, 137, 342–344; timber sale protests, 347–348, 355–356; train wreck and, 106, 124; viability regulations, 108–109; violence, 391; as whiners, 323

Espy, Mike, 113*n*: Alaska timber sales, 85–87, 89; Silver City helicopter crash, 100; Storm King Mountain disaster, 97; Thomas appointment as USFS chief, 71; water rights, 284

Estill, Elizabeth, 157, 284–285, 378

Extinction, 263

Farm Bill, 283, 285–286

Federal Advisory Committee Act, 62, 124, 287, 293, 361

Federal Bureau of Investigation (FBI), 69–70, 172–173, 347, 350, 355

Federal Emergency Management Agency (FEMA), 117

Federal Register, 83, 158

"Feds" as enemy, 158, 172

Feege, Ann, 381

50–11–40 rule, 28

Finley, Mike, 357, 360

Fire. *See* Firefighting; Wildfires

Firefighting: budgeting for, 312–313; course for line officers on, 171; low pay for, 368–369; Silver City (NM) helicopter crash, 97–100; Storm King Mountain disaster, 94–97, 150–151, 171, 397

Fish. *See* Salmon; Trout

Fish habitat: in Alaska, 82; grazing and, 89–90; salmon, 9–10, 49, 82, 100–103; in the Snake River drainage, 110–111

Fish and Wildlife Foundation, 266

Fish and Wildlife Service, U.S.: attitudes of field personnel, 144; bull trout recovery plan, 208; CITES, 126, 128; Columbia River Basin ecosystem planning efforts, 206, 373–374; endangered species listings, 131, 136; environmental laws and forest planning, 181; ESA administrative costs, 299; execution of Rescission Bill, 233; fish habitat, 102–103, 106; forest plans, 75, 252; Mt. Graham observatory expansion, 330, 346; Oregon timber supply crisis, 155–156; owl recovery plan of, 51; petitions for ESA listings, 14; recommendation for owl listing, 43; salvage logging, 153–154, 224, 337, 373; Thomas Report, 21–22; *318* sales, 233, 302; USFS consultation with, 351

Fishing, subsistence in Alaska, 280

Flanigan, Dayle, 203*fig*

Foley, Tom, 57, 206

Food and Agriculture Association (FAO) Committee on Forestry, 183, 392, 395

Foothills fire (ID), 120

Forest Ecosystem Management and Assessment Team (FEMAT): formation of, 7, 55–56; lawsuit against, 62; public lands as focus for ESA compliance, 189; report of, 76

Forest fires. *See* Wildfires

Forest health, 117–120, 123, 170–171, 288, 351–352

Forest History Society, 12

Forest Plan for PNW: budgeting for, 80, 288; court ruling on, 137–138; ecosystem management as basis for, 143–144, 264; Endangered Species Act and, 248; implementation of, 192–194, 220; Option 9, 55–56, 75–77, 250; replacement timber, 296; Republican views of, 176; revisions of, 302; shortfalls in sales under, 250; signing of, 88–89; testimony on, 55, 58–59; *318* sales, 218, 294

Forest and Range Sciences Laboratory, La Grande, OR, 5, 210, 212, 238

Forest and Rangeland Renewable Resources Planning Act. *See* Resources Planning Act (RPA) report

Forest Service, U.S.: actions under salvage rider, 330–332, 342–345, 352–353; assessments, 174–176; bombings in National Forests, 172–174, 222–223; breaking up of, proposed, 305; budget crisis shutdown, 268–270, 274–275; budget dispute with Lyons, 286–287; budget for, 79–80, 185, 306, 327–328; civil rights issues, 317–318; Congressional criticism of, 184; demoralization in, 49–50; DOJ effect on policies of, 12, 130, 252, 280–282; ecosystem management as policy of, 141–144; endangered species regulations, 14, 131; environmental laws and forest management, 178–181, 191–192; fire management at urban/forest interface, 112; fish habitat, 100–103, 107, 110–112; hunting regulations, 83–85; Interior Department relations with, 9, 140; Kaibab Industries case, 141; land exchanges, 184; land-use planning, 174–175; Law Enforcement Investigation Division, 9, 332, 335, 397; leadership meeting, 91–92; lobbying charges by legislators, 217; MOA on salvage logging, 223–225; National Grasslands management, 132–134, 197–198, 217; Northeastern Experiment Station (Amherst, MA), 5; Office of General Council (OGC), 218; Olympic gold medal given to, 350; OMB budget passback, 279; Oregon timber supply crisis, 155–156; overharvesting of timber by, 40; Pacific Northwest Forest and Range Sciences Laboratory (La Grande, OR), 5, 210, 212, 238; Park Service relations with, 359–360; personnel issues, 70, 93–94, 146, 170, 298; planning regulation delays, 314–315; primary purpose of, 44; range program, 286; recreation, 112–115; reductions in force, 264–265; regional office consolidation, 145–146; reinvention of, 129–130, 134–136, 145–147, 165–166, 220; State and Private Forestry Division,

183–184, 194, 243, 266, 376; Timber Bridge Initiative, 196; timber industry relations with, 249; transfer to Indians of land of, 140–141; USDA relations with, 91, 245, 325–327; violence against employees of, 283; volunteers used by, 114–115

Forestry: green, 183; international, 182–183, 196, 243; "New," 26–27, 43; on private lands, 183–184, 242–245, 279–280; sustainable, 49, 125, 392, 394–395; urban, 194–195

Forsman, Eric D., 25*fig*, 44, 53, 57*fig*

Frampton, George, 130; DOJ-Interior relations, 280–281; environmental laws and forest planning, 179; New World Mine proposal, 234; Oregon timber supply crisis, 155; salvage logging, 249–250; *318* sales, 247–248

Francis, Mike, 231

Franklin, Jerry, 36*fig*; biodiversity, 26; Gang-of-Four, 37–38, 40, 385; Makah Tribe, 54; old-growth forests, 384; PNW forest management, 46; Sierra Nevada EIS, 361

Frazier, Greg: Alaska timber sales, 240–241; intervention in USFS personnel selection, 318; roadless areas, 253, 255–256; Thomas resignation offers, 239, 321; Thomas retirement, 378, 398; timber sale protests, 347, 355–356; USDA-USFS relations, 245, 326; USFS actions under salvage rider, 330, 343

Friends of Animals, 83–84

Friends of Animals v. Thomas, 359

Fuel loadings, 171, 324

Fuel reduction, 119–120

Furse, Elizabeth, 148–150, 200–201, 340–341

Gaede, Mark, 294, 346, 372

Gallatin National Forest, 10, 258, 371

Gang-of-Four, 36–38, 40, 42, 88, 384–385

Garber, Dave, 338, 358*fig*

Garcia, Ernesto, 328

Gas leases, 254

Geisinger, Jim, 311–312

General Accounting Office (GAO), 276

General Services Administration (GSA), 246, 272, 306

Geological Survey, U.S., 206

Geringer, Jim, 276

Gila National Forest, 99

Gilliam, Bertha, 205, 389

Gilliland, James: air tankers, 236–237, 272–273; DOJ effect on USDA policies, 281; hunting, 84; salvage logging, 250; *318* sales, 248

Gingrich, Newt, 201, 261, 307

Glacier Bay, 141

Glauthier, T.J., 232, 287

Glickman, Dan: accountability for salvage targets, 222; air tankers, 271; Columbia River Basin ecosystem planning efforts, 375; DOJ-USDA relations, 280; environmental laws and forest planning, 181, 191–192; grazing reform, 210; intervention in USFS personnel selection, 313, 318–319; New World Mine, 338; photos, 182*fig*, 342*fig*; Quincy Library Group, 238–239, 287–288, 291–293; roadless areas, 254–255; salvage logging, 200, 249–251; salvage sale exemptions, 372; as secretary-designate, 158; Sierra Nevada EIS, 360; Thomas resignation offers, 239–240, 257, 319–322; Thomas retirement, 398–399; *318* sales, 248, 250–251, 302; timber harvest targets, 187; timber sale protests, 347, 355; USFS actions under salvage rider, 342–345, 352–353, 370; USFS-USDA relations, 325–326; wildfires, 236

God Squad, 9, 29, 36, 50–53

Goldschmidt, Neil, 290

Gonzalez Rodriguez, Oscar, 392, 393*fig*

Gordon, John, 36*fig*, 37–38, 40, 46, 48, 384

Gordon, Terri, 60

Gore, Albert: Alaskan timber contracts, 340; biodiversity, 7, 189; CITES, 140; Portland timber summit, 52*fig*; salvage logging, 199, 291; streamlining government, 115; Thomas resignation offers, 240, 257, 322; Thomas visit with Clinton and, 66

Gore, Tipper, 240

Gorton, Slade: Columbia River Basin ecosystem planning efforts, 206–207; income from forest activities, 305; owl proposals, 23, 33, 47; testimony before Congress by, 152; timber harvest targets, 182, 186–187; USFS budget reductions, 306; USFS reduction in force, 265
Goshawk, Queen Charlotte, 147
Government Printing Office, 154
Gramm, Phil, 197
Grand fir, 324
Grazing: advisory boards, 214; as chief BLM/USFS function, 197–199; elk in competition with, 31; fish habitat and, 89–90; in Hell's Canyon N.R.A., 308; NEPA assessments on permits, 169–170; permits violations, 156–158, 304; privatization of rights on public lands, 214; reform proposal, 106, 140–141, 197, 208–210, 212–217, 225–226, 333, 355, 389
Greater Yellowstone Coalition, 357–358
Greater Yellowstone ecosystem, 359–360
Greber, Brian, 55, 59
Green, Jerry L, 203*fig*
Green, Nancy, 345
Green forestry, 183
Green labeling, 125–126
GreenStreets program, Chicago, 194
Gunderson, A. Grant, 53, 57*fig*
Gutierrez, Sean, 100

Hall, Doug, 249
Hamilton, Stan, 363
Hansen, James, 262–263, 333–334, 375
Harcharik, David, 125, 392, 393*fig*
Hardy, Yvan, 124, 126, 392, 393*fig*
Hatch, Orrin, 333
Hatfield, Mark O.: Forest Plan for PNW, 59; grazing reform, 213; hearings on owl proposals, 47; Interior Department appropriations hearings, 79; retirement of, 367; Thomas appointment as USFS chief, 69; Thomas Report, 20–21, 33; *318* sales, 162, 289; timber cut levels, 137; USFS appropriations hearings, 80; USFS budget hearing, 185; USFS reinvention, 135

Hayden, Marty, 231
Heissenbuttel, Anne, 375
Helicopter logging, 154, 337
Hells Canyon National Recreation Area, 308–310, 369
Herger, Wally, 152, 202
Heritage Program, 196
Hessel, David, 168, 221, 370
Hickel, Wally, 81
Hired guns, 18, 28–29, 39
Holding, Earl, 333–334
Holthausen, Richard S., 53, 57*fig*
House of Representatives, U.S.: Agriculture Committee, 36–38, 41–42, 55, 58, 135, 285; Appropriations Committee, 160; benefits for Storm King Mountain victims, 97; Forests Subcommittee, 308; hearings on forestry bills, 46–47; Interior Committee, 38, 41–42, 55, 58; Merchant Marine Subcommittee, 55, 58; National Parks and Public Lands Subcommittee, 26, 40; Natural Resources, 135, 262; Rescission Bill, 199
Houston, Doug, 20
Howard, John, 311
Hunter, Malcolm, 27
Hunting, 83–85, 280, 359–360
Hurricane Hugo, 161

Idaho Department of Fish and Game, 111
Illegal immigration, 282–283, 380–384
Impoundments, 283 286
Incident Command System, 235, 368
Indians, 54, 140–141
INFISH, 207
Interagency Arthur Carhart Wilderness Training Center, 130–131
Interagency Scientific Committee to Address the Conservation of the Northern Spotted Owl (ISC): alternatives considered by, 44–45; establishment of, 14, 384; Forest Plan for PNW, 88; Senate attacks on, 20. *See also* Thomas Report
Interior, U.S. Department of: appropriations hearings, 79; budget crisis shutdown, 268, 274–275; Congressional criticism of, 178; DOJ relations with, 280–281; Mt. Gra-

ham observatory expansion, 346; New World Mine, 317, 338, 346–347; Pacific Northwest land management, 75; power in Western states of, 276; *318* sales, 247–248; USFS relations with, 9, 140
Intermountain Experiment Station, 336
Internal Revenue Service (IRS), 69
International Association of Game and Fish Commissioners, 85
International Forestry, 182–183, 196, 243

Jamison, D. Cy: God Squad actions on BLM timber sales, 9, 29, 36, 51–53; Thomas Report approach of, 9, 20, 22, 27, 29, 33
Janik, Phil: as replacement for Reimers, 313, 319, 327; salvage logging, 220; timber harvest targets, 147–148, 227–228; Tongass N.F. land management planning, 340; as USFS chief candidate, 378
Jet boats, 308–310, 370
Jobs as political issue, 253
Johnson, Bennett, 79, 185
Johnson, K. Norman: Forest Plan testimony, 55; Gang-of-Four, 36*fig*, 37–38, 40, 384–385; Makah Tribe, 54
Johnson, Kurt, 393*fig*
Johnson, Tim, 169
Jolly, Dave, 70, 120
Jones, Bob, 393*fig*
Joslin, Robert, 306, 350
Justice, U.S. Department of: air tankers, 272–273; bear baiting, 84, 359; as determiner of legal actions, 132, 280–282; effect on USFS policies, 12, 252; environmental laws and forest planning, 181; hesitation to prosecute by, 215–216; hunting regulations, 84; Kaibab Industries case, 141; Mt. Graham observatory expansion, 330; *318* sales, 247, 249; timber sale protests, 355

Kaibab Industries, 141
Karr, James, 27
Kashdan, Hank, 282
Kempthorne, Dirk, 118–119, 154
Kennamer, James E., 173*fig*

Kennedy, Anne: environmental law and forest management, 191–192; Furse Amendment, 341; law enforcement, 282–283; Mexican Spotted Owl, 251, 253; oversight of USFS activities, 245; Quincy Library Group, 292; salvage logging, 249–250, 372; Sierra Nevada EIS, 360–361; USFS actions under salvage rider, 330–332
Kennedy, Roger, 386*fig*
Kennedy, Ted, 341
Kenya, 129
Kerrey, Bob, 264–265
Ketchikan Pulp Company, 72–73, 227, 340
Kimball, Kate, 155
Kitzhaber, John, 155, 314
Klicker, Robert, 206
Knowles, Donald R., 34–35
Knutson-Vanderberg trust fund, 312
Kootenai National Forest, 120
Kyl, Jon, 198–199, 213, 277, 351–352

Labor unions, 152–153
Lancia, Richard, 27
Land exchanges, 184, 332–336
Land management, 75, 376–378, 396
Land ownership and purchase, 276–277
Land-use planning, 174–175, 252, 281
Lara, Laura, 393*fig*
Larson, Gary, 260
Lavin, Mary Jo, 312, 361–362, 397
Law enforcement: arson, 390–391; bombings, 172–174; illegal immigration, 381–384; Law Enforcement Investigation Division, 9, 332, 335, 397; Level 4 officers, 114–115; management of, 162–165; murder, 114; reduction or elimination of funding for, 196, 282–283; state's rights and, 216; SWAT teams, 350; Warner Creek sale protests, 347–348
Lawrence, James, 324
Leahy, Patrick, 196
Legacy Program, 243
Lennartz, Mike, 393*fig*
Leonard, George, 32, 66, 70, 259
Lewis, John, 307
Lint, Joseph B., 25*fig*

Little Blackwater Demonstration Timber Bridge (MD), 196–197

Logging roads, 341

Long, Jim, 12, 399

Lopez, Johnny, 100

Louisiana-Pacific Corporation, 340

Lowe, John, 77, 100, 206, 212, 309

Lugo, Auriel, 328

Lujan, Manuel: God Squad actions on BLM timber sales, 29, 51–52; Jamison Plan, 29, 33; Thomas Report, 33

Luppold, Bill, 393*fig*

Lynn, Bob, 331

Lyons, James, 362fig; accountability for salvage targets, 222; budgeting for the USFS, 279, 286–287; CITES, 140; Columbia River Basin ecosystem planning efforts, 374–375; Congressional criticism of USFS, 184–186; Forest Plans, 58, 193; Gang-of-four, 36; grazing permit violations, 157; grazing reform, 209–210, 212; hearings on owl proposals, 49; intervention in USFS personnel selection, 298, 313–314, 318–321; Office on Natural Resources and the Environment, 242; Oregon timber supply crisis, 155–156; oversight of USFS activities by, 245, 321, 324, 326–327; PACFISH, 159; performance reviews, 388–389; Portland timber summit, 385; Quincy Library Group, 238–239; reinvention of USFS, 134–135, 145, 165; repealing the salvage rider, 290; Rescission Bill, 199, 218–219, 225; salvage logging, 190–191, 199–201, 224, 229–231, 249–250; salvage sales exemptions, 370, 372; Silver City helicopter crash, 100; Thomas appointment as USFS chief, 6–7, 59–65, 68–70, 257; Thomas relationship with, 259–260, 319–320, 327–328, 383–384; Thomas resignation offers, 238, 319–321; Thomas visit with Clinton, 66; *318* sales, 294, 296; timber harvest targets, 187, 227; timber sales contracts, 73, 85; transfer to Indians of USFS land, 141; USFS actions under salvage rider, 352; USFS leadership meeting,

91; viability regulations, 110; wildfires, 235

Lyons, Tom, 347–348, 355–356

Madigan, Edward, 284

Mahogany, broadleaf, 126–128, 139–140, 183

Makah Tribe, 54

Malheur National Forest, 367–369

Maloney, Kathy, 353–354

Marana Training Center, 171

Marcot, Bruce G., 44, 53

Marijuana, 8, 283

Marion, Al, 164

Marshals Service, U.S., 347

Martinez, Manny, 164, 223, 282

McCleese, Bill, 224, 361

McClure, John, 20

McConnell, Rosalie, 393*fig*

McDonald, Steve, 264

McGee, John, 329–330

McGinty, Katie: endangered species listings, 140; Forest Plan for PNW, 220; grazing reform, 213; NEPA assessment, 379–380; New World Mine, 358*fig*; Oregon timber supply crisis, 155–156; PACFISH, 159; repeal of salvage rider, 289–290; salvage logging, 166–167, 190–191, 224, 249; Sierra Nevada EIS, 361; Thomas visit with Clinton, 66; *318* sales, 296, 301–303

McKelvey, Kevin, 360

Mealey, Steve, 120, 174, 206

Mellon, Kim, 44

Memoranda of agreement (MOA), 219, 223–225, 229

Meslow, E. Charles, 23, 25*fig*, 48

Mexican Spotted Owl, 251–253, 370

Mexico, 394

Militias, 172, 283

Mill conversion, 80–81

Miller, George, 45–46, 49, 106

Miller, Gordon, 393*fig*

Mills, Tom, 70, 187

Mining. *See* New World Mine

Mining Law of *1872*, 235, 338–339

Missoula Smokejumpers, 387

Mitchell, George, 47

Montana: EIS role for New World Mine, 316; environmental regulations, 339; Thomas residence in, 256, 378, 386; wildfires in, 116–117, 120–122. *See also* New World Mine

Moseley, Jim, 23–24, 30–34

Motyka, Conrad "Connie," 242–243

Mount Graham, 329–330, 345–346

Mount Graham red squirrel, 329–330

Mountain Pine Beetle, 171

Muenke, Judge, 370

Mulder, Barry S., 57*fig*

Multiple-use, 249

Multiple Use-Sustained Yield Act, 299

Murders, 114

Murkowski, Frank: environmental laws and forest management, 180, 193; grazing, 198; hearings on Glickman nomination, 168–169, 345; mining, 234; NEPA assessment, 379; review of timber contracts, 72, 86–87, 89; salvage logging, 154, 220, 229–230; timber harvest targets, 147–148, 227–228; Tongass N.F. land management planning, 300–301, 339–340; USFS actions under salvage rider, 353

Murray, Patty, 166, 186, 289–291

Myre, Pauline, 393*fig*

National Association of State Foresters, 242–245, 265–266, 375

National Biological Survey, 130

National Environmental Policy Act (NEPA): assessment of, 378–380; ESA administrative costs, 299; forest management and, 178; grazing, 89–91, 169–170; modification of, 130; requirements of, 123

National Forest Foundation, 174

National Forest Management Act (NFMA): bull trout habitat, 207; changes proposed to, 192; devolution of land management, 376–377, 396; ESA administrative costs, 299; grazing, 91; moratorium on Management Act (NFMA) activities, 252;

national grasslands, 133; requirements of, 123; revamping of planning regulations, 107–109, 270–271; as superseding the Organic Act, 185

National Forest Products Association (NFPA), 42

National Grasslands, 132–134, 197–198, 217

National Interagency Fire Center, Boise, 363, 369

National Labor Relations Board, 272

National Marine Fisheries Service: attitudes of field personnel, 144; CITES, 128; Columbia River Basin ecosystem planning efforts, 206–208, 373–374; endangered species listings, 131, 136; environmental laws and forest planning, 181; Option 9, 75; PACFISH-PLUS, 219; salmon habitat, 10, 101–102, 104–107, 158; salvage logging, 153–154, 224, 337; Snake River salmon stocks, 110; *318* sales, 247–249, 302

National Monuments, 379–380

National Park Service, 10, 141, 258, 274, 359–360

National Press Club, 317

National Program Review (NPR), 315

Natural Resources Conservation Service, 133, 206, 245, 265, 388

Navarro, Pablo, 393*fig*

Nelson, Jim, 203*fig*, 213, 222–223, 304

Nelson, Robert, 211*fig*; endangered species, 136, 224; MOA on salvage logging, 224; public input on salvage sales, 323; salvage sales exemptions, 370; Thomas retirement, 385; timber harvest targets, 136, 221

Nethercutt, George, 206–207, 310–312, 374

Netherlands, 129

"New forestry," 26–27, 43

New World Mine: agreement over, 356–359, 371–372; BLM-USFS dispute over, 316–317; EIS for, 233–235, 258, 338–339, 346–347; inspection of, 348–349; proposal for reopening of, 10

New York Times, 234

News media: on Dale Robertson, 68; on graz-
ing bill, 217; salvage logging ads, 120;
Thomas retirement, 385; on wildfires, 96–
97, 151, 367–369
Noon, Barry R., 23–24, 25*fig*, 44, 360
North American Forestry Commission, 124–
126, 392, 393*fig*, 394–396
North American Free Trade Agreement
(NAFTA), 395
North American Wildlife and Natural
Resources Conference (Minneapolis,
1995), 170–171
Northern Spotted Owl: California Spotted
Owl plan adjustments (CASPO), 315–316;
captive raising of, 24–25; conservation
strategy development for, 14–15; as factor
in *1992* election, 54; managed forests as
home for, 45; in Northern California, 60;
old growth role in preservation of, 17;
recovery team for, 34; viability regulation,
108
Northwest Timber Association (NTA), 28
Nye County, NV, 172, 293–294, 303–305

Obst, Tim, 237
Occupational Safety and Health Administra-
tion (OSHA), 150–151, 171
Office of the Inspector General, 272, 334–335
Office of Management and Budget (OMB),
178, 274–275, 279–280, 287, 289, 351, 372
Office of Personnel Management (OPM), 211,
238–240, 257
Oil leases, 254
Old-growth forest: attitudes toward, 384;
salmon runs and, 49; Scientific Assessment
Team report, 53; spotted owl preservation
and, 17; workshop on management of,
252–253
Olympic Committee, U.S., 350
Olympic Games, *1996*: Snow Basin ski area,
332–335; Whitewater Venue, 307, 349–
351, 370
Olympic Peninsula, 49
O'Neill, Tip, 265, 291
Operation Desert Storm, 272
Operation Gatekeeper, 381–383

Option 9, 55–56, 75–77, 155, 162, 250
Oregon and California (O&C) lands, 14, 51
Oregon Department of Forestry, 364, 367
Oregon timber supply crisis, 155–156
Organic Act of *1897*, 184–185
Ostby, Dan, 365
Overbay, James, 29, 32, 34, 83
Owls. *See* Mexican Spotted Owl; Northern
Spotted Owl

PACFISH-PLUS, 219
PACFISH strategy, 82, 100, 102, 105, 158–159,
207, 373
Pacific Northwest. *See* Columbia River Basin;
Forest Plan for PNW
Pacific Northwest Forest Experiment Station,
192
Pacific Rivers Council, 104–106, 136, 154
Pacific Rivers v. Thomas, 159
Packwood, Bob, 47–48, 154, 278
Panetta, Leon: grazing, 209; National Monu-
ment creation, 379; PACFISH, 159; salvage
logging, 153, 199–201; Sierra Nevada EIS,
360–361, 375
Parrots, 328–329
Patuxent Wildlife Research Center, 188
Payette National Forest, 191
Payne, Jim, 121*fig*
Pence, Carl, 369
Pence, Guy, 222–223
Pence, Linda, 222–223
Penford, Michael, 198
Pennell, Kay, 103
Performance reviews, 259, 388–389
Perot, H. Ross, 54
Perry, James, 237
Peterson, Max, 3, 64, 355, 396
Philpot, Charles, 100
Pinchot, Gifford, 304, 355, 399
Pine, Lodgepole, 171
Pine, Ponderosa, 171, 324, 337, 351
Plenert, Marvin L., 34
Pletscher, Dan, 256
Plumas National Forest, 291
Political correctness, 92–93, 389
Polk, Alan, 190

Pombo, Richard W., 262, 299–300
Pomeroy, Earl, 132, 213
Porter, Doug, 121*fig*
Potts, Janet, 346, 356, 371–372
Power, corrosive influence of, 178–179
Prescribed burns, 364–365, 369
President's Forest Plan for PNW. *See* Forest
    Plan for PNW
Pressler, Larry, 169
Prince of Wales Island, 72, 73*fig*
Prineville Hotshots, 95–96
Privatization, 214, 275
Property rights, 214
Public land ownership and transfer, 140–141,
    184, 202, 276–277
Puerto Rican parrot, 328–329
Puerto Rico, 328–329
Pulaskis, 397

Quincy Library Group, 238–239, 287–288,
    326

Rafting, 110–112
Raines, Michael, 194
Ramey, John, 349–350
Range reform, 106, 140–141
Raphael, Martin G., 44, 53
RARE I & II, 254, 370
Rawls, Charles, 330
Reclamation, U.S. Bureau of, 364
Recreation, 112–115, 274–275, 305
Redds (salmon spawning), 110–111
Reeves, Gordon H., 40, 53
Regula, Ralph, 161, 207, 311
Reimers, Mark: bonus turndown for, 259; per-
    formance review on, 389; Quincy Library
    Group, 292; replacement attempt, 313,
    318–319, 321–322, 327; revocation of sal-
    vage rider, 288–289; Storm King Moun-
    tain disaster, 94, 115–116; USFS-USDA
    relations, 326
Renkes, Greg, 168, 339
Republican Party, 129, 136, 152–153, 176,
    197, 206
Rescission Act of 1995: Clinton signing of,
    225, 260; court interpretations of, 247;

execution of, 176–177, 232–233, 302;
    legislative action on, 160, 162, 186; mem-
    orandum of agreement (MOA) on, 219,
    223–225, 229; problems of, 8; Taylor
    Amendment, 224, 261; White House deal
    over, 199–202. *See also* 318 sales
Research Work Unit, Lincoln, NE, 264–265
Reservoirs, 283–286
Resources Planning Act (RPA) report,
    353–354
Rey, Mark: assessments and land-use plan-
    ning, 174; devolution of federal land
    management, 376–377, 396; environ-
    mental laws and forest management,
    178–181; National Forest Management
    Act, 192; NFMA-driven planning regula-
    tions, 270; as Prince of Darkness, 43;
    salvage logging, 227–228, 230, 352–353;
    Thomas retirement, 396; Thomas' view of,
    28, 168–169; timber harvest targets, 186,
    188; Tongass N.F. land management plan-
    ning, 301; Tongass timber targets, 241–242
Reynolds, Gray, 103*fig*; accountability for
    salvage targets, 221; assessments, 174;
    bombings, 172, 223; budget for USFS, 182;
    Columbia River Basin ecosystem plan-
    ning efforts, 373; fire season projections,
    362; land exchanges, 333; law enforce-
    ment, 282; performance review on, 389;
    Quincy Library Group, 292; replacement
    attempt, 313–314, 318–319, 321–322, 327;
    salvage sales exemptions, 370; Thomas
    appointment as USFS chief, 70; timber
    sale protests, 355; water rights, 285
Rich, Curt, 134
Richmond, Robert, 308, 370
Rios, Bernie, 365
Risbrudt, Christopher, 174, 339, 359–360
Road building, 341
Roadless areas, 74, 253–256, 373
Roberts, Pat, 213
Robertson, F. Dale: challenges of, 35–36;
    clearcutting, 395; congressional hearings,
    43; departure as USFS chief, 3, 7, 61, 66,
    259, 355, 396; ecosystem management,
    141; law enforcement, 163; press criticism

Robertson, F. Dale *(continued)*
of, 68; Scientific Assessment Team, 53; Thomas Report, 20, 22, 26, 32–34; on USFS objectives, 44

Rocky Mountain Elk Foundation, 79

Rocky Mountain Experiment Station, 252, 336

Rodriguez Franco, Carlos, 393*fig*

Rogers, John, 374

Rogers, Mike, 237

Rolienson, Marvin, 231

Romer, Roy, 97

Rominger, Rich: grazing, 209–210; Quincy Library Group, 238–239, 287–288; Thomas resignation offers, 321

Rose, Charlie, 272

Rosenkrance, Lester, 115–116

Roussopoulos, Peter, 378, 393*fig*

Rutzick, Mark, 52, 106, 229, 232

Salmon: in coastal streams, 49; Columbia River Basin habitat, 9–10, 100–103, 158–159, 208; habitat protection in coastal streams, 49; listing as endangered, 206; logging and, 82, 249, 337; and old-growth forest, 49; PACFISH, 100–103; Snake River stocks, 110–111

Salvage logging: accelerated, 176–178, 199; accountability for targets, 218–219, 221–222, 226; carrying out legislation on, 153–155, 160–161, 219–221, 232–233; demonstrations against, 118; exemptions on sales, 370–371; lack of specialists for assessments for, 166; molding public opinion on, 118–120; public input on, 323–325; reasons for, 193; revocation of rider, 288–291; status of sales for, 249–250; Thunderbolt sale, 190–191, 337–338; timber salvage task force, 260–261; USFS actions under rider, 330–332, 342–344. *See also* Rescission Act of *1995*

Salvage rider. *See* Rescission Act of 1995

Salwasser, Hal, 206, 338

San Bernardino National Forest, 112

Sanchez, Charlie, 100

Santa Ana winds, 381

Satterfield, Steve: bonus for, 259; budget crisis shutdown, 274; performance review on, 389; Quincy Library Group, 292; salvage logging, 200; timber harvest targets, 160

Saudi Arabia, 272

Sayer, Gary, 203*fig*

Schaefer, Mark, 359

Schiffler, Lois, 216, 293–295

Schmitten, Rollie: environmental laws and forest management, 179; PACFISH, 102, 105; salmon vs. rafting, 111; salvage logging, 153–154, 229

Schwalbach, Monica, 203*fig*

Schwarzkopf, Norman, 277, 278*n*

Science: power in USFS of, 50; values vs., 18–19, 76

Scientific Assessment Team, 53, 88

Scientific Panel on Late-Successional Forest Ecosystems (Gang-of-Four), 36–38, 40, 42, 88, 384–385

Secret Service, 350

Sedell, James R., 36*fig*; Forest Plan testimony, 55; PNW forestry legislation, 40, 46; reduction of owl range, 48–49; Scientific Assessment Team, 53

Senate, U.S.: Agriculture, Nutrition and Forestry Committee, 118–119, 151, 169; Appropriations Committee, 79, 135, 182; benefits for Storm King Mountain victims, 97; Budget Committee, 245; Energy and Natural Resources, 135, 151, 168–169, 179, 192, 197, 212, 220, 227, 276; grazing as issue in, 90; Public Lands Subcommittee, 28

Senior Executive Service, 3, 11, 61–62, 68–69, 238, 257, 319

Sesco, Jerry, 125, 393*fig*

Settlement fund, 289

Sharpe, Maitland, 202

Sheep, 308

Sheng, Sherry, 92

Sierra Legal Defense Fund, 229, 232

Sierra Nevada Evaluation Program (SNEP), 315–316, 361

Silver City helicopter crash, 97–100

Silviculture; sustaining owls in managed forests, 45

Simpson, Alan, 156

Six Rivers National Forest, 371

Skeen, Joe, 202

Ski areas, 274–275

Smith, Bob, 58, 70

Smith, Curt, 233

Smith, Gordon, 278, 290, 367

Smith, Sammy, 100

Snake River, 110, 308–310

Snow Basin ski area, 332–335

Society of American Foresters, 333, 384, 390

Society for Range Management, 389–390

Solis, David M., 53

Sosa Cedillo, Victor, 124, 393*fig*

South Africa, 129

South Canyon Fire Investigation report; 115–116. *See also* Canyon Creek Fire; Occupational Safety and Health Administration (OSHA)

Southern Pine Beetles, 371

Space, James, 174

Species diversity, 109

Spotted owls. *See* Mexican Spotted Owl; Northern Spotted Owl

Sprague, Lynn, 360, 381

Spruce, 171

Spruce Budworm, 171, 324

Squirrels, 329–330

Stanislaus National Forest, 324

State, U.S. Department of, 394

States rights, 84–85, 216

Stelle, Will, 249, 374

Stevens, Ted: criticisms of USFS, 184–185; income from forest activities, 305–306; review of timber contracts, 72, 85–87, 89; salvage logging, 154; timber harvest targets, 147–148, 227–228, 241; timber harvesting as USFS priority, 220; Tongass N.F. land management planning, 300–301

Stewardship Incentives Program, 196, 243, 279–280, 286–287

Stewart, Ron, 264

Storm King Mountain disaster, 94–97, 150–151, 171, 397

Subsistence hunting and fishing, 280

Sufficiency language, 154

Sullivan, Mike, 157

Sununu, John, 30–34

Supulski, Bill, 331

Sustainable forestry, 49, 125, 392, 394–395

Sustained yield, 49–59

SWAT teams, 350

Sweeny, James, 42

Symms, Steve, 47

Takings, 214

Taylor, Charles, 160–161, 222

Taylor Amendment, 224, 261

Teer, James G., 27

Telescopes, 329–330

Tennessee Park Police, 350

Tennessee State Police, 350

Tennessee Valley Authority, 349–350

Term limits, 152

Terrorism, 350, 391

Texas Parks and Wildlife Department, 5

Thinning, 119–120

Thomas, Craig: assessments, 174; environmental laws and forest management, 179–180; grazing, 156, 198; mining, 234; NEPA assessment, 379; public land ownership and transfer, 276

Thomas, Jack Ward: appointment as ISC head, 15; appointment as USFS chief, 6–7, 59–66, 68–70, 257, 384; arthritis of, 258–259; Atlanta visit, 306–307; as Boone and Crockett Club Professor, 256–257, 378, 383, 388; Clinton's visit with, 66, 67*fig*; Congressional hearings on forestry bills, 46; conversion of appointment to Senior Executive Service, 3, 11, 61–62, 68–69, 238, 257; corrosive influence of power, 178–179; defense of field staff, 263; disaffection with D.C. politics, 210–211; on duty of USFS professionals, 136; early days of, 4–6; on ecosystem management, 267–268; Gang-of-Four report, 37–38, 40, 384–385; grazing reform editorials, 209; interagency meeting on the environment and resource management, 122–124; keeping of journal

Thomas, Jack Ward *(continued)*
by, 6; Lyons relationship with, 259–260,
319–320, 327–328, 383–384; Makah Tribe,
54; memo on accountability for salvage
targets, 218–219, 221–222, 225; performance evaluation, 259; in photos, 25*fig*,
36*fig*, 57*fig*, 67*fig*, 93*fig*, 103*fig*, 121*fig*,
173*fig*, 182*fig*, 203*fig*, 211*fig*, 236*fig*, 240*fig*,
342*fig*, 362*fig*, 386*fig*, 393*fig*, 398*fig*; on
political correctness, 92–93, 389; on politicians vs. resource professionals, 282–283,
286–287, 297–298; qualifications of, 3;
resignation offers, 211–212, 238–240, 257,
294, 296–297, 319–323; retirement as
USFS chief, 12, 378, 383, 396–399; Russian
trip cancellation, 288; salmon habitat protection, 10; Scientific Assessment Team,
53; Society for Range Management membership, 389–390; South Canyon Fire
Investigation report, 115–116; on taking
responsibility, 277–278; wife's relationship with, 17–18
Thomas, Margaret: Clinton's visit with, 66,
67*fig*; husband's relationship with, 17–18;
illness and death of, 6, 8, 54, 77–78, 103,
384; JWT as USFS chief, proposed, 59–60,
63–64, 385
Thomas Report: Bush administration
response to, 23–25, 30–33; completion of,
16, 19; House hearings on, 26–27; Senate
hearings on, 20–22; task force working
group study of, 23–25, 27–29. *See also*
Interagency Scientific Committee to
Address the Conservation of the Northern
Spotted Owl (ISC)
318 sales: appeal of court decision on,
247–251; buyouts of, 289–291, 301–303;
effect on salmon of, 249; First and Last
timber sales, 294–296; Fish and Wildlife
role, 233, 302; lawsuits on, 229, 232; NW
Forest Plan affect from, 218, 294; replacement timber for, 233, 295–296; Rescission
Bill directive to proceed with, 162
Thursday Night Massacre, 389
Timber Bridge Initiative, 196
Timber harvest targets: Alaskan legislators'

threats over, 147–148; departure, 49–50;
industry demands, 40–41; mandated by
Congress, 136–137, 176–177, 219–221;
Option 9, 77; problems in execution of,
160–161; scenarios for, 160–161; in the
Tongass National Forest, 227–228, 240–
242; USFS attacked over, 182–187
Timber industry, 40–41, 249, 270
Timber production income, 305
Timber sales: budgeting for, 279–280; fifty-
year contracts, 72–73, 80–83, 85–89; First
and Last, OR, 294–296; political benefits
from, 251; Prince of Wales Island, AK, 72,
73*fig*; protests over, 347–348, 355–356;
ten-year contracts, 87; Thunderbolt Sale
Complex, ID, 190–191, 337–338; Warner
Creek, OR, 347–348, 355. *See also 318* sales
Timber salvage task force, 260–261
Timber stand thinning, 313, 324
Timber summit, Portland, OR, 7, 52*fig*, 54, 56,
385
Timber theft, 8, 163–165, 282–283
Timber Theft Investigation Branch (TTIB),
165, 282
Toiyabe National Forest, 172, 174, 210
Tongass National Forest, 85, 227–228, 240–
242, 300–301, 339–340
Train wrecks, 39, 41, 55–56, 58, 76, 106, 124
Trilateral Commission, 206
Trout: Bull, 101, 105 106, 207–208; West
Slope Cutthroat, 207
Trout Unlimited, 285
Turner, John, 20–22
Tussock moth, Douglas-fir, 171
Twiss, John, 103*fig*, 130–131

Umatilla National Forest, 107, 111, 367
Umpqua National Forest, 294–296, 365–366
Unger, Dave, 69, 222, 326–327, 360–361, 378
Uniform Grazing Management on Federal
Land bill, 197–198, 202–205
United Nations Heritage Program, 234
University of Arizona, 329
University of Montana, 256–257, 378, 383,
386, 388
Unsoeld, Jolene, 43, 59

Urban/forest interface, 112, 313, 366–367
Urban forestry, 194–195

Valdes, Tony, 203*fig*
Values vs. science, 18–19, 76
Vegetation mapping, 359–360
Vento, Bruce: elimination of programs of, 196; grazing, 202, 204, 213; as hearing participant, 262, 264; PNW forestry legislation, 46, 49, 56; Thomas appointment as USFS chief, 64; Thomas report hearings, 26; train wrecks, 40–41
Verner, Jared, 25*fig*, 360
Viability regulations, 107–110
Volkmer, Harold L., 37, 49, 59, 384
Volunteers, 114–115

Walcott, Rob, 29
Wallop, Malcolm: Forest Plan for PNW, 59; grazing permit violations, 157; Interior Department appropriations hearings, 79; Thomas appointment as USFS chief, 69; Thomas Report, 20
Wallowa-Whitman National Forest, 107, 111, 308, 323–324
"War on the West," 9–10, 79, 90, 104, 158, 172
Wasatch-Cache National Forest, 332, 335
Washington Post, 32, 68, 140
Water rights, 214, 283–286
Watt, James, 68
Weatherly, Debbie, 161
Weingardt, Bernie, 332, 334
Wellstone, Paul, 198, 213
Wes and Helen Show. *See* Chenoweth, Helen; Cooley, Wes
West Slope Cutthroat Trout, 207
Weyerhaeuser Company, 184, 335
Whistleblowers, 164, 282, 334–336
Whitmore, Les, 393*fig*
Wilcove, Dave, 19
Wilderness: attacks by Congress on, 131; classification of de facto areas, 254; de facto, 154; eliminations of programs for, 196; in Idaho, 154
Wilderness Conference (*1995*), 130–131
Wilderness Society, 136
Wildfires: Boundary (OR), 323; Canyon Creek (CO), 94–97; Foothills (ID), 120; in Idaho and Montana, 116–117, 120–122; illegal immigration and, 381–382; increased danger of, 206; insect damage and, 171; Long Island (NY), 235–237; management of, 363–365; media reporting of, 367–369; planning for, 351–352; prescribed burns, 364–365, 369; projections for, 361–363; Silver City (NM) helicopter crash, 97–100; South Fork Salmon drainage (ID), 337; Storm King Mountain (CO) disaster, 94–97, 150–151, 171, 397; Summit (OR), 367; Tower (OR), 367; in the urban/forest interface, 112, 313, 366–367
Willamette National Forest, 390–392
Williams, Jerry, 369
Williams, Pat, 213, 254–256
Williams, Robert, 314, 364
Wirth, Timothy, 20
Wise use coalition, 172
Wold, Johanna, 324
Wolf, Alexander Archipelago, 147
Wyden, Ron: Bull Run watershed logging ban, 148–150; election to Senate, 278, 290, 367; timber sale protests, 356

Yates, Sid, 196
Yellowstone ecosystem, 359–360
Yellowstone National Park, 234, 258, 356
Yeutter, Clayton K., 30–31, 33–34
Young, Don: ESA administrative costs, 298–301; mining, 234; review of timber contracts, 72, 89; salvage logging, 200–201; timber harvest targets, 147–148, 241; Tongass N.F. land management planning, 301

Zielinski, Elaine, 100, 200

Lightning Source UK Ltd.
Milton Keynes UK
UKOW02n0931280116

267295UK00008B/130/P